A Brain for Numbers

The Biology of the Number Instinct

Andreas Nieder

The MIT Press
Cambridge, Massachusetts
London, England

© 2019 Massachusetts Institute of Technology

All rights reserved. No part of this book may be reproduced in any form by any electronic or mechanical means (including photocopying, recording, or information storage and retrieval) without permission in writing from the publisher.

This book was set in ITC Stone Serif Std and ITC Stone Sans Std by Toppan Best-set Premedia Limited. Printed and bound in the United States of America.

Library of Congress Cataloging-in-Publication Data

Names: Nieder, Andreas, author.
Title: A brain for numbers : the biology of the number instinct / Andreas Nieder.
Description: Cambridge, MA : The MIT Press, [2019] | Includes bibliographical references and index.
Identifiers: LCCN 2018056262 | ISBN 9780262042789 (hardcover : alk. paper)
Subjects: LCSH: Number concept in animals--History. | Mathematical ability. | Cognitive neuroscience.
Classification: LCC QL785.24 .N54 2019 | DDC 156/.3--dc23 LC record available at https://lccn.loc.gov/2018056262

10 9 8 7 6 5 4 3 2 1

Dedicated to my father and my mother

Contents

Preface xi
Introduction xiii

Part I Conceptual Foundations 1

1 Thinking about Numbers 3

1.1 Mathematical Reality 3
1.2 Cardinal Numbers as Objective Properties of a Set 5
1.3 Knowledge of Numbers 8

2 Numerical Concepts, Representations, and Systems 11

2.1 Numerical Concepts 11
2.2 Mental Numerical Representations and Mental Systems 12

Part II Numbers Deeply Rooted in Our Ancestry 19

3 Understanding Numbers across the Animal Tree of Life 21

3.1 The Diversification of Animal Life 21
3.2 The Theory of Evolution 25
3.3 Classic Studies on Animal "Counting" 28
3.4 How to Test Animals on Numerical Cognition 32
3.5 The Phylogeny of Numerical Competence 36
3.6 Signatures of Animal Number Discrimination 49

4 The Utility of Number for Animals 63

4.1 A Matter of Fitness 63
4.2 Staying Alive 64
4.3 Benefits for Reproduction 72

5   Biological Heritage in the Human Brain   77

5.1 Baby Steps   77
5.2 Approximate Number System versus Object Tracking System   81
5.3 Number Discrimination in Humans Lacking Number Words   82
5.4 The Ancient Logarithmic Number Line   85

Part III   Numerical Quantity in the Brain   89

6   Localizing Numerical Quantity Representations in the Human Brain   91

6.1 The Building Plan of the Cerebral Cortex   91
6.2 At a Loss for Numerical Quantity after Brain Injury   101
6.3 Mapping Numerical Quantity on the Healthy Human Brain   104

7   Number Neurons   115

7.1 The Language of Neurons   115
7.2 The Discovery of Number Neurons in the Monkey Brain   119
7.3 The Neuronal Code for Number   127
7.4 Number Neurons Are Necessary for Number Judgments   129
7.5 Number Neurons Represent Different Presentation Types and Modalities   132
7.6 Convergent Evolution of Number Neurons: Lessons from Crows   137
7.7 Number Neurons in the Human Brain   140
7.8 An Innate Number Instinct   145
7.9 Number Models and Networks   148
7.10 Numerical Working Memory   151

Part IV   Number Symbols   157

8   Signs for Numbers   159

8.1 Evolution Pushed *Homo sapiens* toward Symbolic Thinking   159
8.2 Number Signs: Icons, Indices, and Symbols   161
8.3 Invention of Number Symbols in Human History   163
8.4 How Children Learn to Deal with Number Signs   168
8.5 Teaching Number Signs to Animals   170

9   Neural Foundation of Counting and Number Symbols   177

9.1 The Patient Who Lost All Numbers beyond Four   177
9.2 Imaging in the Human Brain During Symbolic Numerical Tasks   180

9.3 Numerical Association Neurons in Monkeys   182
9.4 Symbolic Number Neurons in the Human Brain   186
9.5 A Brain Area Dedicated to Numerals   187

10 The Calculating Brain   193

10.1 Non-symbolic Calculation in Indigenous People, Infants, and Animals   193
10.2 Single-Neuron Arithmetic   200
10.3 Cortical Location of Calculation   206
10.4 Dissociation of Calculation Types: Procedure versus Facts   212
10.5 Left versus Right Brain   216
10.6 Dissociated Brain Networks for Calculation and Language   219
10.7 Professional Mathematicians and Mathematical Prodigies   224

11 Space and Number   233

11.1 Small Numbers on the Left, Large Numbers on the Right   233
11.2 Carried along the Number Line During Calculation   237
11.3 Space and Number in the Brain   240

Part V Development   245

12 The Developing Number Brain   247

12.1 Counting in Children   247
12.2 Startup Tools for a Symbolic Number System   249
12.3 Out of Approximate Quantity and into Symbolic Number   251
12.4 Brain Activity in the Developing Brains of Children   254
12.5 Abstractness of Number Representations in the Brain   259

13 Developmental Dyscalculia   265

13.1 Developmental Dyscalculia and How It Affects Life   265
13.2 Domain-General and Domain-Specific Impairment in Dyscalculia   267
13.3 Tracing Dyscalculia Back to Brain Anatomy   270
13.4 Functional Differences in Brain Activation of Dyscalculic Children   273
13.5 It's (Partly) in the Genes   274

Part VI The Brain Departing from Empirical Reality   281

14 The Magical Number Zero   283

14.1 A Special Number   283
14.2 Zero in Human History   285

14.3 Development of Zero-like Concepts in Children   292
14.4 Zero-like Concepts in Animals   295
14.5 Neuronal Representations of "Nothing" and Empty Sets   299

Epilogue   307
Notes   309
Index   365

Preface

The story of how numbers are represented by neurons in the brain experienced a breakthrough at the turn of the new millennium. Around the time when I had to decide on how to continue my scientific career after my PhD, the seminal behavioral study by Elizabeth Brannon and Herbert Terrace in monkeys was published in the renowned journal *Science* in 1998. The paper showed in an unprecedentedly controlled way that rhesus monkeys understood the concept of numerical quantity. If monkeys could crunch numbers, they could only do so because their brains endowed them with such skills. To me, finding out how the brain could give rise to numerical competence seemed like a true scientific frontier worth investigating.

So I approached Earl K. Miller in his laboratory at the Massachusetts Institute of Technology in Cambridge, Massachusetts, in the United States. At that time, Earl was already a rising star in the field of primate cognitive neuroscience and had just gotten tenure at MIT after only three years of working as an assistant professor. Because of his expertise and his newly established laboratory, he had assembled a group of brilliant students to work with him, using the most modern and efficient scientific equipment at that time. Fortunately, Earl invited me to join his laboratory in order to discover how neurons enabled monkeys to grasp numerical quantity.

Thus, in 2000, my wife and I moved as a newly married couple from Germany to the United States. In my bag was a research grant awarded by the German Research Foundation to study "Preverbal representation of numerosity in prefrontal cortex of macaque monkeys." However, little did I know how important and influential my postdoctoral years would become for my career. Surely, this book would not exist without my fruitful time abroad.

I am indebted to many people who helped me along my scientific path. First and foremost, I am most grateful to my advisors and mentors, Georg M. Klump, Hermann Wagner, and Earl K. Miller, for sharing their knowledge in biology and neuroscience with me. I owe special thanks to Earl for his invaluable support in launching my research program about numbers in monkeys; I could not have wished for a better advisor and host laboratory. My gratitude extends to the many supportive colleagues I was lucky to meet in Earl's laboratory, most of them now leading their own flourishing research groups.

When I established my own monkey laboratory in Tübingen, I could not have done it without the help of Peter Thier, who generously shared his knowledge and research infrastructure with me. During this period, practical support from Uwe Ilg was invaluable. Of course, possessing a laboratory in itself does not produce any new data; it requires the hard and dedicated work of one's colleagues. I was privileged to work with many smart students who over the years embarked with me on this journey into the number brain. Without their essential contributions, most of the neurophysiological data on number neurons simply would not exist.

Of the many colleagues and friends in the "number business" who I was fortunate enough to meet over the past two decades, I am particularly grateful to Stanislas Dehaene. His continuing enthusiasm and interest in the single-cell correlates of numerical competence is a lasting source of insight and motivation. After many years of collecting neurophysiological data from non-human primates, I am most grateful to Florian Mormann for sharing his unique resources to record single-cell activity in the human brain. In some sense, my research agenda is coming full circle thanks to the human data collected with his help.

None of this research would have been possible without the generous assistance of several funding organizations, such as the Human Frontiers Science Program, the German Research Foundation, the Volkswagen Foundation, and the Federal Ministry of Education and Research in Germany.

I am also grateful to Michael Liapin; the preparation of this book greatly benefited from his close scrutiny. Warm thanks go also to Robert V. Prior, my editor at MIT Press, for his precious trust and support.

Last but not least, I am forever grateful to my family, my wife Bärbel, and my children Claudius, Philipp, and Vera, who have supported me unconditionally during the many months of writing this book.

A. N.
Tübingen, Germany
September 2018

Introduction

Life without numbers is inconceivable for us. How else would we count objects, tell time, calculate prices, and so on? Our scientifically and technically advanced culture simply would not exist without numbers. Numbers are fascinating because they constitute very abstract units of thoughts. When assessing the number of items in a set, the sensory appearance of the elements is meaningless. Three fingers, three calls, and three hand movements can all be classified by the cardinal number "three." Numbers are also intriguing to examine because, once extracted from sensory input, we use them to calculate and transform numerical information according to abstract principles. This is what arithmetic and mathematics are all about. Numerical operations therefore provide a "window of opportunity" to study the biology of numerical cognition. Ultimately, it gives rise to our number theory, a full-blown symbol system that parallels the language faculty in humans.

Comprehending and processing numbers is a form of behavior, and questions of behavior—how and why it emerges—can be addressed from different biological angles and at different levels. This was most vividly recognized by Nobel laureate Nikolaas Tinbergen, who formulated the four fundamental different types of problems in biology in his influential article, "On aims and methods of ethology," from 1963: the problems of "causation," "survival value," "ontogeny," and "evolution." With respect to number processing, these problems can be expressed as four questions: How does the brain give rise to our understanding of numbers and arithmetic? What is the utility to possessing number skills? How does numerical competence develop during the lifetime of an individual? And, finally, how did numerical competence evolve over the history of life on Earth? My aim in writing this book was to elucidate and attempt to answer all four of these questions from a biologist's point of view. However, the emphasis of this book is on how the brain and its neurons process numbers. This field of research has made significant progress over the past two decades and has yielded insights that gave rise to a new understanding of the evolution and the neural foundation underlying our grasp on numbers.

Naturally, each of the different disciplines contributing to our understanding of numbers has different research methods at its disposal and provides scientific knowledge at different, yet complementary, levels of explanation. In order for the reader to be able to appreciate the scope and limits of these various methods, as well as the different levels of explanations, these different methods will be briefly introduced along the way.

At the most microscopic scale, this book will show how studies with twins have begun to decipher the genetic basis of mathematical functions and disorders. To understand the workings of the brain, the book explores the neuronal code for numbers by recording the electrical activity of single neurons. These recordings, mainly in experimental animals, are complemented by brain imaging scans of human brains using positron emission tomography (PET) and magnetic resonance imaging (MRI). Much is to be learned about the causal role of neurons and brain areas when they are inactivated. While chemical inactivation and transcranial magnetic stimulation (TMS) transiently shut down neurons to influence calculation behavior, neuropsychological research studying patients with brain injuries historically also provided rich insights into the location and workings of number-processing brain areas.

An arsenal of behavioral methods is available to investigate how the minds of adults, infants, and animals represent numbers. These can broadly be classifiesd into methods exploring spontaneous behavior and those exploring behavior in trained individuals. Developmental psychologists mainly rely on explorations of the spontaneous behaviors in newborns and infants when they discover what the human mind is capable of without number experience. Exploitations of spontaneous behaviors are also important when learning about the ecological relevance of numbers for animals in the wild. Training experiments, on the other hand, investigate the behaviors of animals under controlled laboratory conditions to show the full range of numerical skills. As will be shown, working with trained animals offers the advantage of combining behavioral research with brain research. Comparing behavioral findings across the animal kingdom reveals how numerical competence emerged during the course of biological evolution. It turns out that processing numbers offers a significant benefit for survival, which is why this behavioral trait is present in many animal populations in the first place.

One of the key findings over the past decades is that our number faculty is deeply rooted in our biological ancestry, and not based on our ability to use language. This number instinct in its non-symbolic humble beginnings can therefore be traced through evolution and development. Numerical competence does not emerge *de novo* in humans, but builds on ancient biological precursors. Therefore, understanding the biological mechanisms of this evolutionarily and developmentally fundamental number instinct is instrumental to understanding our symbolic mathematical capabilities. Any scientific theory of number competence requires a mechanistic and functional, and therefore biological, explanation. The quest for this explanation is exactly the topic of this book.

Part I Conceptual Foundations

1 Thinking about Numbers

1.1 Mathematical Reality

Where do numbers come from? What sounds like a simple enough question is a fundamental controversy in the philosophy of mathematics, one that was first pondered by the Ancient Greeks. Since then, this question more than ever remains unresolved. There is overwhelming scholarly work devoted to the topic, but this is not the place to touch upon its philosophical underpinnings and challenges. The overall "big picture" of this question will, suffice to say, be just enough to cover the conceptual basis for this book.

Numbers have something to do with mathematics; that much is obvious to anyone. For the purpose of this book, we will be using Phillip J. Davis's and his co-authors' definition of mathematics (a definition that they call "naïve, but adequate for the dictionary and for an initial understanding"):

> Mathematics is the science of quantity and space, plus the symbolism relating to quantity and to space.[1]

The sciences of quantity and of space are generally known as *arithmetic* (from the Greek *arithmos*, "number") and *geometry*. They continue:

> Arithmetic, as taught in grade school, is concerned with numbers of various sorts, and the rules for operations with numbers—addition, subtraction, and so forth. And it deals with situations in daily life where these operations are used.

Easily said, arithmetic, a major branch of mathematics, involves numbers. In this vein, this book is about how the brain deals with arithmetic, or simply numerical quantities and the operations involving them.

But what kind of entities are numbers? To start, we must ask ourselves, do we *discover* numbers as objective realities, or do we *invent* them as products of our mind? This simple question has deeply troubled mathematicians to the core. As Mario Livio, author of *Is God a Mathematician?* expresses:

> If you think that understanding whether mathematics was invented or discovered is not that important, consider how loaded the difference between "invented" and "discovered" becomes in the question: Was God invented or discovered? Or even more provocatively: Did God create humans in his own image, or did humans invent God in their own image?[2]

Livio's comparison concerning the relationship between God and humans illustrates how charged this question is, and how entire believe systems depend on how it is answered. Mathematicians are torn on how to respond to this question and, subsequently, adopt two radically different philosophical positions[3]: The first position, termed *mathematical platonism* or *mathematical realism*, is the view that abstract mathematical objects, including numbers and sets, exist independently of us and our thoughts. Numbers have objective properties. We therefore *discover* numbers, just as we discovered the law of gravity and other laws of physics. The British mathematician Godfrey Harold Hardy (1877–1947), a proponent of mathematical realism, writes:

> I believe that mathematical reality lies outside us, that our function is to discover or observe it, and that the theorems which we prove, and which we describe grandiloquently as our "creations," are simply our notes of our observations. This view has been held, in one form or another, by many philosophers of high reputation from Plato onwards, and I shall use the language which is natural to a man who holds it.[4]

The second position, called *non-platonism* or *anti-realism*, comprises all opponents of mathematical platonism that include *formalism, fictionalism,* and *logicism*. Each is united by the overall idea that numbers and other mathematical objects don't exist as real entities and are independent from our mind. According to this position, we therefore *invent* numbers, just as we invent arbitrary games. American mathematicians Edward Kasner (1878–1955) and James Roy Newman (1907–1966) express this position when they write:

> Mathematics is man's own handiwork, subject only to the limitations imposed by the laws of thought. ... We have overcome the notion that mathematical truths have an existence independent and apart from our own minds. It is even strange to us that such a notion could ever have existed.[5]

Despite the metaphysical significance of the question and the heated debates among mathematical figureheads, a solution to this problem is not in sight. Even individual mathematicians have a hard time deciding which side they are on. They are doomed to a somewhat contradictory attitude, which is revealed in this excerpt from Davis and co-authors:

> The typical working mathematician is a Platonist on weekdays and a formalist on Sundays. That is, when he is doing mathematics he is convinced that he is dealing with an objective reality whose properties he is attempting to determine. But then, when challenged to give a philosophical account of this reality, he finds it easiest to pretend that he does not believe in it after all.[6]

One would most certainly not expect such a vague stance from mathematicians devoting their professional lives to exact sciences. But be that as it may, one thing is certain: Our survival depends on whether we successfully interact with our environment, which in turn is determined by how well we perceive the outside world, including the number of objects and everything around us. If we ignore the physical reality with numbers out there, we will definitely pay for it. Just as our ancestors might have fallen prey to the notorious saber-tooth tiger, today we just as well might get hit by a car or experience some other terrible accident. Surely, there is some sort of physical reality out there that we shouldn't ignore. And the organ that is responsible for preventing such accidents and ensuring our species' survival is the brain. The brain, in turn, is connected to sensory organs that provide us with perceptions as input, as well as bones and muscles that allow us to respond appropriately as output.

This book tries to portray the fascinating feats of a working brain that has had to deal with an objective reality during the course of hundreds of millions of years of biological evolution. Even if much of pure mathematics may just be a "meaningless game,"[7] my fundamental conviction and the basis of this book are that number is a property of real objects and events. By representing numerical quantity and arithmetic operations, the brain is gathering and processing information about the outside world to help its carrier survive in a hostile and competitive world.

1.2 Cardinal Numbers as Objective Properties of a Set

Numbers come in many varieties: natural numbers, rational numbers, real numbers, complex numbers, and others. This book is dedicated only to natural numbers. Just as Platonists and their opponents cannot settle the question of mathematical reality, philosophers of mathematics also can't agree on what numbers actually are. A number may be conceived as "multitudes of units,"[8] "nothing but names" or numerals,[9] a mental entity or projection,[10] and as several other ideas.[11] All of these interpretations have their respective advantages and problems.

In this book, however, I am adopting yet another position, namely the set-size view of numbers. It posits that a cardinal number is a real and *objective property* of a set. This view offers the most consistency with a biological conception of numbers. However, this view hardly originates with me. In fact, the English philosopher and empiricist John Locke (1632–1704) included numbers in his list of real (or primary) properties in his influential book *An Essay Concerning Human Understanding* (II.VIII.17)[12] dating back to the year 1690. As such, a cardinal number is independent of any observer, in contrast to subjective properties or sensations such as color or pain. For instance, the cardinal number *two* of protons and neutrons in the nucleus of elemental helium is a real property of its set; it determines what the element helium *is*, irrespective of our thinking about helium. Similarly, spiders and insects can be distinguished by the former having eight legs, whereas the latter possess six legs. And we know about cardinal numbers because of their instances—the number of siblings in the family, the number of syllables in a song—that we perceive. And from these instances, we infer an abstract category: "number of items."

This view of cardinal number as a real and objective property of a set presupposes that there is an external world that is independent of us (a philosophical position called *ontological realism*), and that the cardinal number is part of it. I, for one, consider this to be true. The physical world we inhabit is an objective reality. The universe, according to current knowledge, is ca. 13 billion years old; the planet earth only ca. 5 billion years old. In contrast, our species, *Homo sapiens*, has only existed on this planet since about 300,000 years ago. The physical nature we interact with has therefore existed long before man, and it will continue long after man has disappeared. If we consider the temporal and spatial dimensions of cosmic history relative to our own evolutionary history, it would be absurd to deny the existence and properties of physical nature as independent of human existence and experience.

Physical facts are objective and support realism in relation to an external world. It is the same with numbers: if a set contains three items, this set

Thinking about Numbers

contains three items irrespective of whether I watch it or whether I judge it to contain two items. My subjective experience of set size may be erroneous or illusory, but the set size exists objectively.

The notion of cardinal numbers as an objective and perceptually accessible property is eloquently expressed by the Austrian-born mathematician and philosopher Kurt Gödel (1906–1978), one of the greatest mathematicians of the twentieth century and probably the most important logician since Aristotle.[13] Gödel shook the world of logic when in 1931 he published his two *incompleteness theorems* (*Unvollständigkeitssätze*), which demonstrate the inherent limitations of formal axiomatic systems such as arithmetic. His intellectual skills deeply impressed another titan of mathematics, Albert Einstein. The two geniuses became close friends during their time at the Institute for Advanced Sciences at Princeton University after World War II. One anecdote states that the elderly Einstein told people he went to work "just to have the privilege of walking home with Kurt Gödel." The following excerpt exhibits Gödel's ideas on humankind's interaction with mathematical objects:

> But, despite their remoteness from sense experience, we do have something like a perception also of the objects of set theory, as is seen from the fact that the axioms force themselves upon us as being true. I don't see any reason why we should have less confidence in this kind of perception, i.e., in mathematical intuition, than in sense perception, which induces us to build up physical theories and to expect that future sense perceptions will agree with them. .....[14]

One can distill three important ideas from Gödel. First, he assumes that sets are real objects, thus expressing mathematical realism. Second, he assumes that we have an intuitive understanding of mathematical objects, a flair for mathematics, just as we have an intuition for the nature of sensory objects. Such a number instinct was also recognized by mathematician Tobias Dantzig (1884–1956) almost two decades earlier. In his book *Number: The Language of Science*, first published in 1930, he referred to this instinct as a "number sense." According to Dantzig, this number sense, which will play a dominant role in this book, is already present in some "brute species." Dantzig emphasized that this number sense should not be confused with counting ("an attribute exclusively human"), but is an evolutionary precursor for counting. Right at the beginning of his book, he explains:

> Man, even in the lower stages of development, possesses a faculty which, for want of a better name, I shall call *Number Sense*. This faculty permits him to recognize

that something has changed in a small collection when, without his direct knowledge, an object has been removed from or added to the collection.[15]

Today, the idea of an intuitive ability to assess numerical quantity is inextricably associated with the work of Stanislas Dehaene, a French mathematician and neuroscientist. Dehaene, whose studies will be mentioned throughout this book, is professor of experimental cognitive psychology at the Collège de France and director of the Neurospin facility near Paris. In his book from 1997 *The Number Sense: How the Mind Creates Mathematics*, Dehaene substantiated and popularized the idea of a number sense with an unprecedented wealth of empirical evidence both from his own work and the newly emerging field of mathematical cognition. In the introduction to his book, Dehaene posits: "This 'number sense' provides animals and humans alike with a direct intuition of what numbers mean."[16] Similar to our perception of time and space, he argues, humans also perceive numbers as a spontaneously accessible feature of the world. We don't have to learn to estimate quantity; we come born with a fundamental understanding, an innate instinct, of what numerical quantity *is*. This innate intuition of numbers is one of several systems of "core knowledge" human cognition is built on.[17] Human numerical cognition is built on this innate intuition both in ontogeny—that is, during the development of an individual—and in phylogeny, the evolutionary history of humans.

The third profound insight in Gödel's statement is his reference to perception. He calls mathematical intuition "a kind of perception." Nowadays, the study of perception is indeed firmly anchored in experimental neuroscience; much of what we know about behavior and the brain stems from an analysis of perception. With this reference, Gödel therefore predicts that the experimental sciences will have dominant authority regarding the nature of numbers. This idea reverberates throughout the chapters of this book, in which insights from evolutionary biology, neurobiology, and psychology are woven together to elucidate the biology of numbers.

1.3 Knowledge of Numbers

Earlier in the book, I claimed that the natural world and cardinal numbers are independent of us, thereby confessing *ontological realism*. But how can we even know about the world and numbers? We can only know that the world and numbers exist if we have the capacity to experience objective facts. A philosophical position that postulates such a capacity to experience objective facts is called *epistemological realism*, and I admit to this idea.

How can we be certain to know objective facts? This question leads us to the philosophical discipline of "theory of knowledge," also termed "epistemology." Its fundamental question is "What can I know?" While epistemology was done from a philosopher's armchair in ancient times, today any noteworthy epistemological framework has to reflect on what science tells us about our relation to the world. A theory of knowledge that emphasizes the role of natural scientific methods is called "naturalized epistemology," a term coined by American philosopher Willard V. O. Quine (1908–2000).[18] For Quine, similar to Descartes, the sciences first involved in epistemology were psychology and the physiology of perception.

In the following years, the Darwinian theory of evolution became another cornerstone of naturalistic epistemologies. This resulted in the rise of "biological" or "evolutionary epistemology."[19] Evolutionary epistemology is the attempt to address questions of "how can I know" from an evolutionary point of view. Philosopher Paul Thomson at John Carroll University even claimed that "Darwin's theory is potentially much more important for epistemology than psychology, information theory, or cognitive science."[20] The main theses of evolutionary epistemology are summarized by German philosopher Gerhard Vollmer:

> Thinking and knowing are capabilities of the human brain, and this brain arose during biological evolution. Our cognitive structures fit (at least partly) to the world because—phylogenetically—they emerged through adapting to the real world and because—ontogenetically—they have to cope with the environment of each individual.[21]

The biologist George Gaylord Simpson (1902–1984) used the following, more radical and provocative statement to summarize this stance:

> The monkey that had no realistic perception of the branch he was jumping for was soon a dead monkey—and did not belong to our ancestors.[22]

In other words, we owe our superb three-dimensional vision to our arboreal ancestors for whom this constituted a survival advantage. Since then, we are not only able to exploit three-dimensional vision for seeing branches, but also for monitoring manipulations of objects with our dexterous hands. In the same way, we owe our symbolic mathematical abilities to the non-symbolic number instinct that provided a survival advantage for our primate ancestors. Applying Simpson's logic to the concept of knowing numbers, it is clear that the monkey that had no realistic grasp of the number of food items was soon a starving monkey—and did not belong to our ancestors. Being able to assess the number of objects in a set was always a survival benefit for animals, and I will provide evidence for this claim in a later chapter. Animals with a realistic assessment of set size had a better chance to survive and were therefore selected for in a competitive environment, which in turn led to our ability to assess quantities in the course of evolution.

The same principle continues in our lineage. We are certain that $1 + 1 = 2$ because those previous hominids who believed in $1 + 1 = 2$, rather than $1 + 1 = 1$, survived and reproduced, and those who did not, were not evolutionarily successful. A hominid ancestor who had a ready grasp of elementary mathematics would be better suited for life's struggles than one who did not. Consider, for example, two prehistoric men, one of whom assesses number and the other who does not. They both sneak through a dense forest outside of their territory. Suddenly, they hear the war cries of several men. One prehistoric man exclaims, "Ah, it looks as though the inhabitants of this forest are here, but it is unclear how many." The other says nothing, but rapidly makes off. Which one of these two prehistoric men was more likely to be our ancestor?

Adaptations to the sense organs and the brain occur in the natural world, and in response to it. If the theory of evolution is correct (which is undisputed among biological scientists based on a plethora of evidence) and we possess innate and heritable "organs of knowledge," then these organs of knowledge are subject to the same major driving forces—genetic variation and natural selection—as all other organs and their resulting faculties. Like an instinct, the principles of mathematics are reflections of the innate dispositions that are wired into the brain of every mature, healthy human being.[23] This is why understanding the evolution and the physiology of the brain is the key to unlocking the secrets of arithmetic and mathematics.

2 Numerical Concepts, Representations, and Systems

2.1 Numerical Concepts

We use numbers in our lives for a myriad of reasons. From counting objects to encoding credit cards, numbers are incredibly flexible and can be applied to virtually everything that is imaginable, be it fact or fiction. So before delving into their biology, it is important to clarify some numerical terminology. The following sentence from a sports game nicely illustrates that numerical cognition encompasses different numerical concepts:

> Despite a *seventy-eight yard run* by *number thirty-four* the Bears lost by *two touchdowns* and dropped into *sixth place*.[1]

Three different concepts are expressed by the number assignments in the previous example[2]: The first concept is the cardinal number. It refers to a quantitative number assignment and, thus, applies to the empirical property "number of elements in a set" (figure 2.1). Cardinal numbers concern the questions "How many?" or "How much?" They refer to a discrete set size, such as "two touchdowns" in the sentence above, and continuous measures, as evidenced by "seventy-eight yard run."

The second concept is the ordinal number. It applies to the empirical property "serial order of an element in a sequence," or, simply, its rank (figure 2.1). For instance, the expression "sixth place" in the example sentence refers to an ordinal number. It concerns the question "Which one?" Ordinal numbers are inherently linked to cardinal numbers, because cardinal numbers are necessarily ordered according to their values: two precedes three, which in turn is followed by four, and so on. Cardinality therefore encompasses ordinality. As a result, cardinality is at the core of the number concept, and the prime concern of this book.

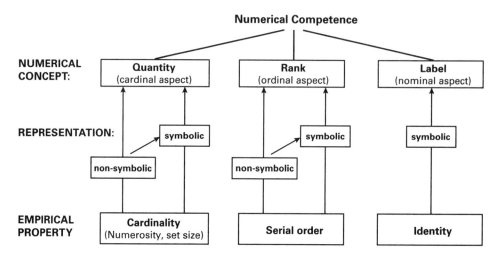

Figure 2.1
We represent object properties according to the three numerical concepts—quantity, rank, and label. The empirical properties "cardinality" and "serial order" are represented non-symbolically (non-verbally) and—uniquely in humans—also symbolically (verbally) via numerals and number words. The symbolic representations (partly) originate from the non-symbolic representation, as indicated by the connecting arrow. Numerical labels are mere linguistic identifiers and thus not "numerical" in a strict sense.

The third and final concept is the nominal number (figure 2.1). It identifies objects within a set. Nominal numbers are used like proper names, as expressed in identifying the player "number thirty-four." In a way, nominal number assignments are atypical because they are not really numerical, but rather used as adjectives. Therefore, nominal numbers are confined to the linguistics humans use.

In addition to these concepts, it is equally important to distinguish between the *empirical properties* that numerical concepts refer to, and the *representations* of these properties (figure 2.1). The next section clarifies what a representation is and discusses the different classes of representations that can be distinguished.

2.2 Mental Numerical Representations and Mental Systems

The term "representation" will be used a lot throughout this book, so it is important to clarify what I mean when I use this expression. Let's begin with mental representation first. A broad definition of "mental representation" is that of a "semantic mental object." In other words, this includes our meaningful thoughts, ideas, perceptions, judgments, and so on, which can be measured as behavioral output. The most common measurements of behavioral outputs are reaction times and the proportion of correct responses. Because of this measurable output of mental processes, mental representations can also be depicted graphically. For instance, if I ask you to tell me as quickly as possible which of two numbers is the smaller one, your reaction time will be faster or slower as a function of the numerical distance between the numbers, and these reaction time differences can be graphically depicted. Similarly, subjects can be correct most of the time when forced to decide which number of two very different numbers is smaller, but often err when the numbers only differ by one or two; again, the emerging percent of correct responses can be plotted as a function of numerical distance. Such functions constitute graphical reflections of mental representations, and I will use such graphical depictions of numerical representations later when discussing the signatures of animals' and humans' number discrimination.

When discussing the mental representations of numbers, the most fundamental distinction is the one between non-symbolic and symbolic number representations. They emerge from non-symbolic (e.g., arrays of dots) and symbolic (e.g., Arabic numerals) stimulus formats, respectively. The stimulus format refers to the way a number is presented, either as percepts of the actual number of elements or mediated by symbols. Parts II and III of this book will be concerned with non-symbolic representations in the mind and brain of non-human animals and humans, whereas Parts IV and V focus on uniquely human symbolic representations. How these two different number formats are represented, either format-dependent or independently, will be one of the focal points in the respective chapters.

Non-symbolic number representations are perceptual-like representations of set size, similar to an estimate of the number of dots in a dot array made without counting. Sometimes the adjective "analogical" is used to denote non-symbolic representations. In the number domain, this refers to a one-to-one correspondence between the represented number and the empirical number of items in a display or an event.

The number of items in a set that give rise to its set size, or numerosity, can be presented in different ways. When items are scattered in space (i.e., ∴), or across space, they can be assessed in parallel at one glance. In contrast, if items are presented one after the other in a temporal succession (i.e., ●- ● - ●, etc.), or across time, they need to be evaluated in sequence. These two simultaneous and sequential presentation ways differ fundamentally, and one of the questions of this book will be whether they address the same or different brain processes. In addition, the items of a set can address different senses. For instance, we can see dots, hear sounds, or sense touches on our skin. The concept of cardinality applies to all of these events, irrespective of sensory modality. Likewise, this book will also focus on whether number representations are modality-dependent or independent.

Non-symbolic number representations can be considered primordial number representations, both from an evolutionary point of view in the animal kingdom, as well as from a human developmental one. This type of number representation is all that animals have and will ever have available. Similarly, we humans only have non-symbolic number representations at our disposal as infants, before we master counting around age 4. Non-symbolic number representations also remain our only number representations if no one teaches us to count, or if we live in a culture that has not developed symbolic counting. The latter is shown by fascinating studies with indigenous people that demonstrate this idea. In numerate human adults, the approximate non-symbolic number representations are eventually overshadowed and replaced by exact symbolic number representations. However, non-symbolic number representations still exist under the surface of numerical symbols, and they can be brought to light with certain tricks.

Cognitive scientists discuss two separate mental systems that may allow animals, infants, and adults to represent number without symbols: the object tracking system and the approximate number system. The evidence for both of these mental systems will be discussed in more detail in Part II of the book. The object tracking system (OTS), also known as object file system, allows grasping only small set sizes, from one to about four, in an unconscious but relatively precise way. The workings of the OTS have been deduced from the behavioral characteristics that emerge when preverbal infants are confronted with small sets of objects. The idea is that the visual system singles out (or individuates) objects that it wishes to track by assigning "pointers" to them,[3] or by storing objects in "files." Each file can store precisely one object, and all files can store their items in parallel, or simultaneously, which is why the reaction time during discrimination from one to four items hardly increases. In addition, this process works without attending to individual objects; the objects are stored automatically and unconsciously.[4,5] However, because it is thought that the visual system only contains three or four of such files, a maximum of only four objects can be stored. This is referred to as the "set-size limit" of the OTS. As a result of this, larger numbers cannot be tracked. Importantly, however, the OTS is not dedicated to number representations; numbers are only implicitly, or unconsciously, represented because of filled-up files.

The OTS is thought to be responsible for a "subitizing" effect, from the Latin word *subitus* ("suddenly"). This effect describes an effortless, fast, and accurate process to judge a small number of items.[6] While real counting of larger sets of items takes about 200–350 ms per item,[7,8,9] one to four items can be judged with reaction times of only about 40–100 ms per item. This rather mild increase in reaction time per additional item is taken as evidence that the items of a set are assessed more or less at the same time, and with a mechanism that processes each item in a parallel manner. The rather sudden increase in reaction time beyond numerosity four has been interpreted as a switch from simultaneous non-verbal judgments to a serial symbolic counting strategy for higher numbers.

Originally, subitizing has been interpreted as a recognition of sets of dots as reoccurring figural patterns: one is a singleton, two a line, three a triangle, four a square. This idea is supported by evidence that rapid assessment could be extended up to seven or eight items if the patterns were familiar.[10,11,12] Moreover, participants rate different patterns of the same numerosity as more similar within the subitizing range (≤4 items) than outside of it (>5 items).[13] However, when all points are arranged collinearly or when objects are not simple dots but complex household items, the subitizing effect is still present. Because of this contradiction, the pattern-recognition account for subitizing has largely fallen out of favor. More recently, subitizing is explained by the OTS mechanism.

The second mental system for non-symbolic number judgments is termed the *approximate number system* (ANS; formerly known as analog magnitude system). The ANS is a quantity estimation system for an unlimited amount of set sizes. It represents the number of items consciously. However, as the name already indicates, it can only represent the number of items inexactly, or approximately. The ANS has two important behavioral characteristics, called the "numerical distance effect" and the "numerical size effect". Numerical distance effect refers to the finding that numerically distant numbers are easier to discriminate than numerically closer numbers. For instance, if the discrimination between 5 and 6 is very defective, the discrimination between 5 and 7 is more accurate and the discrimination between 5 and 8 is even better. On the other hand, the numerical size effect says that, at a given numerical distance, it is easier to discriminate numbers with low values than numbers with high values. For example, it is easier to discriminate 2 versus 3 than 8 versus 9, even though the numerical distance is 1 on both cases. In order to reach a similarly good discrimination performance as present in 2 versus 3, the numerical distance would have to increase in proportion with the magnitudes of the numbers, in this example to 8 versus 12. In other words, the ability to discriminate between quantities varies as a function of ratio, and is therefore considered "ratio-dependent."

It is easy to see that the discrimination of numerical values of sets follows systematic relationships. The German physician Ernst Heinrich Weber (1795–1878) was the first to discover such systematic relationships, not for number judgments, but for the sensation of weights.[14] He realized that we do not perceive the absolute weight difference between two objects, but the ratio of the weight difference. If one object weighs 100 g, the comparison object needs to weigh 110 g for us to just be able to discriminate the weights; the so-called *just-noticeable difference* is 10 g. However, for an object of 200 g, the comparison object has to weigh 220 g to tell the difference in weight. In this second case, the just-noticeable difference is 20 g. Weber realized that this systematic relationship could be captured mathematically for all possible weights by a constant: the just-noticeable difference, ΔI, divided by the reference weight, I, is a constant, c. More generally, the Weber fraction reads $\Delta I/I = c$. For the weight example above, the constant is 0.1, because 100 g / 10 g = 0.1 and 200 g / 20 g is also 0.1. This relationship is nowadays called Weber's law, and it not only applies to weight discrimination or other sensory intensities, but also to numerical discrimination. For example, if an animal is just able to discriminate 5 from 10 in 50% of the cases ($\Delta I = 5$; I = 10), then we can predict it will just be able to discriminate 10 from 20 ($\Delta I = 10$; I = 20). In both cases, the Weber fraction is 0.5 and therefore constant, because 5 / 10 = 0.5, and 10 / 20 = 0.5. In fact, Weber's law is a hallmark of the ANS.[15,16] The lack of a ratio effect is a distinctive signature that allows experimental differentiation of the ANS from the OTS. This pattern of a constant Weber fraction value across a broad range of numerosities emerges in studies with humans, non-human primates, and other animals, suggesting that the ANS is a basic mechanism available for representing numerosity.[17,18,19]

The precise value of the Weber fraction differs between species, and to some extent also between individuals. Small Weber fractions indicate the capacity to discriminate small numerical differences, whereas large Weber fractions suggest inferior number discrimination ability. The Weber fraction therefore is an objective measure for the precision with which numbers can be discriminated from other numbers.

If we continue to increase the number of visual items within a given area, we will reach a point where we can no longer resolve the items, and they will merge into what is commonly termed texture. Texture-density estimation is also discussed by some as a potential third mechanism of numerosity assessment. David Burr, professor of physiological psychology at the University of Florence in Italy, and coworkers[20] suggest that texture-density estimation becomes active if a visual scene becomes too crowded, for example as with a large flock of birds. In this case, the density becomes too great to segment the items from each other, and the scene is perceived as a texture rather than an array of elements. However, it is unclear whether this process deals with the number of items in a set or rather relates to sensory discriminations independent of number.

In contrast to non-symbolic, or analogical, number representations, symbolic number representation is characterized by the reflection of cardinality via different number notation symbols, such as Arabic numerals ("8") or numerical words ("eight"). Number symbols are signs that refer to cardinality, and importantly are part of a symbol system that has a compositional syntax and semantics. As part of a symbol system, number symbols are processed and transformed in a logical way based on rules in arithmetic. Because we usually use the language system to express number symbols, symbolic number representations are often called linguistic or verbal number representations. However, since the linguistic and the number systems are not the same, as will be emphasized in this book, I prefer the term symbolic.

Symbolic number representations can be based on different number notations, such as the written Arabic numeral "8" or the numerical name "eight." At the same time, symbolic number notations can address different sensory modalities, such as seeing a written numeral versus hearing a spoken numerical name. How these sensory modalities and number notations are represented will also be one of the topics of this book.

The transition from non-symbolic to symbolic number representations is one of the big mysteries in cognitive neuroscience. Even though hotly debated, I will later present several lines of evidence suggesting that our symbolic number representations—both ontogenetically and phylogenetically—actually build on non-symbolic number representations. Comparative psychologists have showed that animals can discriminate numerical information, and developmental psychology experienced a breakthrough when tapping into numerical cognition in human infants of only few months or even days of age. These developments indicated that numerical competence does not emerge *de novo* in humans, but builds up on a biological precursor system.[21,22,23] It is also obvious that our language faculty seems to play a role early in development, as it is conspicuous that children learn to count once they have mastered speaking. I will discuss some of the ideas in depth in Part V, which deals with development specifically.

Whatever the foundations, one thing is clear: once children do arrive at this symbolic stage, number representations are qualitatively transformed and massively enhanced to become the constituents of a full-blown and uniquely human number theory. Symbolic number representations provide us with the most exact quantity representations. Five is precisely five all the time, never four or six. Number symbols become part of a combinatorial symbol system and can be processed and transformed according to arithmetic rules. Symbolic number representations are the constituents of our science and technology. Number theory can be considered the other symbol system in addition to natural language. Without number theory, we would all still live in hunter-and-gatherer societies.

To sum up, the empirical property of "number of items" is represented in the mind by (non-symbolic or symbolic) *mental representation*. Evidently, all mental events are caused by the workings of neurons, or in other words observable physical processes in the brain. Therefore, a different type of cardinality representation needs to be addressed, one that tells us how cardinality is reflected at the neuronal level. This is indeed the fascinating "neuronal representation." Because neuronal representations give rise to the mind, there are of course no mental representations without neuronal representations, but mental and neuronal representations are far from the same. As I will show in the later parts of the book, the neural machinery giving rise to number representations has been deciphered over the past two decades down to the level of single neurons.

Part II Numbers Deeply Rooted in Our Ancestry

3 Understanding Numbers across the Animal Tree of Life

3.1 The Diversification of Animal Life

Even though we humans are rightly proud of our symbolic number and mathematical skills, we all start out with a non-symbolic ability to assess a given number of items without symbols. Where does this non-symbolic capacity stem from? Is it a uniquely human capability, or could it be an evolutionary heirloom from the past that was somehow genetically transmitted from our animal ancestors to be part of our brains?

In order to find out, we should go back in time and investigate our direct ancestors. Unfortunately, this is not possible. However, we can study modern animal species and work our way back based on the degree of relatedness between different species and ourselves. This evolutionary descent of animals and their corresponding relatedness leads to a phylogeny of animals, often depicted as a "tree of life." I will sketch out the most important steps during the evolution of major animal groups, or taxa, to explain the diversity and relatedness among animal life forms on Earth.

Let's start at the beginning in the truest sense of the word and consider some key events in Earth's history.[1] The origin of the solar system and planet Earth date back 4.6 billion years. The first primitive and single-celled living organism (a microbe) emerged relatively shortly thereafter, around 4.0 billion years ago. This postulated first life form that resembled a bacterium is called the last universal common ancestor, or LUCA, for it is apparently the one and only ancestor of all subsequent living species, including plants, fungi, animals, and yes, also us. For the next 2.5 billion years or so, the only living organisms were single cells, and all life was only present in the oceans. During the longest time of this early phase of Earth's history, there was almost no oxygen in the atmosphere, so life outside of water was not really feasible.

Only about 2.5 billion years ago, the amount of atmospheric oxygen shot up to about 10% of its present level. This so-called "oxygen revolution" had an enormous impact on life. Certain organisms took advantage of this abundant oxygen, because it allowed them to develop new and much more efficient ways to produce the chemical energy that all living beings need to survive. The main actors of this key evolutionary event were a special group of bacteria, called the cyanobacteria. They invented photosynthesis, a physiological process that uses the abundant energy of sunlight to convert carbon dioxide and water into sugar and oxygen. Photosynthesis is responsible for producing the oxygen content of Earth's atmosphere, and supplies all of the organic compounds and the energy necessary for life on Earth. In short, we humans and other animals breathe and eat the products of photosynthesis. We entirely depend on it.

Until about 1.5 billion years ago, only single-celled life forms existed. Then, some single cells somehow decided to join forces and cluster together to form a single organism that consisted of multiple specialized and interdependent cells. The current descendants of such multicellular life forms include plants, fungi, and animals. This division of labor at the cellular level opened up new possibilities.

But even with the advent of multicellularity and the first animal species, such as sponges, life forms were still primitive and simple. This was about to change roughly 600 million years ago, at the beginning of the geological period called the Cambrian.[2] All of a sudden, a huge variety of new life forms appeared and filled the seas with an astonishing diversity of animals. The number of animal groups and species seemingly exploded, which is why this key evolutionary event is termed the *Cambrian explosion*. All major animal groups that we see today—such as the arthropods, the mollusks, the chordates—emerged during the Cambrian explosion. The reasons for this huge radiation of animal species are intensively discussed.[3] It might have been a rise in temperature on Earth, an increase in the concentration of oxygen, or the advent of newly emerging predatory animals.[4] Some of the new, evolving animal species conquered the land, in another milestone in the history of life. And with the new lifestyle that divided animals into prey and predator, new behavioral and cognitive skills were necessary. As we will see, numerical competence is one of them, which is why it is spread throughout the animal kingdom.

Most animal species are characterized by a front end and a rear end (among other traits), and therefore are grouped as bilateral (two-sided) animals, or *bilaterians*. Bilaterians have diversified into two major lineages. On the one side are the *protostomes*, which maintain the embryonic primordial mouth as actual mouth in the adult animal. On the other side, we find the *deuterostomes*, which form a second, new mouth during embryonic development. Both protostomes and deuterostomes gave rise to big groups of animals, each called a *phylum*, a big limb on the tree of life.

Among the protostomes, the arthropods (insects, spiders, crustaceans), with over one million species, are by far the largest animal phylum of all animals. All arthropods have a segmented exoskeleton and jointed appendages. The central nervous systems of protostomians usually lack a pronounced centralized organization in favor of a more segmented one. Later on, this book discusses the numerical capabilities of several insects, such as honeybees and beetles. Keep in mind that they are only very distantly related to us, for they belong to the protostomes.

The deuterostomes contain the phylum Chordata (chordates), to which vertebrates belong. Among those vertebrates are humans and other animals. Vertebrates have a backbone and a more complex nervous system, enabling them to become more efficient at two essential tasks: capturing food and avoiding being eaten. The first jawed vertebrates, fishes, radiated into diverse vertebrate groups, or classes, classically (but taxonomically incorrectly) known as fishes, amphibians, reptiles, birds, and mammals.

In general, to identify different species, biologists refer to them by their Latin scientific names. The Swedish natural scientist Carl von Linné (Carl Linnaeus, 1707–1778) invented the modern system of naming organisms, called binary nomenclature. This means that each species is given a scientific name that is composed of two parts. The first part of the name identifies the genus, or smallest group of animals, to which the species belongs. The second part identifies the species within the genus. For example, the scientific name for us humans in *Homo sapiens*. This indicates that humans belong to the genus *Homo* (of which all other species are extinct by now) and within this genus to the species *Homo sapiens*.

Linné also grouped organisms into a hierarchy of increasingly inclusive categories, a hierarchical taxonomic system. The first grouping is built into the binary nomenclature: Species that appear to be closely related are grouped into the same genus. Beyond genera, taxonomists employ progressively more comprehensive categories of classification. Related genera are placed in the same family, families into orders, orders into classes, classes into phyla (singular: phylum). Finally, the kingdom of animals comprises all animal phyla. The resulting biological classification of an animal is somewhat like a postal address for species, able to home in on a particular apartment, or more widely on a building with many apartments, or even more widely to encompass their city, and so forth.

The evolutionary history of a group of organisms can be represented in a branching diagram called a *phylogenetic tree* ("tree of life"; see figure 3.1) The branching pattern indicates how taxonomists have classified groups of organisms nested within more inclusive groups. Biologists reconstruct the evolutionary history of a species or group of species, their "phylogenies," by classifying organisms based on their evolutionary relatedness. Organisms share many characteristics because of common ancestry. Common ancestry means that different species of animals all evolved from the same animal species millions of years ago. For instance, all insects evolved from a six-legged ancestor, and therefore the possession of six legs identifies an insect. Such a "derived" trait that emerged newly in one common ancestor and has then been passed on to all its descendents is called a "homologous" trait; it establishes common descent. There is much to learn about a species if we know its common evolutionary history. For example, an organism is likely to share many of its genes, brain structures, and even behavioral traits with its close relatives. The study of numerical cognition benefits from a phylogenetic perspective, because many of the capacities we see in modern species have not evolved in these species originally, but have already evolved in their ancestors and have been passed on to the modern species from there.

However, specific traits may also emerge independently in animals, without a last common ancestor that possessed this trait and passed it on. Flight, for example, was invented independently in insects, birds, and bats. This is called an "analogous" trait; it does not indicate relatedness and only occurred by means of "convergent evolution" because it provided a survival advantage for these animals. Numerical competence also seems to have emerged multiple times in different, only remotely related animal groups.

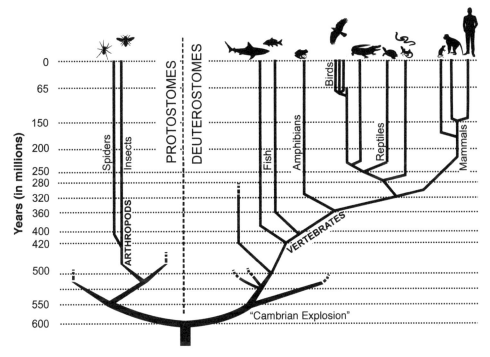

Figure 3.1
Simplified tree of animal life displaying animal groups discussed in this book.
The left side of the tree shows insects and spiders (of the phylum Arthropoda) as representatives of the protostomes that diverged from the deuterostomes on the right side of the tree about 600 million years ago. The right side of the tree depicts the traditional five vertebrate groups: the fish (shark and carp), the amphibians (frog), the birds (crow), the reptiles (crocodile, turtle, lizard, and snake), and the mammals (rat, monkey, human).

As we will see, this will tell us something about the importance of numerical competence for diverse groups of animals.

3.2 The Theory of Evolution

What causes the evolutionary history of species or group of species—in other words, their phylogeny? The answer about the mechanisms behind biological evolution are provided by the theory of evolution, first developed by Charles Darwin (1809–1882) in his landmark work from 1859 *On the Origin of Species*.[5] The theory of evolution is a grand unifying theory in biology that is supported by an overwhelming amount of scientific evidence. It explains the existence of diverse species, their relatedness in phylogeny, their adaptedness to their environments, and their continued evolution. The theory of evolution places humans within the context of phylogeny. Our human brain and the fascinating numerical capabilities it endows us with is a current complex endpoint (not a goal!) of the evolutionary history of species.

In essence, biological evolution posits that species accumulate heritable differences, called "genetic variations," generation by generation. In the long run, this results in descendants that differ from their ancestors in many features. New species with different bodies and brains arise from existing ones over many generations. Darwin used the term "descent with modification" to describe this evolutionary phenomenon. Importantly, individuals themselves do not change and evolve during their lifetime, but populations of individuals change through time, from one generation to the next.

In his time, Darwin could not know how this heritable information was laid out in organisms. However, today we know that each cell of the body contains the genetic information for the entire building plan of an animal, called "genome." The specific implementation of the genome in each organism, the "genotype," determines its "phenotype," which is how an organism looks, works, and largely behaves. Like letters of the alphabet, the genotype is written down in a sequence of four molecular bases called adenine (A), thymine (T), cytosine (C), and guanine (G), on a molecule called deoxyribonucleic acid (DNA). This universal genetic code is found in all living beings. Based on this code, DNA contains units of information, called "genes." Put simply, a **gene** is a snippet of DNA that contains the instructions to make a specific protein. Different genes are needed to make different proteins, and such proteins are needed to build and maintain a body and its organs. Usually a specific gene comes in two different modes, called "alleles," one provided by the mother, the other by the father. For instance, a flower may have a gene that codes for the color of its blossom, with the two alleles each coding for either white or red color. Depending on whether an individual flower carries the alleles for white or red color, its blossom will look accordingly.

The question now is what causes genetic variation, and thus heritable differences from one generation to the next. The ultimate source of genetic variations are mutations of the genetic code that give rise to new alleles. Usually, such mutations are detrimental and cause diseases or genetic incompatibilities, but sometimes they can give rise to functioning new features.

But variation alone does not explain why certain alleles and genes are more frequent in one population than in another. In addition to genetic variation, "natural selection" is a second key process in biological evolution. Natural selection determines how frequently, and if at all, certain alleles and genes are passed on to the next generation. This is because the environments these species inhabit are also constantly changing, which requires an adaptation of the phenotype by means of the genotype. When a particular kind of phenotypic trait (such as an anatomical feature, a physiological process, or a behavior) allows its carrier to be more successful in life than others and therefore is of "adaptive value," such traits will be selected for and their underlying genes will become more frequent in the next generation.

Natural selection is the only mechanism that produces adaptive evolutionary changes. Importantly, natural selection acts at the level of the living and behaving individual, not its genes. This is significant, because now we have a mechanism that specifically takes cognition and behavior into account when it comes to the question of which traits are passed on to the next generation. Cognition is a starting point for natural selection. Certain cognitive capabilities may be passed on to the next generation precisely because they result in an increased number of offspring surviving in the next generation. For instance, if an animal that is able to discriminate set size is more successful than others in producing offspring (and thus has greater *fitness*), the genetic basis that gives rise to this cognitive capability is selected for. This is how numerical competence can be maintained in a population. The evidence for the adaptive value of numerical competence is presented in more detail in the next chapter, which deals with the utility of number for animals.

The evolution of new cognitive capabilities may all seem very goal-driven from the outside.However, the theory of evolution is emphatically non-teleological. There is no goal toward which evolution is working. It was never a goal to have numerically competent animals and humans on this planet; it just happened because those animals grasped something relevant about the world and thus had an advantage.

Moreover, it is important to keep in mind that evolution is not an optimizing, but only a sufficing process. All it takes for a heritable trait to be selected for is the trait's ability to confer some advantage to its possessor; this trait need not be the best imaginable, or even the best available. In the realm of numbers, precise counting abilities would surely be advantageous compared to the only approximate numerical skills found in non-human animals. However, optimality is a human concept, not an evolutionary one. If approximate estimation is sufficient for an animal to survive, then that is what will be selected for.

As pointed out in the first chapter, the theory of evolution also speaks to epistemology, the question of how we and other animals know anything about the world. If survival depends on how well an animal is able to interact with its environment, and it surely does, then the process of natural selection favors those brains and minds that rightly inform about "how the world is." It is obvious that evolutionary success must correlate with how well reality is perceived, however constrained this percept may seem, and this determines the life and death of an organism. This is the core of evolutionary epistemology, as addressed in the first chapter. We and animals can represent cardinal numbers because they are a significant feature of the world, and assessing number increases fitness and survival. Numerical competence as a way of knowing something important about the world thus arises out of biological evolution.

3.3 Classic Studies on Animal "Counting"

So can animals deal with numbers? This was first postulated more than a hundred years ago. In 1904, Wilhelm von Osten introduced his stallion Hans, later called "Clever Hans," which he had trained to an amazed audience in Berlin, Germany (figure 3.2). This horse seemed to be able to count and calculate. When von Osten wrote down a calculation, say 2 + 3, on a chalk board, Hans would tap with his hoof exactly five times and then stop. And Hans would also solve much more complicated arithmetic problems. Thirteen eminent scientists of that time suspected a circus trick but, despite intensive investigations, could not detect a hoax. So they certified in a much-acclaimed recommendation that Hans was able to do arithmetic. Of course, he could not.

Figure 3.2
Wilhelm von Osten with his horse Clever Hans (around 1908).

Later, the psychologist Oskar Pfungst investigated Hans once more. Pfungst found that the horse could get the correct answer even if von Osten himself did not ask the questions. This ruled out the possibility of fraud. Importantly, however, Hans got the right answer only when the questioner knew what the answer was, and the horse could see the questioner. If the questioner didn't know the result, or was out of sight, Hans constantly failed. Pfungst concluded that Hans used unintentional body cues, slight changes in the posture, facial expression or breathing pattern of the questioner when the correct result was reached, to stop hoof tapings at the right moment. Pfungst wrote[6]

> It was he [the questioner] who gave the directions, and since all that were involved were visual signs, the drama in which Hans appeared as the hero, was nothing but a pantomime.

After Hans was exposed a swindler, he and von Osten retreated from public. No one knows what happened to Hans after 1916, when in the midst of World War I, he was sold to the military. For psychology, however, Hans was responsible for a truly important finding called the "Clever Hans effect." It reminds behavioral scientists how important it is to exclude unintentional cueing in any behavioral experiment with animals and humans alike. The gold standard of behavioral experiments since then is to separate the experimental subjects from the experimenter during the testing.

Even if horses certainly are no mathematicians, in defense of horses's honor, I have to add that more recent experiments showed that they do have a rudimentary capacity to discriminate quantity. When untrained horses (*Equus caballus*) have the option to choose between numbers of apples placed in two buckets, they chose two apples over one, and three apples over two, but have difficulty with four apples versus six.[7]

After the "Clever Hans" failure, the question of numerical competence in animals needed a brave scientist to address it anew. This courageous man was the German zoologist Otto Koehler (1889–1974), who provided the first definitive evidence that animals can be trained to discriminate between the number of objects in a set.[8] While most European zoologists of that time, such as Nobel laureates Konrad Lorenz, Nikolaas Tinbergen, and Karl von Frisch, studied innate behavior, Koehler used operant conditioning to train animals on controlled tasks.

Koehler demonstrated convincingly that birds and mammals can discriminate numerical quantities.[9,10] For example, he tested jackdaws' (*Corvus monedula*) judgment of simultaneously presented items using a matching-to-sample task, in which the sample number was indicated by the number of ink dots (or pebbles and plasticine beads). The subject's task was to find the one of two possible test lids that showed the same number of dots, and lift it to find a food reward in the pot underneath it (figure 3.3). Koehler classified this capability as "simultaneously seeing the number of items" and discriminated it from a second type of numerical competence he termed "successively acting upon the number of items." He tested the second, sequential capability by, for instance, training Amazon parrots (*Amazona spp.*) to peck a certain number of grains from two piles of grains. For example, a parrot trained on "five" could eat all three grains from a small pile and two additional grains from a second, larger pile, before flying off and leaving the rest of the grains untouched.

Of course, Koehler knew about the devastating Clever Hans story. To avoid unconscious cueing of the animals, the factor that proved to be the demise of von Osten and Clever Hans, the experimenter was positioned invisibly to the animal during the sessions. An automatic spring-loaded device shooed the birds if they made errors. In addition, the experimental sessions were videotaped and thoroughly analyzed offline. Most notably,

Understanding Numbers across the Animal Tree of Life

Figure 3.3
One of Otto Koehler's jackdaws being tested for the ability to match numbers of items. The small squares are lids on food pots. If the jackdaw flips over the lid with the same number of dots as on the large card, it will find food underneath (from Koehler, 1941).

Koehler introduced transfer tests in which feedback for the birds' behavior was removed so that the birds could not learn what the correct answer was but had to derive it from their own conceptual knowledge. During such transfer tests, the birds successfully applied the concept of numerical discrimination to novel situations.

Koehler also was very aware of the problem of non-numerical features in displays which the birds could use to solve the tasks. For instance, increasing the number of same-sized ink dots would also increase the amount of blackness on the white display cards. Surely, animals can learn to discriminate more from fewer blackness, but this has nothing to do with number. So Koehler carefully varied and equated the sizes and positions of the dots on the lids for simultaneously presented items, and the temporal sequence of the items in the successive tasks. None of these controls distracted the birds' discrimination abilities, thus demonstrating that they truly relied on numbers to solve the tasks.

Over the years, Koehler and his students tested nine species in the numerical competence project—pigeons, budgerigars, jackdaws, ravens, Gray parrots, Amazon parrots, magpies, squirrels, and humans. His work served as the basis for all following investigations on non-verbal numerical competence. Koehler made it very clear that the capacity he discovered in animals was not counting, but a non-symbolic (or non-verbal) progenitor of it.

> Our birds did not count, for they lack words. They could not name the numbers that they are able to perceive and to act upon, but in actual fact they learn to "think unnamed numbers."[11]

In the wake of Koehler's pioneering studies, a menagerie of different animal species has been investigated for numerical skills to this day. I will skip most of the earlier studies, but classical "animal counting" literature after Koehler includes the review by Davis and Perusse[12] from 1988.

3.4 How to Test Animals on Numerical Cognition

Before tracing numerical skills within the animal tree of life, I want to introduce the basic experimental approaches that researchers have at their disposal to learn about animals' number skills. In behavioral experiments with humans, we can simply talk to the subjects and let them know what we need them to do. With animals it is obviously different, and we cannot instruct them in the same manner. How could they know that a researcher wants them to discriminate the number of objects? This, by the way, is exactly the same problem in behavioral experiments with pre-verbal human infants. Fortunately, behavioral protocols have been designed that can do this without language. In order to learn about animals' numerical competence, two main test protocols have been established that do not require verbal reports from the subjects. One of the protocols is the spontaneous-choice test, which, as described later, also helps developmental psychologists test infants and toddlers. The second approach is behavioral training based on operant conditioning. Both methods have their advantages and limitations,[13] as outlined below.

Spontaneous-choice tests take advantage of the natural tendency of an animal to prefer more or less of something that is of value to it. Such naturally valuable stimuli can, for example, include food or social partners. In a typical experiment that can either take place inside a laboratory or out in the wild, the animal is presented with two sets, each containing different numbers of such relevant stimuli. The assumption underlying these tests is that if animals are able to discriminate between the two quantities of the presented items, they are expected to choose the most advantageous set over the less preferred one. If test sets consist of food items, for example, they should pick the larger quantity because animals usually have a preference for more food.

One obvious advantage of spontaneous-choice tests is that they don't require time-consuming training of the animal. They can also be done in the wild. In addition, spontaneous-choice tests can evaluate the ecological validity of numerical competence: If animals can spontaneously discriminate stimuli that are important to them in an experimental setting, they are likely to also apply this capacity when facing similar situations in their habitat.

However, spontaneous-choice tests also have significant drawbacks. First, when using valuable items, such as food, it is unavoidable that the number of items is confounded with other, non-numerical and continuous magnitudes. Put simply, when, for example, the number of apples increases, the overall appearance of the sets also changes systematically. More apples need more space (or volume), individual apples may be packed more densely, and so on. An animal may discriminate sets by paying attention to volume, surface area, or density to pick "more food" rather than "more food items." Because of problems like these, some authors go so far as to claim that nonsymbolic number discrimination is nothing but a discrimination of continuous quantity, such as space or amount, and not numerical quantity.[14] Spontaneous-choice tests cannot really get rid of the problem of continuous magnitude. However, to conclude that numerical quantity discrimination is an experimental artifact is unjustified based on complementary studies with well-controlled stimuli, which will be shown later.

A second confound that appears in discriminations between items that are of inherent value for the animals is the amount of reward. If we stick with the apple example, it is easy to see that many apples are more desirable to an animal than a few, simply because the animal will get more satisfaction from eating many apples. However, we don't want to know whether an animal can discriminate between magnitudes of reward, or "hedonic value"; we want to know if an animal can discriminate between the numbers of objects.

A third problem that cannot be underestimated is the motivation of animals to choose at all between sets. All spontaneous-choice tests rely on the assumption that the tested animal indeed wants to discriminate between two sets to obtain something. However, this may simply not be the case. An animal that has just fed has no desire to acquire more food. This is a particular problem in the wild, where the nutritional status of animals is unknown.

Finally, there are limits to the kinds of numerical competence that can be tested with spontaneous-choice tests. Numerical discrimination in untrained animals can only address the simplest numerical ability, namely relative numerosity judgments. Relative numerosity judgment means that the animal does not assess the absolute set size, but only one set size relative to another. This can be done exclusively with "more than" or "less than" judgments, respectively; no appreciation of the actual, or absolute, number of items is required. Put another way, relative numerosity judgments are possible without cardinality assessments.

The aforementioned problems in spontaneous-choice tests can be avoided by training animals. Training is based on a very powerful learning mechanism, called operant conditioning, that Otto Koehler applied in his previously mentioned classic studies. Operant conditioning goes back to Edward L. Thorndike (1874–1949) and was elaborated by B. F. Skinner (1904–1990), a professor of psychology at Harvard University. Here, an animal learns by trial and error that a specific stimulus (called the "discriminandum") is associated with a specific action that is required from the animal for it to receive a reward or avoid punishment. For example, a rat can be trained to press a lever (the response) whenever a light flashes (the discriminandum) to get a reward (the consequence).

In operant conditioning, reward is the driving force of performance. Animals need to be motivated to work in order to receive something that is of value to them. Usually food or fluid is used as a reward. As a consequence, food and water intake of the experimental animal needs to be controlled to be of value prior to the actual experiment. Of course, it is ensured that the animal receives the amount of food or water, either during or after the experiment, they need to stay healthy.

In training experiments, animals can perform hundreds of trials per session to result in a statistically solid data set, something that would be impossible in spontaneous-choice tasks. Confounds by non-numerical stimulus features are avoided because animals in the lab usually do not discriminate naturalistic objects, but arbitrary items on a computer screen, such as dots or geometric objects. These arbitrary items are specifically designed by computer software so that non-numerical parameters can be equated and controlled for. As an additional advantage, such arbitrary display items are also of no intrinsic value for the animals. Moreover, animals always receive the same amount of reward for a correct response, which precludes hedonic confounds. Lastly, animals are kept under a controlled feeding or fluid intake schedule prior to the experiment, so an animal's motivation is controlled and constant in each session.

An important advantage of training is that a multitude of numerical concepts can be investigated based on the specific task design. Animals cannot only be trained to perform relative numerosity judgments, which is the only concept that can be addressed in spontaneous-choice tests, but also to perform absolute numerosity judgments, ordinal numerical judgments, or switching between numerical rules. With conditioning, animals can also be trained to associate numerosities with signs, as shown later in this book. This is exactly the same process children have to undergo when learning to understand Arabic numerals and number words.

It is obvious that the two approaches, spontaneous-choice test and training by operant conditioning, have specific advantages and are therefore both necessary for our understanding of numerical cognition in animals. For me, the advantages of the training approach predominate, which is why my research group trains our monkeys and crows to discriminate numerosity. Sometimes it is said that training may recruit other cognitive capabilities to support numerical discriminations that would otherwise not be active under "natural" conditions.[15] However, there is a flaw in this argument's logic. If these "other cognitive capabilities" support numerical competence, then they are numerical capabilities—that is, exactly the type of capabilities that are being investigated. Training approaches can take animals to the limit of what they can accomplish with their numerical competence. It is important to explore these limits because the numerical capabilities an animal may show in certain situations in the wild will only be a fraction of its full cognitive powers.

Leaving the pros and cons of these approaches in purely behavioral research aside, there is yet another distinctive advantage of working with trained animals, and one that will become very important throughout this book. This is the advantage of combining sophisticated behavioral tasks (and the corresponding skilled performances of the animal) with brain research. The controlled behavior that animals exhibit in laboratory experiments can be directly related to simultaneous neurophysiological recordings, which constitutes the most direct way to learn about brain functions. This advantage will be elucidated more when discussing number neurons and how they give rise to numerical competence. But first, let's examine which groups of animals are numerically competent.

3.5 The Phylogeny of Numerical Competence

In the following sections, I will classify the animal protagonists of recent studies on numerical competence based on phylogenetic relationships. In doing so, I will not dissociate between the two testing approaches, spontaneous-choice or behavioral training studies, since the question is simply which groups of animals have been tested successfully. It will be amazing to see how widespread numerical cognition really is (figure 3.4). Let's begin with vertebrates, which humans also belong to.

Fish. Fish belong to the oldest group of vertebrates. While fish were left unexplored for a long time, the past decade has seen a wealth of data showing that fish have rudimentary numerical capabilities.[16] Most of these insights were gathered by Christian Agrillo and Angelo Bisazza from the University of Padua.

Fish have mainly been tested using spontaneous-choice tests, and primarily using social stimuli. Since many fish do not possess any defense mechanisms, the best that they can do to avoid being eaten by predators is to hide among themselves. When individual fish are inserted in an unfamiliar environment, they tend to join with other conspecifics. If two shoals of conspecifics are present, individuals tend to join the larger shoal for better protection.[17] Using such shoal-choice behavior, it was found that several tested fish species, such as mosquitofish (*Gambusia holbrooki*), guppies (*Poecilia reticulata*), angelfish (*Pterophillum scalare*), and swordtails (*Xiphophorus elleri*), can approximately discriminate the number of fish in optional shoals. Usually, the fish can discriminate 1 versus 2, 2 versus 3, and 3 versus 4 fish. As expected of numerosity estimation based on the approximate number system (see chapter 2), the numerical ratio between the quantities needs to be increased for larger numbers, but in doing so fish can discriminate 5 versus 9, or 8 versus 16 fish.

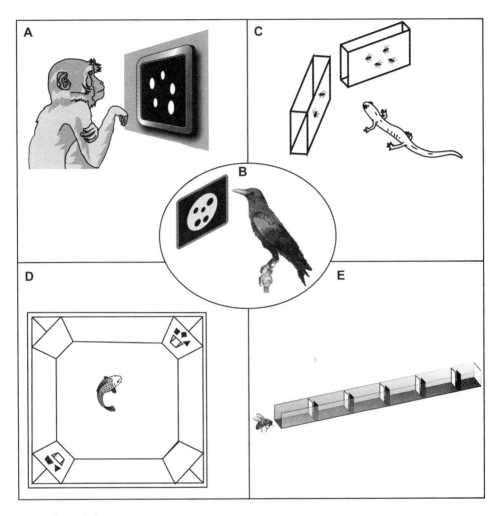

Figure 3.4
Laboratory studies have demonstrated numerical capacities in diverse animal species. A) Monkeys discriminate and memorize the number of dots in computer displays to earn a reward (Nieder et al., 2002). B) Crows assess the number of items in dot arrays on a computer monitor to obtain food (Ditz and Nieder, 2015). C) Salamanders spontaneously select the larger group of potential prey (Uller et al., 2003). D) Fish can be trained to discriminate two from three figures to rejoin their shoal mates (Agrillo et al., 2011). E) Bees are able to estimate the number of landmarks in a flight tunnel to reach a food source (Dacke and Srinivasan, 2008).

Fish are very visual creatures and can also be trained to select artificial items to receive a reward. As outlined earlier, training protocols have the significant advantage of motivated animals learning to discriminate stimuli, which can be controlled for non-numerical factors. Agrillo and coworkers trained mosquitofish to discriminate between different numbers of geometric images.[18] First, the fish were removed from their social group and placed in an unfamiliar fish tank. Since fish don't like to be alone, they try to rejoin their friends, and they could do so by passing through one of two identical tunnels at opposite corners. The entrance of the tunnels was marked with pictures of either two or three geometric objects. Only one of the doors, the one showing the rewarded number, allowed the subject to rejoin its companions (figure 3.4D). For instance, the door displaying three objects allowed access to the social partners, but the door showing two objects did not. After some training, the fish could discriminate between two and three figures based on number alone, when all other non-numerical parameters were controlled for. The fish also managed with larger numbers, such as 4 versus 8.[19] Later studies showed that mosquitofish also learn to discriminate between numbers of objects if they are rewarded with food.[20]

While these experiments tested the fish's discrimination of visual numerosity, a more recent experiment investigated quantitative abilities in blind cavefish of the species *Phreatichthys andruzzii*.[21] They were trained to discriminate between two groups of three-dimensional sticks placed in opposite sides of the experimental tank in order to receive a food reward. When the blind cavefish were trained with stimuli that controlled for continuous quantity, they were able to discriminate between two and four sticks. While visual input can obviously be excluded in blind cavefish, it is not clear which sensory modality they may have used to discriminate stimuli.

Amphibians. After some fish evolved their fins into more limb-like appendages and experienced some other major bodily changes over time, the first tetrapods, the amphibians, conquered the land about 365 million years ago. Amphibians are rather humble vertebrates whose behavioral repertoire basically consists of a set of reflexes. Therefore, amphibians have exclusively been tested using spontaneous-choice tests to discriminate food.

The first study was conducted with salamanders (*Plethodoncinereus spp.*).[22] Salamanders presented with two transparent boxes containing differing numbers of fruit flies were able to select the larger group in 1 versus 2 and 2 versus 3, but not 3 versus 4 choice situations (figure 3.4C). Later, the role of the total activity of prey in quantity discrimination of salamanders was investigated.[23] Live crickets, videos of live crickets, or animated images on a computer monitor were used as stimuli. When no control was made, subjects discriminated the larger group in an 8 versus 16 contrast. However, salamanders chose randomly when the total amount of movement of the stimuli was controlled for. This finding suggests that the total activity of prey, not the number, is a dominant feature in salamanders' foraging behavior. Finally, toads, a small frog (*Bombina orientalis*), have been tested for their capacity to discriminate small food items.[24] In this study, systematic controls for variables such as surface area, volume, weight, and movement were included. Toads showed quantity discrimination in the range of both small (1 vs. 2, 2 vs. 3, but not 3 vs. 4) and large numbers (3 vs. 6, 4 vs. 8, but not 4 vs. 6), even when continuous physical variables were controlled for.

Reptiles. Reptiles were the first vertebrates to truly conquer the dry land about 310 million years ago. Reptiles in a classic and narrow sense comprise lizards, snakes, turtles, and crocodiles. They possess several specializations that ensure neither their bodies nor their eggs dry out on land. Reptiles can be ferocious hunters with acute senses, but they do not have a reputation of being intelligent. One physiological reason for this is probably due to the fact that their body temperature largely fluctuates with the temperature of the environment. When it is cool, they become inactive, and what good is intelligence if it can only be used in certain temperatures? No wonder, then, that the numerical capabilities of reptiles have hardly been investigated. However, there is one exception to this general lack of studies, although it is not flattering for the reptiles, because it suggests reptiles are the black sheep in the animal counting family.

When ruin lizards (*Podarcis sicula*) were tested in a standard free-choice test with food items, they failed to discriminate the number of food items.[25] Whether they were presented with 1 versus 4 (ratio 0.25), 2 versus 4 (ratio 0.5) 2 versus 3 (ratio 0.67), or 3 versus 4 (0.75) food items, they picked any number of items with equal probability. A later study in trained ruin lizards resulted in similarly disillusioning findings.[26]

This failure in a reptile species presented a true exception in numerical cognition studies. Either ruin lizards are the numerically challenged outliers among reptiles, or reptiles in general can't handle numerical quantity. The latter explanation would contradict the idea that all vertebrates share a core number system that is inherited by a common ancestor. The common ancestor of all vertebrates, including reptiles, is a primitive fish-like creature (figure 3.1). Interestingly, however, different species of fishes have passed number tests multiple times. Of course, fish have also evolved over hundreds of millions of years since the last common ancestor of all vertebrates was alive, and one cannot exclude that fish independently acquired the cognitive trait to discriminate numerosity. This, however, would mean that the common ancestor of vertebrates was unable to discriminate between numbers and all successfully tested vertebrate classes such as fishes, amphibians, birds, and mammals invented this ability independently. Hopefully, more reptiles will be tested with different protocols in the future. I think the chances are favorable to find other reptile species that can discriminate numerosity, which would allow us to put reptiles back in the club of numerically competent vertebrates.

Birds. In a strict zoological sense, birds are also reptiles (figure 3.1). Birds descended directly from dinosaurs, and any modern bird can be considered a feathered dinosaur.[27] At the same time, birds evolved many special features that also justify them being considered a separate vertebrate class called Aves. Among these special features are those involved in flying, such as feathers, lightweight bones, and the most efficient lungs of all vertebrates. Birds evolved about 160 million years ago, later than mammals, and are therefore the most recent class of vertebrates. Birds internally maintain a constant high body temperature; they are "endothermic." Endothermy allows birds to maintain a high metabolic rate and a very active lifestyle. Endothermy therefore correlates with unusually large brains and elaborate cognitive behaviors. It is therefore probably not an accident that birds, alongside mammals, belong to the cognitively most advanced vertebrates.

Ever since the pioneering work of Otto Koehler, described above, it is known that birds possess elaborate quantification skills. In their natural habitat, wild American coots (*Fulica americana*) and brown-headed cowbirds (*Molothrus ater*) use numerical information to counteract nest parasitism, which will be covered later and discussed for its adaptive value.[28,29] New Zealand robins (*Petroica longipes*)[30,31] perform relative quantity judgments to go for more during foraging, and the same is true for jungle crows

(*Corvus macrorhynchos*)[32] and jackdaws (*Corvus monedula*).[33] The number of sounds produced is also important to birds. Black-capped chickadees (*Poecile atricapillus*) use the number of "dee" notes in their well-known "chick-a-dee" alarm call to indicate to others the level of danger in predators; the more ferocious the predator, the more "dee"-notes.[34] This behavior of chickadees will also be covered later.

Birds, such as the African grey parrot (*Psittacus erithacus*),[35] pigeons (*Columba livia*),[36] and different species of crows,[37,38,39] have also been trained in laboratories to distinguish arbitrary stimuli based on the absolute number of items, or to order numerosities according to their numerical rank.[40] The numerical competence of crows (figure 3.4B) and their corresponding neural correlates will be touched on later.

One bird in particular to examine is the house chicken (*Gallus gallus domesticus*) and, more precisely, their chicks. The young domestic chick is an extremely precocial species, meaning that they are up and ready to roam around and provide for themselves only a few hours after hatching from the egg. This allows scientists to analyze their quite sophisticated behaviors right after they hatch from their eggs, devoid of sensory experiences. Echoing this concept, chicks also show filial imprinting a few hours after hatching. A chick rapidly learns to follow the first salient, moving object it sees. Usually, the first moving object would be the mother chicken, but chicks have been shown to also follow all sorts of objects, ranging from humans to soccer balls. In addition, chicks naturally prefer to approach larger numbers of social partners (or other objects) that provide protection from predators and necessary warmth. Giorgo Vallortigara, a professor at the University of Trento in Italy, and his team took advantage of this behavior to study numerical cognition. In their experiments, chicks right after hatching are first exposed to different numbers of artificial imprinting objects. By later observing the animals in spontaneous choice tests with sets of different numbers of objects, their numerical abilities can be explored.

In a series of experiments,[41] separate groups of chicks were imprinted on one, two, or three imprinting objects. More often than not, the chicks preferred three objects when tested in choice experiments. The chicks' preferences for the larger number was related to imprinting, because choice in a group of chicks tested without any imprinting did not yield any preference for a larger number of objects.

However, even basic arithmetic capabilities seem to be present from birth in chicks.[42] These tests exploited the fact that chicks seem to prefer a large number of imprinting objects. First, chicks were reared with five identical objects. During the test, a chick was first confined to a specific location, from which it could see different sets of imprinting objects disappear behind an opaque screen. One set of two disappeared behind one screen, and another set of three disappeared behind the other screen. When the chicks were left free, they inspected the screen occluding the larger set of three objects.

This experiment was extended to beyond just the initial disappearance of the two sets behind the two screens. Some of the objects were visibly transferred, one by one, from one screen to the other before the chick was released. Even in this case, chicks picked the screen hiding the larger number of elements. Obviously, chicks can calculate a series of subsequent additions and subtractions of objects to keep track of the larger set of imprinted objects. No wonder Giorgo Vallortigara refers to chicks as "natural-born mathematicians."[43]

Mammals. Mammals, together with birds and reptiles, belong to a group of advanced tetrapods called Amniota, because they all share a special embryonic membrane called the amnion. The first mammals evolved during the Jurassic Period about 200 million years ago from reptilian-like ancestors (figure 3.1). Mammals share many novel (derived) traits, such as mammary glands to produce milk for their young, hair, differentiated teeth, three middle ear bones, and a six-layered cerebral cortex. Mammals are endothermic, like birds, who invented this trait independently. And just like birds, they have large brains and a very active lifestyle.

Mammals diversified into many groups or "orders," in zoological terms, including: Rodentia (rodents; e.g., mice and rats); Cetacea (cetaceans; e.g., whales and dolphins); Carnivora (carnivores, flesh-eating mammals such as cats and dogs); Perissodactyla ("odd-toed" or "odd-hoofed" mammals, such as horses and rhinoceroses); Proboscidea (mammals with long trunks, such as elephants), and, of course, Primates. The order Primates, which humans also belong to, comprises about 350 different species. Lemurs, bush babies, macaques (including the well-known rhesus monkey), chimpanzees, and humans are all primates with opposable thumbs or opposable big toes. Our upright-walking human-like ancestors, the hominids, and the ancestors of chimpanzees diverged from a last common ancestor around 6–7 million years ago. And from the line of hominids, our own species, *Homo sapiens*, originated more than 300,000 years ago.[44,45]

Not surprisingly, the numerical capabilities of mammals have been studied most extensively. Almost all of these studies explored basic relative quantity judgments, primarily using spontaneous-choice tests. Members of the order carnivora that have been tested successfully include dogs,[46] cats,[47] bears,[48] lions,[49] hyenas,[50] sea lions,[51] and raccoons.[52] Other animals in different orders, such as dolphins[53,54] and whales,[55] horses,[56] and elephants[57] have also been studied. Some of the earliest publications testing rodents demonstrated numerical competence in trained rats.[58,59]

In a classic study from 1958, Francis Mechner[60] trained rats to press a lever a certain number of times—either four, eight, 12, or 16 times—before switching over to press another lever in order to obtain a reward. After hundreds of trial repetitions of the same target number, he plotted the distribution of presses the rats performed on the first lever. He found that the number of presses formed a bell-shaped curve around the respective target numbers (figure 3.5). The rats most frequently pressed the correct target number, the peak of the function, but also pressed fewer or more times, thus making systematic errors that progressively disappeared with the numerical distance to the target number. This is the well-known numerical distance effect previously alluded to in chapter 2. Mechner graphically depicted the rats' behavior to create what can be called the mental representations of specific non-symbolic numbers. The bell-shaped functions indicate how a given number is distinguished from all other numbers in the animal's behavior.

Besides rodents, most mammalian species that have been studied belong to the order Primates (figure 3.4A). When ordered from phylogenetically primitive to advanced primates, the list of investigations starts with lemurs[61] from Madagascar that belong to the most basal primates. Among the more advanced monkeys in a narrow sense are the squirrel monkeys[62] and the capuchin monkeys[63,64] originating from South America. Old-world monkeys from Africa include rhesus monkeys[65,66,67] and the baboons.[68] Apes are the most advanced non-human primates, with three species examined: orangutans,[69] gorillas,[70] and chimpanzees.[71,72] Of course, by no means is this an exhaustive list of species that have been investigated to assess quantitative abilities. However, the above do represent the major zoological

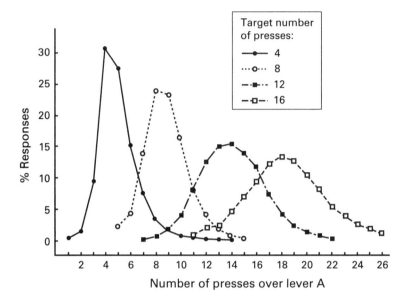

Figure 3.5
Rats learn to press a lever an instructed number of times. Groups of rats were trained to press lever A either four, eight, 12, or 16 times. The rats approximately match the respective target number, which results in bell-shaped performance curves centered around the target number (adapted from Mechner, 1958).

groups necessary to grasp the phylogenetic breadth of research in this area of number study. Monkey behavior will be examined more deeply later on in descriptions of number discriminations in animals and the neurophysiological basis of numerosity.

Arthropods (insects and spiders). While numerical cognition traditionally has been studied almost exclusively in vertebrates, several insect and spider species have also been investigated in more recent years. Insects and spiders are particularly interesting, for they are the "aliens" in the animal kingdom. Namely, they possess an external shell, all sorts of weird appendages, and strange sensory organs compared to other animals. Humans and all other vertebrates are very distantly related to insects and spiders, which belong to the by far largest animal phylum—arthropods (i.e., animals with jointed appendages).

The last common ancestor of vertebrates and arthropods lived at least 600 million years ago, some time before the Cambrian Explosion, which can be considered an eternity from an evolutionary point of view (figure 3.1). This extremely long time of separation provided ample opportunities for the evolution of independent body features. Besides being bilateral animals—in other words animals with a front and a rear end—we really do not have a lot in common with arthropods. The relatively simple nervous system of arthropods has a very different building plan, and it arose independently of the vertebrate nervous system. In contrast to vertebrates, which have the central nervous system at the back (dorsal), arthropods have it toward the belly (ventral). Due to their small body size, arthropods also possess only a small number of neurons. A honeybee, for example, has fewer than one million neurons in its brain, compared to the 86 billion neurons that make up our human brain. Nonetheless, all these differences did not prevent bees and other arthropods from developing numerical competence.

Researching arthropods offers another, more psychological advantage that has to do with our own prejudices. When working with monkeys and apes, we are at great risk of overinterpreting and anthropomorphizing other primates' skills. We tend to think that just because they look familiar to us, their minds must also operate like ours.[73] However, insects and spiders obviously look nothing like us. Because we humans hardly see ourselves in insects and spiders, we don't automatically assume that what is going on in their "minds" is identical to what is happening in ours. We therefore can analyze arthropod behavior much more objectively than other animals' behavior.

By now, there is evidence that spiders spontaneously use numerical cues during foraging. Communal spider-eating spiders (*Portia africana*) base their decision of settling near a prey spider's nest on the number of conspecifics already present; they prefer one conspecific spider over zero, two, or three spiders.[74] This peculiar behavior will be described in more detail in a later chapter about the utility of numbers. Besides spider-eating spiders, sit-and-wait spiders also seem to rely on numerical cues. Golden orb-web spiders (*Nephila clavipes*) accumulate prey pantries on their webs. If they lose their pantries, the spiders search for them, and their search time increases proportionally to the prey counts they lost (but also to the prey mass they lost).[75] This is suggestive evidence that these spiders have some rudimentary memory of prey numerosity.

Most of the evidence for numerical competence stems not from spiders, but from different species of insects. The 17-year periodical cicadas (*Magicicada spp.*) live up to their name because their larvae assess the seasonal cycles of host trees from which they sap sugar underground to determine the hatching time point after precisely 17 years. Miraculously, after these 17 years have passed, all individuals at any location emerge simultaneously within several days of each other. Experiments show that these cicadas assess the number of years rather than the passage of absolute time.[76]

In beetles, chemical number discrimination has been demonstrated. Mealworm beetles (*Tenebrio molitor*) can discriminate between odor bouquets containing the scents of different numbers of female beetles in a two-choice situation.[77] Ants, a group of hymenopteran insects, also seem to show aspects of numerical competence. Red wood ants (*Formica sp.*) were reported to estimate the number of turns in a maze, and even to communicate this information to nest mates.[78] Unfortunately, this experimental design was not able to disentangle whether wood ants really used numerical information, or rather distance information, which ants are known to measure quite accurately. The desert ant (*Cataglyphis*) uses some sort of "step counter" mechanism to measure distance in order to locate its hive after returning from foraging trips.[79] Experiments showed that the ant overestimated travel distance after an artificial elongation of the ant's legs (via stilts), whereas shortening the legs (causing stumps) made them underestimate the distance. However, it is not known whether the number of strides is measured directly or indirectly by integrating the magnitude of leg-position–sensor (proprioceptive) feedback.

Nonetheless, the most amazing insect of all from a cognitive point of view is the honeybee. Honeybees collect nectar and pollen from near and far and return these goods to their hives. In order to measure the distance from the hive and later return, bees seem to memorize the number of landmarks they pass on the way to a feeding site. In 1995, Lars Chittka and Karl Geiger from Queen Mary University of London[80] established a row of four yellow tents and trained bees to collect sugar water at a feeder between the third and fourth tent. When the number of the tents and the distances between them were changed, bees used the absolute distance flown, but also, additionally, the number of landmarks passed to navigate. Later in 2008, Marie Dacke and Mandyam Srinivasan from the Australian National University[81] provided clear evidence that honeybees can sequentially enumerate landmarks. The authors trained bees to forage near one of five landmarks in a 4-meter-long tunnel (figure 3.4E). The bees indeed searched in the vicinity of the learned landmark number, even if the appearance of the landmark was changed. These findings suggest that bees sequentially enumerate the number of variable landmarks, and in doing so show signs of abstraction. As expected of approximate number representations, the bees' performance became more imprecise when more landmarks had to be memorized. Importantly, however, their numerical abilities reached a limit of around four landmarks; they failed if the number of landmarks was larger than four.

If bees can assess the number of sequential stimuli, can they also judge the number of items at a glance? In 2009, a group from the University of Würzburg in Germany[82] trained bees to perform a working memory task in which they had to memorize and match displays that contained the same number of dots. During the baseline training, bees saw a certain number of dots on a card at the entrance of a Y-maze. They had to memorize the numerosity while flying through the maze toward the fork of the corridor. At the bifurcation, they had to choose the maze arm that showed the same numerosity as displayed at the entrance in order to receive a sugar reward. During training, only numerosity 2 and 3 were displayed, and the bees eventually performed up to 75% correctly. Confirming hypotheses of abstract assessments, variations of the displays had no impact. No matter whether the position, the color (from blue to yellow), or the shape of the items (from dots to stars) was changed, or whether the edge length and cumulative area of the items was controlled, the bees still mastered the task. Only when four or more items had to be discriminated did the bees fail and performance drop to chance level. This limitation mirrored the finding by Dacke and Srinivasan that bees' sequential landmark enumeration was also limited to four. Both studies seem to show that bees have a set size limit of about four items. However, new research demonstrating that bees can also assess larger numerosities is in progress.

A very recent study by Scarlett Howard and Adrian G. Dyer from Monash University in Australia demonstrated even more astonishing number skills in honeybees.[83] Honeybees can rank numerical quantities according to the rules of "greater than" and "less than," and they extrapolate the "less than" rule to place empty sets next to a set of one at the lower end of their "mental number line." For their experiments, Howard and her team lured free-flying honeybees from maintained hives to their testing apparatus and marked the insects with color for identification. The bees were rewarded for discriminating displays on a vertical rotating screen that showed different numbers of items, ranging from zero to five. The systematic changes in appearance of the changing number displays were controlled for, thus ensuring that the bees actually discriminated numbers rather than low-level visual cues. First, the bees were trained to rank two numerosity displays at a time. One group of bees was rewarded with a sweet sugar solution whenever they flew to the display showing more items, thereby following a "greater than" rule. The other group of bees was trained on the "less than" rule and was rewarded for landing at the display presenting fewer items. The bees learned to master this task with displays consisting between one and four items, not only with familiar numerosity displays, but also for novel displays. Surprisingly, the bees obeying the "less than" rule spontaneously landed upon the occasionally inserted and unrewarded displays showing no item (i.e., an empty set). Bees' competence in dealing with empty sets will be explained in the last chapter, which focuses on the number zero.

Honeybees possess amazing numerical skills that rival those of many vertebrates. Honeybees have a reputation of being insect geniuses: not only can they enumerate and order numbers, but they also possess elaborate working memory to ponder about upcoming decisions,[84] understand abstract concepts such as "sameness" and "difference,"[85] and learn intricate skills from other bees.[86] And they achieve all of this with fewer than one million neurons. Even outside of the number domain, it is impossible not to be fascinated by these little creatures on the other side of the animal tree of life.

To sum up, numerical competence, at least in its most rudimentary form, is omnipresent in the animal kingdom. Perhaps with the exception of reptiles (but I doubt it), number skills are found in all vertebrate groups: fish, amphibians, birds, and mammals. This suggests that all vertebrates inherited basic numerical competence from some primordial fish-like creature that constitutes the last common ancestor of all vertebrates as a homologous trait. Alternatively, the last common ancestor might not have possessed number skills, but rather the different vertebrate groups acquired it independently of each other through convergent evolution over hundreds of millions of years. One also has to be aware that the cognitive brain structures, most notably the endbrain, that represent numerical information evolved dramatically and quite distinctly in some of these vertebrate classes. More on this will be said in part III of this book. Whether neuronal numerical representations shifted from one brain area to another or whether newly evolving brain areas endowed these species with numerical competence in the first place is still an open question.

3.6 Signatures of Animal Number Discrimination

Although Otto Koehler could show that animals can discriminate the number of items, he never characterized discrimination performance in detail. Only the classic work by Francis Mechner[87] mentioned previously provided more detailed insights about numerical representations in lever-pressing rats (figure 3.5). It showed how well rats discriminate a target number of lever presses from more or less presses. The rats' behavior is representative of and can be seen in all tested animal species. The following section demonstrates that animals, most notably non-human primates, cannot only produce one specific number, but that they can also discriminate between various numbers of items, even in visual and auditory sets. With this behavior, some animals show a surprisingly abstract understanding of numerical quantity.

When we hear "numerical quantity," we most likely picture visual items that make up a set. In comparative cognition, arrays of visual items, such as black dots, are therefore the most common type of numerosity stimuli. In this presentation format, numerosity can be estimated at a single glance in a direct, perceptual-like way from a spatial arrangement of stimuli.

In fact, this was the kind of stimuli Herbert S. Terrace and his then-PhD student Elizabeth M. Brannon from Columbia University used in a pioneering study from 1998 with rhesus monkeys.[88,89] Brannon and Terrace unequivocally demonstrated that monkeys can discriminate between computer-generated images that contained as many as nine objects and can respond to them in ascending order. With this experiment, they also showed that some behavior in animals clearly goes beyond the simple stimulus-response associations that famous behaviorist B. F. Skinner, Terrace's mentor at Harvard, though animals were confined to.

The stars of this experiment were two male rhesus monkeys named Rosencrantz and Macduff. Brannon and Terrace trained them on a math quiz by presenting them with 35 sets of images on a touch-sensitive video screen (figure 3.6A). Each picture contained a different number of distinct objects

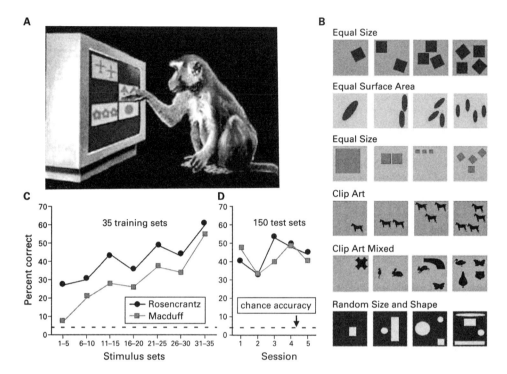

Figure 3.6
Monkeys order numerosities displayed on a touch screen. A) A drawing of a monkey responding on the touch screen. **B)** Exemplars of different types of stimulus sets used by Brannon and Terrace (1998). *Equal size:* elements were of same size and shape. *Equal area:* cumulative area of elements was equal. *Random size:* element size varied randomly across stimuli. *Clip art:* identical non-geometric elements selected from clip art software. *Clip art mixed:* clip art elements of variable shape. *Random size and shape:* elements within a stimulus were varied randomly in size and shape. **C)** First, monkeys learned with fixed sets or stimuli to order numerosities one to four in ascending order. The graph shows the percent correct performance for 35 training sets. Each was presented for 60 trials and each data point reflects the average of five sessions (300 trials). **D)** After learning, the monkeys were also immediately able to order new sets of test stimuli. The graph shows percent-correct performance for 150 trial unique test sets tested in five test sessions with 30 trials each. Chance accuracy is less than 4% in this task (from Brannon and Terrace, 1998).

from one through four, such as one triangle, two bananas, three hearts, and four apples (figure 3.6B). The stimuli appeared in random positions on the screen in order to prevent the monkeys from learning the required sequence as just a series of fixed motor movements. Other features of the pictures unrelated to number, such as size, surface area, shape, and color, were also varied randomly. Such feature variations and controls made it impossible for the monkeys to use non-numerical cues for stimulus discrimination; they were left to only judge the number of objects. The task for the monkeys was to touch the pictures in ascending numerical order— for instance, one square, two trees, three ovals, and four flowers. If successfully accomplished, they received a banana-flavored food pellet as reward. If they made an error, the screen turned black for several seconds, and a new trial then began with different pictures.

Over the course of learning 35 different training sets, the monkeys got better and better at responding in the ascending numerical order (figure 3.6C). Of course, one may say: they have learned the sequence of specific pictures. Would the two monkeys really understand the numerical rule for ordering the number of items in the pictures? To find out, Rosencrantz and Macduff were confronted with 150 new and trial unique stimulus sets, showing numbers of objects from one to four. Indeed, the monkeys could transfer the numerical ordering rule to novel stimuli (figure 3.6D).

But could the monkeys also apply the ordering rule to novel numerosities they had not experienced during training? To find out, Rosencrantz and Macduff were faced with pictures containing up to nine items. Again, the task was to first touch the picture containing the smaller number of objects, then the one with the larger number of objects. For example, if a monkey was shown one picture with five large circles and another containing seven small circles, the correct order was five, then seven. Rosencrantz and Macduff responded correctly, even though the number of objects in the pictures was never shown before and exceeded those they had been trained on. They could transfer the numerical ordering rule to novel set sizes. There was no way Rosencrantz and Macduff could have done this, unless they had learned some numerical rule for ordering the contents of the pictures. It was clear that the monkeys knew things about numbers that Brannon and Terrace hadn't taught them.

Elizabeth Brannon continued and extended this line of research in her own lab at Duke University. Together with her PhD student, Jessica

F. Cantlon, she characterized monkeys' number discrimination behavior and compared it to humans' ability performing the same visual numerical ordering task. When adult humans and monkeys were given a task in which they have to rapidly compare two visual arrays and touch the array with the smaller numerical value (without counting the dots), their performance was surprisingly similar and shows the ratio-dependency expected from Weber's law. If the ratio between the numerical values is low, for instance comparing two to 32 dots (ratio 0.06), the performance is perfect. However, when 27 and 32 dots need to be compared (ratio 0.84), the resulting performance is close to chance. The explanation for this performance pattern is that both humans and monkeys are representing the numerical values in an approximate way using the approximate number system.

When I joined Earl K. Miller's laboratory at the Massachusetts Institute of Technology (MIT) in 2000 as a postdoctoral researcher, my main intention was to study the neuronal correlates of number judgments in monkeys. For this purpose, the monkeys first had to be trained to discriminate between numbers of items. However, we wanted to take a different behavioral approach. Rather than training monkeys to compare the relative numerosity of two or more displays, as Brannon and Terrace did, we wanted them to assess and memorize the absolute numerical values in visual dot displays. This gave us the opportunity to analyze detailed numerical representations of randomly selected numerosities.

We did this with a special type of delayed response task known as a delayed matching-to-sample task; its temporal layout can be seen in figure 3.7A. We showed monkeys one set of dots on a computer monitor, and then, after a one-second interval, showed another set of dots. The monkeys were taught over many weeks of training that if the second set was equal to the first set, releasing a lever that they had grasped to start a trial would earn them a reward in the form of a sip of apple juice. If the second set was not equal to the first, which randomly happened in half of the trials, then the monkeys were not allowed to respond; instead, they had to wait for a third set which always matched the first before releasing the bar to earn apple juice. So the odds of being rewarded were 50% if the monkeys did not pay attention. However, since the monkeys really wanted the apple juice, they did their best to discriminate the wrong, non-matching numerosity from the correct, matching numerosity. Whenever one such trial was finished, a new trial started with a different set of dots.

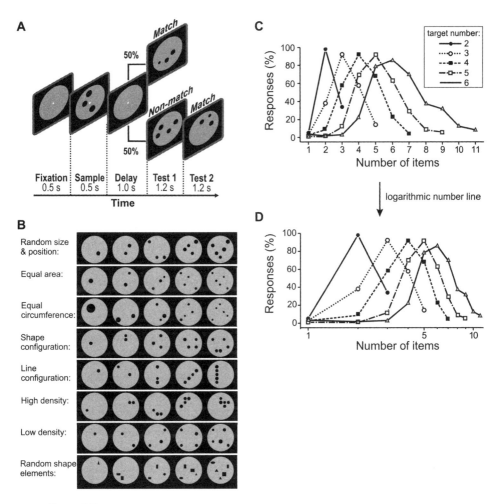

Figure 3.7
A typical experiment in which a monkey must decide whether two sets of dots sequentially displayed on a computer monitor are matched in quantity or not. A) Delayed match-to-sample task layout with numerosity three as example sample stimulus. B) Examples of different types of stimulus sets used by Nieder et al. (2002). C) Behavioral performance functions of two monkeys averaged. The individual functions represent the probability with which the monkeys judged numerosities as being equal to the respective target numerosity from two to six. The resulting bell-shaped performance functions are skewed on a linear number axis. D) When the same behavioral performance functions are plotted on a logarithmically compressed number axis, the bell-shaped curves become symmetric (A, B from Nieder et al., 2002; C, D from Nieder and Miller, 2003).

Just like Brannon and Terrace before us, we took extra care to ensure the monkeys could not use the visual appearance of the item arrays to solve the task. Thus, every numerosity was presented with 50 different dot arrays per session so that the monkeys could not learn these patterns, and the arrays were made anew for each behavioral session using a specific computer software. In addition, the sample and test displays were never the same to prevent the monkeys from matching visual patterns. Moreover, non-numerical visual features, such as the total area, the total circumference of dots, or the overall density or geometric arrangement of the dots, were kept constant in control trials (figure 3.7B). None of these changes caused problems for the monkeys. This showed that they abstracted the quantity of items from visual displays that varied widely in appearance, and then held that information in their memory over a short delay. The monkeys did their best to estimate the number of dots they had seen. And so, after hundreds of computer-controlled trials and random repetitions of target numbers, we could explore how monkeys discriminated the cardinal values of numerosities.

We found that the monkeys' responses formed a bell-shaped curve around the respective target numbers (figure 3.7C), the monkeys' mental representation of that non-symbolic number, similar to the result found by Mechner for lever-pressing rats (figure 3.5). In contrast to Mechner's approach, however, our monkeys were not trained to discriminate only one specific numerosity from all others, but any numerosity between one and five from all others. The monkeys most frequently detected the correct target number at the peak of the function. However, they also responded to the non-match numbers that had similar values as the target numbers; only when the numerical distance between matching and non-matching numbers increased did they make fewer and fewer errors.

However, one property of the bell-shaped performance curves was conspicuous: the functions were not symmetric bell curves; instead, they were slightly and systematically distorted when plotted on a linear number axis (figure 3.7C). Clearly, the slopes were shallower for numerosities higher than the target numerosity (i.e., the center of each distribution) than for numerosities lower than the target. The ideal bell-shaped curve, the Gauss function, did not fit the performance functions well, at least not in the way the number line was scaled. It is known that similar asymmetric performance functions arise for the discrimination of simple sensory magnitudes. The German Gustav Theodor Fechner (1801–1887), when elaborating on

Weber's insights, discovered that our sensation of physical magnitude scales not linearly with physical magnitude, but with the logarithm of physical magnitude.[90] Fechner realized that our subjective sensation of stimuli, S, is proportional to the logarithm of the physical stimulus magnitude, I. This relationship ($S = k \cdot log(I)$), is known as Fechner's law. This means that, in order to be able to judge that a numerosity is just noticeably larger than a preceding one, increasing numerical distances, or just noticeable differences, are required. If true, this would result in skewed performance functions with steep slopes toward smaller numerosities, and shallow slopes toward larger numerosities.

That is precisely what we observed in the monkeys. We took exactly the same performance data of the monkeys and plotted them on a logarithmically compressed number axis (figure 3.7D). And, sure enough, the functions became nicely symmetric, and they matched a Gauss function significantly better.[91] Because the logarithm of the responses is normally distributed, numerosity representations are also called "log-normal" distributions. Abstract numerical representations shared the same fundamental principles that Fechner had postulated more than 150 years earlier for sensory magnitudes. The mental number line for non-symbolic numbers is logarithmic. These deep similarities indicate that the brain applies similar strategies when encoding sensory and cognitive magnitudes.

Humans can apply two strategies to assess the number of dots in arrays. We can either count them serially, one by one, which would mean that we enumerate individual items, which in turn would take us more time for each additional item. Alternatively, we can estimate set size at one glance, which would uniformly take about the same amount of time across sets of different sizes. But how did the monkeys solve this issue? Interestingly, the discrimination of higher numerosities did not take them longer compared to smaller ones.[92] In addition to this, the monkeys did not look at each individual items in the displays before making a decision. Because we monitored their eye movements with an infrared-camera tracking system throughout the trial, we could see that they consistently only made an eye movement to one of the dots in each set before releasing the lever to indicate a matching numerosity. Both findings indicated that they assessed numerosity at one glance. They were not sequentially enumerating the items, but encoded all individual items in parallel, which explained why more items in a set did not cause longer reaction times or more eye movements.

In the initial set of experiments at MIT, we had the monkeys work with only small numbers of one to five. Of course, we wanted to see if they could also manage novel and much larger numerosities. Years later, in my laboratory in Tübingen, my PhD student Katharina Merten and I extended the delayed-matching-to-sample task.[93] Initially, we trained the new monkeys to discriminate numerosities of one to five. After proficient performance with those, one monkey was confronted with novel numerosities six, seven, and eight in transfer trials, in which it was randomly rewarded and therefore could not learn the correct response; the monkey had to infer how to treat the novel numerosities from what it had learned with the smaller set sizes. Much to our delight, the monkey continued to discriminate these novel numerosities with a similar accuracy to the previously learned small set sizes. The monkey understood what numerical quantity meant. To further demonstrate an abstract knowledge of quantity, both monkeys in this study only trained on small numerosities were abruptly confronted (i.e., from one day to the other) with numerosities ranging up to thirty. Once again, both monkeys showed spontaneous generalization to novel large numerosities; they had a concept of non-symbolic number and knew what cardinality means. In addition, we found precisely the same logarithmic scaling for the broad range of numbers—and not only in monkeys, but also in humans that we tested with the same protocol.[94]

Journalists often ask "How high can a monkey count?" With this question, they are asking about the limits of an animals' discrimination ability, or, simply, which numerical value they can tell apart from the immediately adjacent one. For the very small numerosities one and two, the average discrimination accuracy in monkeys is close to perfect. However, performance declines rapidly with increasing values. In our laboratory, the discrimination threshold was found to be between four and five items in monkeys. Thus, with a minimum numerical distance of one , the animals failed for four versus fives and higher.

This, however, is not the whole story, because animals can also discriminate larger numerosities provided the numerical distance between them is large enough. In this case, their performance recovers systematically. The performance functions for both small and large numerosities showed the numerical distance effect. In addition, the performance curves also became broader with increasing target values, a reflection of the numerical size effect. Clearly, when assessing non-symbolic number, both monkeys and humans follow Weber's law.[95]

The concept of numerosity applies to two very different ways of presenting the individual items. I can put three apples together on the table, or I can show them one by one. The first instance is a simultaneous presentation, discussed in the previous section; the second is a sequential presentation. Even though in both cases we are dealing with the number three, how the individual items are perceived is very different. In the simultaneous presentation, the items are gathered in space, whereas in the sequential presentation they need to be collected individually over time. What makes sequential enumeration particularly interesting is that real counting is a sequential process.

Species from different animal classes, such as fishes, birds, and mammals, can enumerate small numbers of sequentially presented items. As discussed earlier, even honeybees could enumerate sequentially when they kept track of a small number of landmarks encountered one after the other during flight. In most of these studies, the animal had to detect only one specific number of sequential items—for instance, they had to always respond to three light flashes. Very few studies have tested whether animals can represent different absolute numbers of sequential events.

In preparation for electrophysiological recordings, my laboratory trained two rhesus monkeys in a delayed matching-to-sample task to judge one to four sequentially presented sample dots on a computer monitor, and then match them after a brief delay period to the number of elements in dot arrays (see the behavioral protocol in figure 7.4A).[96] If the two sets matched, the animals had to release a lever; if they did not match, they were not allowed to respond. We were very aware of the fact that the succession of single dots gives rise to temporal parameters that the monkeys could potentially learn to solve the task, and we wanted to avoid this. For example, if the presentation duration of each single dot and the pauses between them are constant, then the presentation of four dots lasts twice as long as the presentation of two dots. However, when the total presentation duration is kept constant across numerosities, then the duration of single dots decreases systematically, which again would be a temporal feature the monkeys could have learned. We therefore applied different sequential layouts of the sample dots, each controlling for a specific temporal problem. The only aspect all of these protocols had in common was the number of sample items. Since the monkeys could not predict which control was being presented, they had to rely on number to solve the task. And this is exactly what they did.

We first trained monkeys to only discriminate two sequential items from four (and vice versa), because we wanted to know if they could apply the concept of sequential numbers to sets that they had not been trained on. Indeed, when we occasionally inserted three sequential items as sample numerosities, they were immediately able to transfer their discrimination to the novel sequential numerosity without further training. This indicates that monkeys can understand the concept of sequential numerosity.

In addition to this sequential presentation of number, we also had monkeys discriminate arrays of dots—just as described in the previous section—in the same session. Therefore, in one trial they had to assess the number of dots presented in a sequence, whereas in the next trial they had to extract the numerosity from simultaneous dot arrays. This did not cause them any problems. The performance data did show that differentiating the number of sequentially presented items was more difficult for both monkeys, but this is understandable, given that sequential enumeration is cognitively more demanding; it incorporates multiple encoding, memory, and updating stages, and imposes an ordinal aspect (numerical rank) in addition to a cardinal component (numerical quantity). Sequential enumeration may even be regarded as a form of addition. Nevertheless, the result from this study indicated that the monkeys perceived both sequential and simultaneous item presentations as renditions of numerical quantity.

As an abstract quantity category, numerosity is not only expected to be represented irrespective of the presentation of items in space and time, but also across different sensory modalities. For example, three beer cans on the table or three claps of thunder in a row are both instances of "three," despite the elements being visual items in the first and acoustic events in the second case. If animal truly understands the concept of number, they are expected to enumerate seen and heard items. In order to reliably answer this question, the same individuals have to be confronted with more than one sensory modality of number presentation to see whether they treat these as the same or different types of stimuli. Only a few studies, all done with monkeys, have looked into this issue, but the results are rewarding.

To find out if monkeys can perform such cross-modal numerosity judgments, Kelly Jordan and Elizabeth Brannon presented rhesus monkeys with a sample sequence of either one to nine visual items on a touch screen, or acoustic sounds played through a speaker.[97] After the sample presentation, the monkeys correctly chose the matching visual test arrays that contained the same number for both visual and auditory sample numerosities. Moreover, the monkeys were just as accurate in matching across sensory modalities (auditory–visual) as they were within a single modality (visual–visual). For example, if they heard five sounds, they selected an array containing five elements most of the time. As expected for the approximate number system (ANS), performance was dependent on the ratio between the displayed match and the non-match numerosities. For both visual and acoustic numerosities, the monkeys' performance obeyed Weber's law. My laboratory has reported similar cross-modal numerosity discriminations in an electrophysiological study with monkeys trained to discriminate one to four visual or acoustic items.[98] This study will come up again in descriptions of neuronal correlates of multimodal numerosity judgments (see the behavioral protocol in figures 7.4A and 7.4B).

Jordan and Brannon went an important step further and also presented samples that consisted of interspersed visual *and* acoustic items. For example, monkeys had to match two visual items and two sounds to the test numerosity four. In the first 150-trial test session, the monkeys already performed with above-chance accuracy for these bimodal visual-auditory samples. This demonstrates that non-human primates can cross-modally enumerate the number of visual objects they see and the number of sounds they hear. They show this capacity over a relatively large range of numerosities and use the ANS to solve the task. Non-human primates, and possibly also other animal species, grasp that number is a concept that applies to elements across sensory modalities.

So far, the discussions about numerical capabilities have centered on primates. This may give the impression that primates are somehow extraordinary in numerical competence, while in fact this impression primarily results from the abundance of behavioral data from this group of animals. We simply know more about primates, and it therefore seems that they are special in this respect. Instead, approximately equivalent numerical skills are present in birds. This is interesting to note because, as mentioned earlier, the last common ancestor between mammals and birds lived about 320 million years ago (figure 3.1). Since then, primates and birds have evolved in parallel.

Many years after Brannon and Terrace performed their seminal work with rhesus macaques, Michael Colombo, professor in the department of psychology at the University of Otago in New Zealand, knocked on Herb Terrace's door and asked whether he could use the monkeys' stimulus set to try the same task in pigeons. Colombo refused to believe that monkeys were the only animal to have such a grasp on numbers. So he tested pigeons with the monkeys' stimulus set and behavioral protocol.[99] The only difference, of course, was that the pigeons, instead of touching with a finger, pecked with their beaks at the pictures in the trained ascending numerical order. And sure enough, the pigeons were just as able to solve the task with novel numerosities as the macaques Rosencrantz and Macduff years earlier. This and other studies performed in his lab made Mike Colombo an advocate of the belief that avian cognitive capabilities should not be considered inferior to those of mammals.

One group of birds is particularly renowned for its sophisticated intelligence: corvid songbirds (crows and relatives). The cognitive flexibility of corvids is, in many respects, on par with those of advanced non-human primates. In agreement with their intelligence, the endbrain of corvids is particularly large when compared to other birds of equal size, and it therefore contains an astonishingly high count of densely packed neurons.

Starting with the previously mentioned pioneering work of Otto Koehler, corvids have been shown to judge absolute numerical values in conditioning experiments. Hooded crows (Corvus cornix) trained to differentiate a restricted set of one to four numerosities successfully transferred this behavior to stimulus sets with novel cardinalities of five to eight, which demonstrates a conceptual understanding of numbers in this group of birds.[100] Moreover, several species of corvids have been shown to spontaneously discriminate the relative quantity of food items without previous training.[101,102,103]

The characteristics of absolute numerosity representations are also not exclusive to non-human primates. My PhD student, Helen M. Ditz, and I trained carrion crows to perform exactly the same delayed match to numerosity task with one to 30 items as monkeys did earlier.[104,105] After training, the crows knew that they had to peck with their beaks at the test image that contained the same number of dots as the sample images (figure 3.8). And as always, we controlled non-numerical parameters. The crows also

Figure 3.8
The carrion crow (*Corvus corone*). The corvid songbird recognizes and discriminates the number of dots shown in computer displays (photo by Andreas Nieder).

could distinguish absolute cardinal values, albeit with less precision than the monkeys did. The crows' performance functions looked almost identical to those of monkeys, showing a numerical distance and size effect. Also, the crows' performance exhibited a logarithmic number line, just as seen in monkeys and humans. Weber's and Fechner's laws clearly hold for non-symbolic number judgments of corvids, perhaps even for most other animal species. This is a fundamental similarity we humans share with animals, at least as long as we are not symbolically counting.

All these behavioral characteristics of non-symbolic number discrimination in different animal species (including ourselves) answer the question of which system is at work. Is discrimination based on the set-size limited object tracking system (OTS) that shows no ratio dependence, the ANS exhibiting ratio dependence and Weber law signature, or both? Whenever animals are investigated in detail with stimuli that are controlled for non-numerical parameters, discrimination of numerosities shows a numerical distance (or numerical ratio) effect, the classical signature of the ANS that obeys Weber's law. When both small and large numerosities have been investigated, the exclusive workings of the ANS have been demonstrated repeatedly in mammals such as primates,[106,107,108] as well as birds[109,110] and also fish. The rare support for the presence of an OTS comes exclusively from spontaneous-choice tasks (mammals,[111,112] birds,[113] fish[114]). However, it is questionable whether spontaneous-choice tasks are suitable to discover the systems at work due to the methodological problems summarized earlier. In addition, failure to discriminate larger numerosities is no evidence for the absence of the ANS. Early studies often reported that certain species were constrained to small numerosities in support of the OTS. However, subsequent studies investigated further and found large number discriminations in agreement with the ANS.

Besides the set-size limit to about four items, the absence of ratio dependence in differentiating small numerosities is usually taken as evidence for the OTS. However, some researchers have argued that a different ratio dependence does not necessary imply the existence of two separate systems.[115,116] Charles Ransom Gallistel and Rochel Gelman from Rutgers University,[117] for example, pointed out that the representation of larger numbers would be more variable than those of small numbers. As a consequence, representations of nearest values may overlap only in the large number range, leading to a ratio effect only in that case. And, of course, recognizable geometric patterns may emerge for small numbers, with one item being a dot, two items a line, three items a triangle, and four items a quadrangle.[118] Besides, it is unclear how implicit object files filled by the OTS are turned into explicit number representations. Collectively, the data gathered in animals support the presence of a single enumeration system, namely the ANS, as an evolutionarily ancient non-verbal system we inherited from our animal ancestors.

4 The Utility of Number for Animals

4.1 A Matter of Fitness

Considering the multitude of situations in which we humans use numerical information, life without numbers would be difficult to imagine for us. But what was the benefit of numerical competence for our ancestors, before they became *Homo sapiens*? In other words, why would animals crunch numbers in the first place? This question seeks to understand the benefit of numerical competence for individual animals and entire populations of species. Only if numerical capabilities are beneficial (or at least not harmful) for the individuals will this faculty be maintained in a population across generations and sometimes conserved over millions of years in large taxonomic groups of animals.

As outlined in the previous section about the theory of evolution, beneficial traits from a biological point of view are those that help their carrier stay alive and reproduce. An individual must survive long enough to pass on its genes directly to the next generation by successfully mating. An animal that meets this demand has high "individual fitness." But there is a certain degree of sophistication in this process. An individual can also increase its fitness indirectly by helping relatives to reproduce, because relatives share a substantial proportion of the same genes. Taking this opportunity into account results in "inclusive fitness," the ultimate type of fitness.

How can an animal increase its overall fitness and become a winner in this respect? Well, by possessing genetically inherited traits that ensure its success in this evolutionary race against the ever-changing environment. An animal that masters these challenges well is adapted to its overall environment. Beneficial inherited traits that serve this need are therefore said to be of "adaptive value."

It is important to remember one aspect: While the genotype determines which capabilities are available, it is the phenotype that is under close and life-long scrutiny. The phenotype is the access point of selection pressures, and it is at this level that beneficial traits have to stand the test to be passed on via genes into the next generation.

To accomplish this goal of staying alive and passing on genes, two classes of behaviors are necessary. First, behaviors are needed for an individual to survive across the lifespan until reproduction is possible in a mature adult, and for some species, to later pamper offspring to make sure they survive long enough. For an individual, this means first and foremost to find food and avoid becoming food, but also to pick one's way through a cluttered environment and to count on help from friends in daily business. The following section of this chapter demonstrates that numerical cognition serves this purpose. Several studies examining animals in their ecological environments suggest that representing number enhances an animal's ability to exploit food sources, hunt prey, avoid predation, navigate in its habitat, and persist in social interactions.

In addition, a second set of skills are necessary that directly increase the chances of passing on the right genes during reproducing. Animals always compete for mating partners, and even when such a partner is found, the next challenge is to make sure that their own offspring succeed rather than a competitor's progeny. In the last section of this chapter, I will demonstrate that numerical competence plays a major role in this endeavor, from monopolizing a receptive mate to increasing the chances of fertilizing an egg and finally promoting the survival chances of offspring.

4.2 Staying Alive

Quorum sensing. Before numerically competent animals evolved on the planet, single-celled microscopic bacteria—the oldest living organisms on earth—already exploited quantitative information. The way bacteria make a living is through their consumption of nutrients from their environment. Mostly, they grow and divide themselves to multiply. However, in recent years, microbiologists have discovered they also have a social life and are able to sense the presence or absence of other bacteria; in other words, they can sense the number of bacteria.

Take, for example, the marine bacterium called *Vibrio fischeri*. It has a special property that allows it to produce light through a process called bioluminescence, similar to how fireflies give off light. If these bacteria are in dilute water solutions, in other words when they are alone, they make no light. But when they grow to a certain cell number of bacteria, all of them produce light simultaneously. Therefore, these bacteria can distinguish when they are alone and when they are together. Somehow they have to communicate cell number, and they do this using a chemical language. They secrete communication molecules, and the concentration of these molecules in the water increases in proportion to the cell number. And when this molecule hits a certain amount, called a *quorum*, it tells the other bacteria how many neighbors there are, and all bacteria glow. This behavior is called "quorum sensing"[1]: the bacteria vote with signaling molecules, the vote gets counted, and if a certain threshold (the quorum) is reached, every bacterium responds. This behavior is not just an anomaly of *Vibrio fischeri*; all bacteria use this sort of quorum sensing to communicate their cell number in an indirect way via signaling molecules.

Quorum sensing is not confined to bacteria; animals are using it to get around, too. Japanese ants (*Myrmecina nipponica*), for example, decide to move their colony to a new location if they sense a quorum.[2] In this form of consensus decision making, ants start to transport their brood together with the entire colony to a new site only if a defined number of ants are present at the destination site. Only then is it safe to move the colony. Quorum threshold increases with colony size—that is, the larger the colony, the more individuals are required at the new site to reach the quorum. With larger colonies, the quorum threshold becomes slightly larger, but not proportionally. Despite some ratio dependency, the discrimination behavior of the ants does not fully obey Weber's law. Of course, one also has to keep in mind that it is virtually impossible in free-ranging ant colonies to control for non-numerical factors that could supporting the ants' decision process.

Navigation. Enumerating landmarks can play an important role for animals to find their way in their daily goal to survive. The honeybee, for example, relies on landmarks to measure the distance of a food source to the hive. In an early experiment by Lars Chittka and Karl Geiger mentioned above, bees were trained to collect sugar water at a feeder between the third and fourth tent in a row of four tents.[3] Changing the number of tents and distances between them caused a performance pattern that indicated a compromise between conflicting distance information, namely the absolute distance flown and the number of landmarks passed. Even though the number of landmarks cannot explain entirely how honeybees measure distance, numerosity is nonetheless an important factor.

In an elegant and thoroughly controlled experiment using flying tunnels, Marie Dacke and Mandyam Srinivasan[4] provided clear evidence for bees' aptitude to sequentially enumerate landmarks. Bees were trained to forage in a 4-meter-long tunnel equipped with five yellow stripes distributed on the tunnel wall as landmarks (figure 3.4E). Different cohorts of bees were trained to find the feeder at landmark 1, 2, 3, 4, or 5. During training, bees were prevented from relying on distance information by varying the separation between the landmarks and thus the position of the rewarded landmarks. When the bees mastered this task, they were tested in new tunnels without being rewarded. The bees indeed searched in the vicinity of the learned landmark number. In other words, bees trained to the first landmark searched mainly near landmark 1, bees trained to the second landmark hovered close to landmark 2, and so on. Trained bees found the correct landmark even when the spatial layout was changed; that is, when landmarks were closer together or more spaced apart, or even irregularly spaced. Correct performance even remained when the landmarks changed from the original yellow stripes to disks or even overlapping baffles they had to fly through. Bees sequentially enumerate the number of variable landmarks, and in doing so, show signs of abstraction. For honeybees, assessing numbers is vital to finding their way between a nectar source and the hive.

Choosing between food patches. Numerical cognition helps animals develop efficient foraging strategies. The theory of optimal foraging[5] states that animals, when faced with two or more food options, would benefit if they can choose the one that provides the greatest energetic gain, and more food items obviously are more nutritious than just a few. Since food is a natural incentive, it allows for testing even cognitively less advanced vertebrates, such as amphibians, which can hardly be trained on complex numerosity discrimination tasks. Motivated animals are expected to spontaneously approach the larger patch of food items during such tasks. And indeed, amphibians go for more. When showing red-backed salamanders (*Plethodon cinereus*) in forced-choice situations two transparent test tubes that contain different numbers of flies (figure 3.4C), they reliably choose the tube containing more flies by moving toward it and touching it with their snouts.[6] They were able to select two flies over one, and three over two, but failed to discriminate three versus four and four versus six. A later study,[7] however, showed that salamanders are also able to discriminate larger numerosities, such as 16 versus eight crickets. All they need is a sufficiently large numerical ratio between the test sets. Similarly, frogs[8] (*Bombina orientalis*) preferred more mealworm larvae in free-choice experiments and discriminated one versus two, two versus three, three versus six, and four versus eight prey items.

"Going for more" is a good rule of thumb in most cases, but sometimes the opposite strategy is favorable. The field mouse (*Apodemus agrarius*) loves live ants, but ants are dangerous prey because they bite when threatened. When a field mouse is placed into an arena together with two ant groups of different quantities, it surprisingly "goes for less."[9] Mice that could choose between five versus 15, five versus 30, and 10 versus 30 ants always preferred the smaller quantity of ants. The field mice seem to pick the smaller ant group in order to ensure comfortable hunting and to avoid getting bitten frequently.

Food items are probably the most popular stimuli to test the spontaneous numerical capabilities of animals. Thus, a variety of animal species, such as robins, crows, coyotes, dogs, elephants, and different species of monkeys and apes, have been shown to differentiate number of food items. As mentioned earlier, food items are particularly hard to control for non-numerical parameters, such as the amount of food items, the overall space many food items will cover, or hedonic value. It can therefore never be excluded that the animals pay attention to such parameters rather than the actual number. Still, many studies suggest that animals in the wild are sensitive to number of food items, and foraging is a particularly important domain in which numerical competence pays off.

Hunting prey. Numerical cues play a role in communal spider-eating spiders, known as araneophagic spiders. One such spider-eating spider is

Portia africana in Kenya (hereafter Portia). Small Portia juveniles adopted an especially intricate predatory strategy when preying on *Oecobius amboseli* (hereafter oecobiid), a small spider species that builds tent-like silk nests on boulders, tree trunks, and the walls of buildings.[10] Portia often practice communal predation, especially when the prey is an oecobiid. In typical sequences, two Portia settle alongside each other at an oecobiid's nest, and when one Portia captures an oecobiid, it is joined by the other to feed alongside it. Portia base their decision of settling near a prey spider nest on the number of conspecifics already present. They prefer one spider being present over zero, two, or three spiders. The reason why these spiders prefer to hunt in pairs rather than in larger groups may have to do with increasing numbers of freeloaders in large groups. The more members a hunting party has, the more likely it becomes that some members won't cooperate. Therefore, larger groups are often worse at capturing prey than smaller groups.

The probability that wolves capture elk or bison varies with the group size of a hunting party. This is at least what research with wild wolves at Yellowstone National Park suggests. Wolves often hunt large prey, such as elk and bison, and large prey can kick, gore, and stomp wolves to death. Therefore, there is a lot of incentive to "hold back" and let others go in for the

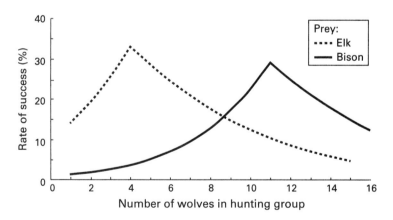

Figure 4.1
Wolves have an optimal group size for hunting different prey. The lines are population-averaged fitted values to hundreds of wolf-elk and wolf-bison encounters (after MacNulty et al., 2014).

kill, particularly in larger hunting parties. As a consequence, wolves have an optimal group size for hunting different prey.[11] For elks, capture success levels off at two to six wolves (figure 4.1). However, for bisons, the most formidable prey, nine to 13 wolves are the best guarantor of success. Therefore, for wolves, there is "strength in numbers" during hunting, but only up to a certain number that is dependent on the toughness of their prey.

Avoiding predation. Animals that are more or less defenseless often seek shelter among large groups of social companions. By joining large groups, the probability of becoming prey for each individual member is decreased. For many fish, becoming social is therefore the main anti-predatory strategy. The bigger the shoal in which to seek shelter, the better for the fish. Indeed, individual fish that are inserted in an unfamiliar and potentially dangerous environment tend to join other conspecifics. If two shoals are present, they usually join the larger shoal,[12,13,14] which means that they are able to distinguish the larger shoal from the smaller one. The ability to compare the number of conspecifics can therefore be vital in such cases as a life or death situation.

Biologists recognize at least three distinct advantages of such behavior. First, an individual fish reduces the risk of being caught as the quantity of individuals in the group increases, a phenomenon termed the "dilution effect."[15] Second, a predator has a harder time singling out an individual that lives in larger shoals, a phenomenon known as the "confusion effect."[16] And finally, many individuals together have a higher chance of detecting a predator, a consequence termed the "many eyes effect."[17] Evidently, hiding within larger groups has undeniable advantages.

But hiding out is not the only anti-predation strategy involving numerical competence. Black-capped chickadees (*Poecile atricapilla*) in Europe found a way to announce the presence and dangerousness of a predator.[18] Like many other animals, chickadees produce alarm calls when they detect a potential predator, such as a hawk, to warn their fellow chickadees. For stationary predators, these little songbirds use their namesake "chick-a-dee" alarm call. It has been shown that the number of "dee" notes at the end of this alarm call indicates the danger level of a predator. A call such as "chick-a-dee-dee" with only two "dee" notes may indicate a rather harmless great gray owl. Great gray owls are too big to maneuver and follow the agile chickadees in woodland, so they aren't a serious threat. In contrast, maneuvering between trees is no problem for the small pygmy owl, which is why it is one of the most dangerous predators for these small birds. When chickadees see a pygmy owl, they increases the number of "dee" notes and call "chick-a-dee-dee-dee-dee." Here, the number of sounds serves as an active anti-predation strategy.

Social territory defense. Groups and group size also matter a lot if resources cannot be defended by individuals alone. Many animals therefore live in social groups that claim and defend territories against intruders. Defending a territory usually means that grave and potentially lethal conflicts with rivaling groups can occur. The ability to assess the number of individuals in one's own group relative to the opponent party is therefore of clear adaptive value. The doctrine "strength in numbers" applies in particular for social territory defense. It serves as the basis for smart decisions on whether to attack and risk damaging fights, or retreat and lose territory.

Several mammalian species have been investigated in the wild, and the common finding is that numerical advantage determines the outcome of such fights. In a pioneering study, Karen McComb and coworkers at the University of Sussex[19] investigated the spontaneous behavior of female lions (*Panthera leo*) at the Serengeti National Park when facing intruders. The authors exploited the fact that wild animals respond to vocalizations played through a speaker as though real individuals were present. If the playback sounds like a foreign lion that imposes a threat, the lionesses would aggressively approach the speaker as the source of the enemy. In this acoustic playback study, the authors mimicked hostile intrusion by playing the roaring of unfamiliar lionesses to residents. Two conditions were presented to subjects: either the recordings of single female lions roaring, or of groups of three females roaring together. The researchers were curious to see if the number of attackers and the number of defenders would have an impact on the defender's strategy. Interestingly, a single defending female was very hesitant to approach the playbacks of a single or three intruders. However, three defenders readily approached the roaring of a single intruder, but not the roaring of three intruders together. Obviously, the risk of getting beaten and hurt when entering a fight with three opponents was foreboding. Only if the number of the residents was five or more did the lionesses approach the roars of three intruders. In other words, lionesses decide to approach intruders aggressively only if they outnumber the latter. This clearly shows their ability to take quantitative information into account.

Our closest cousins in the animal kingdom, the chimpanzees (*Pan troglodytes*), show a very similar pattern of behavior. Chimpanzees live in groups of 20–150 individuals. Males defend the group's territories and sometimes kill members of neighboring groups. Chimpanzee inter-group fights include both "gang attacks," in which multiple fighters concentrate attack on a single victim, and "battles," with multiple opponents on each side. Primatologists think that this aggressive behavior, which reduces the strength of the victim's group, enhances the attacker's chances of success in future battles. This, in turn, results in increases in territory size, better access to food and females, as well as reduced danger from neighboring groups. Gang attacks are a nasty but readily understood case in which some of the attackers immobilize the victim while others beat, bite, and otherwise injure the victim. From a numerical cognition point of view, "battles" with multiple fighters on each side are much more interesting, because the numbers of individuals in the confronting parties matter a lot.

Michael Wilson and coworkers from Harvard University used a similar playback approach for chimpanzees as previously described for the lionesses.[20] In each trial, a single vocalization of a single foreign male was played through a speaker. These playbacks elicited cooperative responses, with the nature of the response depending on the number of adult males in the resident party. Parties with three or more males consistently joined in a chorus of loud vocalizations and approached the speaker together. Parties with fewer adult males usually stayed silent, approached the speaker less often, and traveled more slowly if they did approach. The chimpanzees behaved like military strategists. They intuitively follow equations used by military forces to calculate the relative strengths of opponent parties. In particular, chimpanzees follow predictions made in Lanchester's "square law" model of combat.[21] This model predicts that, in contests with multiple individuals on each side, chimpanzees in this population should be willing to enter a contest only if they outnumber the opposing side by a factor of at least 1.5. And this is what wild chimps do. Parties with one male do not approach a single caller until joined by allies, while parties with two males approached a single caller in four out of seven cases.

Using a similar playback technique, numerical assessment of hostile callers was also demonstrated in wild spotted hyenas (*Crocuta crocuta*).[22] Just like lions and chimpanzees, hyenas live in social clans that contain up to 90 individuals, and they cooperatively and fiercely defend their territories. The playback experiments demonstrated that they respond with increasing levels of vigilance to calls produced by one, two, and three unknown intruders played back via speakers. Hyenas also took more risks approaching the speaker when they outnumbered the calling intruders. Numerical comparison thus guides their decisions to enter social contests, and they only do so when their group outnumbers their opponents. This has nothing to do with cowardliness. On the contrary, this calculated behavior saves the lives of clan members.

4.3 Benefits for Reproduction

Mating. Staying alive—from a biological stance—is a means to an end, and the aim is the transmission of genes. In mealworm beetles (*Tenebrio molitor*), many males mate with many females, and competition is intense. Therefore, a male beetle will always go for more females in order to maximize his mating opportunities. To find female mating partners, the male beetles can discriminate between odor bouquets containing the scents of up to four different female beetles.[23] After mating, males even guard females for some time to prevent further mating acts from other males. The more rivals a male has encountered before mating, the longer he will guard the female after mating.[24] It is obvious that such a behavior plays an important role in reproduction and therefore has a high adaptive value. Being able to estimate quantity has improved males' sexual competitiveness. This may in turn be a driving force for more sophisticated cognitive quantity estimation throughout evolution.

One may think that everything is won by successful copulation. But that is far from the truth for some animals, because the real prize is fertilizing an egg. Once the individual male mating partners have accomplished their part in the play, the sperm continues to compete for the fertilization of the egg. Since reproduction is of paramount importance in biology, sperm competition causes a variety of adaptations at the behavioral level. In both insects and vertebrates, the males' ability to estimate the magnitude of competition determines the size and composition of the ejaculate.[25] In the pseudo-scorpion *Cordylochernes scorpioides*, for example, it is common that several males copulate with a single female.[26] Obviously, the first male has the best chances of fertilizing this female's egg, whereas the following males face slimmer and slimmer chances of fathering offspring. However, the production of sperm is costly, so the allocation of sperm should be weighed considering the chances of fertilizing an egg. Males smell the number of competitor males that have copulated with a female and adjust by progressively decreasing sperm allocation as the number of different male olfactory cues increases from zero to three. Because males can assess the number of competitors based on the amount of olfactory cues, sperm competition marks the transition from processing amount in an automatic and noncognitive way to processing discrete quantity in a more cognitive way.

Brood parasitism. Breeding a clutch and raising young are costly endeavors. Some bird species have invented a whole arsenal of trickery to get rid of the burden of parenthood and let others do the job. They become brood parasites by laying their eggs in other birds' nests and letting the host do all the hard work of incubating eggs and feeding hatchlings. Naturally, the potential hosts are not pleased and do everything to avoid being exploited. And one of the defense strategies the potential host has at its disposal is the usage of numerical cues.

This has been shown in American coots (*Fulica americana*), a water bird. Unlike the famous cuckoo, which lays its eggs in nests belonging to other bird species, coots target their own species. Coot hens sneak eggs into their neighbors' nests and hope to trick them into raising the chicks. Of course, their neighbors try to avoid being exploited. A study in the coots' natural habitat suggests that potential coot hosts can enumerate their own eggs, which helps them to reject parasitic eggs.[27] They typically lay an average-sized clutch of their own eggs, and later reject any surplus parasitic egg. Coots therefore seem to assess the number of their own eggs and ignore any others.

An even more sophisticated type of brood parasitism is found in cowbirds (*Molothrus ater*), a songbird species that lives in North America. In this species, females also deposit their eggs in the nests of a variety of host species, from birds as small as kinglets to those as large as meadowlarks, and they have to be smart in order to guarantee that their future young have a bright future. Cowbird eggs hatch after exactly 12 days of incubation; if incubation is only 11 days, the chicks do not hatch and are lost. It is therefore not an accident that the incubation times for the eggs of the most common hosts range from 11 to 16 days, with an average of 12 days. Host birds usually lay one egg per day; once one day elapses with no egg added by the host to the nest, the host has begun incubation. This means the chicks start to develop in the eggs, and the clock begins ticking. For a cowbird female, it is therefore not only important to find a suitable host, but also to precisely time their egg laying appropriately. If the cowbird lays her egg too early in the host nest, she risks her egg being discovered and destroyed. But if she lays her egg too late, incubation time will have expired before her cowbird chick can hatch. Clever experiments by David J. White and Grace Freed-Brown from the University of Pennsylvania suggest that cowbird females carefully monitor the host's clutch to synchronize their parasitism with a potential host's incubation.[28] The cowbird females watch out for host nests in which the number of eggs has increased since her first visit. This guarantees that the host is still in the laying process and incubation has not yet started. In addition, the cowbird is looking out for nests that contain exactly one additional egg per number of days that have elapsed since her initial visit. For instance, if the cowbird female visited a nest on the first day and found one host egg in the nest, she will only deposit her own egg if the host nest contains three eggs on the third day. If the nest contains fewer additional eggs than the number of days that have passed since the last visit, she knows that incubation has already started and it is useless for her to lay her own egg. It is incredibly cognitively demanding, since the female cowbird needs to visit a nest over multiple days, remember the clutch size from one day to the next, evaluate the change in the number of eggs in the nest from a past visit to the present, assess the number of days that have passed, and then compare these values to make a decision to lay her egg or not.

But this is not all. Cowbird mothers also have sinister reinforcement strategies. They keep watch on the nests where they've laid their eggs. In an attempt to protect their egg, the cowbirds act like mafia gangsters.[29] If the cowbird finds that her egg has been destroyed or removed from the host's nest, she retaliates by destroying the host bird's eggs, pecking holes in them or carrying them out of the nest and dropping them on the ground. The host birds better raise the cowbird nestling, or else they have to pay dearly. For the host parents, it may therefore be worth to go through all the trouble of raising a foster chick from an adaptive point of view.

The cowbird is an astounding example of how far evolution has driven some species to stay in the business of passing on their genes. The existing selection pressures, whether imposed by the inanimate environment or by other animals, force populations of species to maintain or increase adaptive traits caused by specific genes. If assessing number helps in this struggle to survive and reproduce, it surely is appreciated and relied on. This explains why numerical competence is so widespread in the animal kingdom: it evolved either because it was discovered by a previous common ancestor and passed on to all descendants, or because it was invented across different branches of the animal tree of life. Irrespective of its evolutionary origin, one thing is certain: numerical competence is most certainly an adaptive trait.

5 Biological Heritage in the Human Brain

5.1 Baby Steps

What would it be like if we were to be ignorant of the culturally transmitted number symbols, such as number words or Arabic numerals, that we use to count elements? Would we still be able to say something about numerical quantity? And if so, how would we do it? The answers to these intriguing questions would reveal a lot about our biological heritage and the predispositions of our brain with respect to quantity assessments.

Newborns and infants represent such an interesting status of counting nescience. Our brain has to mature over the first years of life to be ready to grasp symbols. During this initial phase, the brain's premier job is to maintain the body and ensure such vital functions as breathing, feeding, and temperature regulation. So this is what the brain is doing exclusively at or shortly after birth, while our fancy cognitive functions can develop later in life. However, the human mind is not a blank slate when we are born, and this applies also to the realm of numerical quantity.

Developmental psychologists seek to understand the origins of human conceptual abilities during the course of development of an individual. In order to better understand how certain traits and abilities develop throughout a lifetime, infants are studied from birth throughout the first years of their lives. But a problem emerges: how is it possible to say anything about whether infants can discriminate numbers? After all, infants cannot speak yet. How can infants report how many objects they see? They share this fate with non-human animals, but animals can at least tell us which set contains more food items by actively going for more. Young infants, on the other hand, cannot even crawl. However, developmental psychologists have found a way to work around this problem. They resort to looking time as the infants' way to indicate differences in set sizes. And since looking time can be measured with stopwatches, it is possible to collect quantitative data from infants who are barely able to sit straight, let alone crawl.

One of the standard protocols to test if infants can discriminate numbers is the "habituation protocol." Infants sit on their parent's lap or in a chair, depending on their age, and watch a puppet show or a computer monitor on which a sequence of displays is shown. The infants are monitored by a secret camera so that their eye fixation on the displays can later be analyzed by unbiased raters who are ignorant to the experiment. The habituation protocol is based on the finding that repeatedly presented displays containing the same number of items become boring for the little ones. The babies tend to look less frequently at these displays because they get used (or habituated) to the displays. As a consequence, fixation time drops. However, new displays elicit their increased attention, they become sensitized, and looking time increases. The rationale, therefore, is that if infants can discriminate displays with a different number of items, they should stare longer at this display showing a new numerosity than one that was used during habituation. In addition, all of this works spontaneously and doesn't require any training.

The first and seminal study that used the habituation protocol with five- to six-month-old infants was done by Prentice Starkey and Robert G. Cooper at the University of Pennsylvania in 1980.[1] The infants were habituated to visual displays of either two or three dots arranged in a line. To discourage infants from attending to the length of the lines, the spacing between the dots varied. After the babies were habituated to either two or three dots, they were shown test displays of three and two dots on different trials. Infants habituated to two dots looked significantly longer at the displays that contained three dots, and vice versa. This indicated that infants discriminated between the two numbers.

If five-month-old infants are sensitive to the number of visual dots in an array, maybe even younger babies are as well. As a result, researchers proceeded to study younger and younger infants. And perhaps these younger infants could not only discriminate numerosity within one modality, such as vision, but also abstractly across different sensory modalities?

Biological Heritage in the Human Brain

Veronique Izard, Elizabeth S. Spelke, and Arlette Streri from Harvard University took on this challenge and investigated newborns who were just 50 hours old.[2] To study cross-modal numerosity, they combined the numerosity of visual dot arrays and the number of sequentially played sounds. The newborns were tested right in the maternity hospital where they were born. They were seated in an infant seat in front of a monitor and surrounded by speakers. During the familiarization phase, the screen remained black and sounds were played through speakers for two minutes (figure 5.1). In one experiment, newborns were familiarized either to the acoustic numerosity 12 in half of the trials by repeatedly playing a 12-count sequence, and to acoustic numerosity four in the other half of the trials by presenting a four-count sequence that was equally long in duration to the larger count one. After familiarization with auditory sequences, the newborns were presented with four visual numerosity test displays while the auditory stimulus continued to play in the background. The displays showed four and 12 items in alternation, i.e., 4-12-4-12 numerosities that had equal sizes of each item

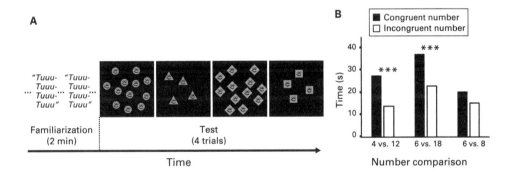

Figure 5.1
Newborn infants perceive abstract numbers. A) Behavioral protocol. Newborn infants heard auditory sequences containing a fixed number of syllables (here four) during initial familiarization, and were then tested with four images of the same or a different number of items (here four or 12). B) Newborns looked consistently longer at the displays that contained the same number (i.e., were congruent in number) as the auditory sequences presented during familiarization. This was true when the numbers to discriminate were separated by a large ratio of 3:1 (4 vs. 12, and 6 vs. 18), but not for a smaller ratio of 2:1 (4 vs. 8). Asterisks indicate statistically significant discrimination between congruent and incongruent numbers (from Izard et al., 2009).

and equal density across the arrays. It turned out that the newborns looked significantly longer at sets of four items if they had been familiarized with four sounds. Conversely, the infants looked longer at sets of 12 items if they had been familiarized with 12 sounds. This indicates that newborns can recognize the matching number of visual items and acoustic sounds. Not only do the little ones abstract across sensory modality, but also across sequential and simultaneous presentation formats!

In subsequent experiments, Izard and coworkers could show that this discrimination was not specific to four versus 12, but equally present for six versus 18. The newborns thus generalized to other numerosities, at least as long as the numerical difference between the numerosities was large enough. However, as expected with the estimation of large numerosities, the newborns failed the comparison between four and eight. Newborn infants, nonetheless, showed the ratio-dependent numerical processing that is characteristic of older children and adults. The main finding of this study with 50-hour-old newborns, however, is that we seem to be born with the neuronal machinery for numerical quantity.

Since then, many similar studies have been performed with infants. It is now known that by the age of five months, infants are able to discriminate between numbers differing in a 1:2 ratio (e.g., 16 vs. 32, 8 vs. 16, 4 vs. 8), when presented with arrays of dots,[3,4] sequences of sounds,[5] or sequences of actions.[6] These findings are fully explained by the approximate number system (ANS) and its Weber's law signature that are frequently encountered in animal studies.

An additional effect emerges when infants grow older. With age, the precision of numerical discrimination in infants and children improves. Newborns are only able to discriminate a 1:3 ratio (4 vs. 12) but not a 1:2 ratio (4 vs. 8). However, 6-month-old infants can already discriminate numerosities with a ratio of 1:2 (such as 8 vs. 16) but still not with a 2:3 ratio.[7] Ten-month-old infants are already able to discriminate numerosities with a 2:3, but not yet a 4:5, ratio.[8] And the quality of number discrimination continues to enhance throughout childhood. Six-year-olds are able to discriminate a 5:6 ratio and adults even a ratio of 9:10.[9] Exactly why this enhancement occurs is difficult to say. It could be a simple brain maturation process, or an improvement driven by learning and experience with number, or a combination of both.

5.2 Approximate Number System versus Object Tracking System

The ratio dependence of number discrimination found in the many studies described above is characteristic of Weber's law, and thus of the ANS that is also frequently found in non-human animals. But many developmental psychologists suggest that there is a second source of numerical representation, called the object tracking system (OTS), which is geared toward smaller numerosities.[10,11] The reason for this assumption is that the performance pattern with smaller numerosities sometimes differs notably from that observed with larger numerosities.

The first experiment to report such a difference was that by Prentice Starkey and Robert G. Cooper mentioned earlier.[12] Using a habituation protocol, they found that 5- to 6-month-old infants discriminated two from three dots and vice versa. In a second condition, however, infants of the same age failed to discriminate between four and six dots. Note that the ratio between the larger numbers was still 2:3. According to Weber's law, following the ANS, there was no reason of why the infants would fail; still, they were unable to tell four and six apart.

In a later experiment, 10- and 12-month-old infants chose between two quantities of hidden crackers.[13] Infants watched an experimenter sequentially hide, for example, one cracker in a bucket on the left, and two crackers in a bucket on the right. With choices of 1 versus 2 and 2 versus 3 equal-sized crackers, infants spontaneously chose the larger quantity. However, with choices of 3 versus 4, 2 versus 4, 3 versus 6, and even 1 versus 4 crackers, infants chose randomly despite the highly conspicuous ratio between quantities. Surprisingly, the infants' success depended not on the numerical ratio, as expected by Weber's law for the ANS, but on the absolute number of up to three items presented. This is called the set-size signature of the OTS. This limit of three or four on infants' small number quantification also appeared in other studies using different test protocols.[14]

The strongest evidence supporting the existence of an OTS comes from spontaneous-choice studies in infants. However, whether this system is widespread, or actually a separate system at all, remains debated. One reason for this doubt is that it is rarely seen in animals performing spontaneous-choice tests analogous to the tests applied to infants, and never in studies with trained animals. In addition, the OTS faces conceptual challenges. The most severe issue is how cardinality might be derived at all within an object tracking framework. After all, the OTS only stores items; it does not enumerate them. Zenon Pylyshyn from the University of Western Ontario suggests that judging the numerosity of a set of items might involve two different processing stages[15]: an individuation stage that might derive object entities of a (visual) scene in parallel, and a (serial) enumeration stage in order to judge the numerosity of individuated items. According to this explanation, however, the first, individuation stage (object tracking) is only a preprocessing stage that does not provide access to cardinality, and thus might not be regarded as a system to assess numerical information at all. This is probably one reason why brain imaging of the neural substrates of object tracking has remained inconclusive.[16,17] Moreover, this parallel object individuation effect only holds for simultaneously displayed item arrays; the subitizing effect disappears when items must be enumerated serially one after the other. Finally, if object pointers exist only in the visual domain, how would small numbers of sounds or touches be represented?

If the OTS is a general mechanism, it is never seen in trained animals; in training studies, all effects are in agreement with the ANS. In addition, infants who originally were thought to be restricted to assessing up to three or four items succeeded later in studies applying larger numerosities. From a conceptual point of view, it is difficult to understand why, in addition to the ANS, which covers both small and large numerosities, humans would need a second system that redundantly covers only small numerosities. The proponents of the OTS usually state that it represents small set sizes more precise than the ANS, but the original performance data of infants are far from perfect. Finally, as pointed out above, the OTS alone does not have the power to represent cardinality explicitly. Because of all these problems, the OTS is intensively debated and called into question even among developmental psychologists.[18]

5.3 Number Discrimination in Humans Lacking Number Words

The study of newborns and infants represent a first instance of counting nescience. However, there is a second condition in which counting may escape us: if we were to live in a culture that never developed number words. If there were no one around to teach us, symbolic counting would never be revealed to us. Interestingly, such cultures still exist. They are indigenous people who still live in hunter-and-gatherer societies. Usually, they have

Biological Heritage in the Human Brain

not attended school or have done so only briefly. Anthropologists studying the numerical skills in these cultures can tell us a great deal about how the fully mature and adult human brain processes numerical quantity in the absence of spoken or written number symbols.

In 2004, Peter Gordon from Columbia University reported an utmost reduced number system among the indigenous people of the Pirahã tribe in the state of Amazonas in Brazil.[19] The Pirahã only have number words for one ("hói"), approximately two ("hoí"), and many ("baágiso" or "aibaagi"). These people show poor discrimination of any numerical quantities larger than three. Gordon used a variety of number tasks to test Pirahã participants. For example, the participants were presented with an array of AA batteries and had to recreate the array on their side of a table with their own set of batteries. In a more difficult version of this task, they were asked to draw a single line for each battery displayed on the other side of the table. The Pirahã were accurate for sets of one or two, but their performance systematically deteriorated from three to 10, particularly when they had to memorize the target number. However, their performance for larger numbers was not random. With increasing target numbers, the average answers increased as a rough approximate of the correct number. And in accordance with Weber's law, the distribution of the answers became broader with increasing target numbers.[20]

Similar findings were reported in the Munduruku, another group of indigenous people who also live in a rather isolated manner in the Amazon rainforest of Brazil. The Munduruku know only very few number words, and they use them rather like estimates than as number words. For almost two decades, the French linguist Pierre Pica has regularly visited the Munduruku.

Initially, Pica showed the Munduruku varying numbers of dots on a notebook screen and ask them to count in their language.[21] The Munduruku numbers for one, two, three, four, and five are "pug," "xep xep," "ebapug," "ebadipdip," and "pug pogbi," respectively. So for one dot on the screen, the Munduruku said "pug." Two dots were denoted with "xep xep." Beyond two, however, they started to become imprecise (figure 5.2). At the sight of three dots, they said "ebapug" only about 80% of the time. Four dots were responded to with "ebadipdip" in only 70% of the cases. When five dots showed up, they reacted with "pug pogbi" only 30% of the time, with "ebadipdip," the word for four, given in 15% of the cases. Thus, already beyond

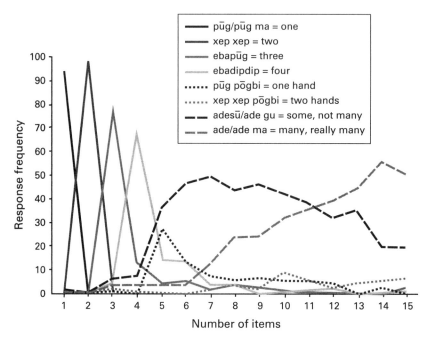

Figure 5.2
The imprecise number naming system of the Munduruku in Amazonia. Participants were shown sets of one to 15 dots in random order and were asked to name the quantity. For each quantity on the x axis, the graph shows the rate that it was named with a given word (from Pica et al., 2004).

two, the Munduruku number words were just estimates. Even for the small range of numbers for which they had words, performance still resulted in a distribution. One could say they counted "one," "two," "three-ish," "four-ish," "five-ish." If a set contains more than five items, the Munduruku label it as "adesu," which means "some," or "ade," the word for "many." That's it. Some linguists even doubt that the words for one to five are number words at all; rather, these words could describe the property of a set just as adjectives do. Interestingly, however, and despite such rudimentary counting abilities, the Munduruku are still able to perform approximate calculations with set sizes. This will be examined in chapter 10 on calculation.

Exactly the same performance pattern surfaces in our own Western society if we are forced to abandon symbolic counting and rely on the ANS. Bringing people into situations in which they estimate set size rather than count is easier said than done. Counting is so well trained and provides us with such unprecedented precision that we can't simply decide to turn it off for an experiment. It has to be outwitted. One way of discouraging counting is to present the items too rapidly to be counted. This is taking advantage of the fact that it takes us around 200 ms to count each additional element. We did this with science students and showed them displays with up to 30 dots for only 500 ms.[22] In this situation, humans are left with the ANS and make noisy estimates of the target numbers, just like the Pirahã who never learned to count.

A second way to avoid counting is to flood the symbolic system with a non-numerical language task. We can only perform one symbolic task at a time, so if the symbolic system is already busy, it cannot represent other symbolic items.[23,24] In one experiment, adult subjects were forced to enumerate non-symbolically by pressing a key a certain number of times, while saying "the" at every press. Obviously, we can't keep track of counting if a different word, like the article "the," is already occupying the slot for a symbol. In this situation again, the ANS enters the stage. Moreover, humans are equally good at estimating numerosities across as opposed to within visual and auditory stimulus sets, or across compared to within simultaneous and sequential sets.[25] We can, if we care to, bring the ancient ANS to light. It is always there, never absent—only hidden.

5.4 The Ancient Logarithmic Number Line

After the initial study was published in 2004, Pierre Pica returned a few years later to the Munduruku and examined their spatial understanding of numbers.[26] How do they visualize numbers when spread out on a line? We Westerners do this all the time—on rulers, folding meter sticks, or graphs.

Pica tested the Munduruku with an unmarked line that connected two sets of dots at their ends (figure 5.3A). To the left side of the line was one dot; to the right, 10 dots. Each volunteer was then shown a third set of dots with random numbers between one and 10 dots. For each dot set, the participants had to point at where on the line he or she thought the number of dots should be located. This point was then noted down. This procedure was repeated several times, with different set sizes, to explore how the Munduruku spaced numbers between one and 10. For comparison, he gave exactly the same test to American adults.

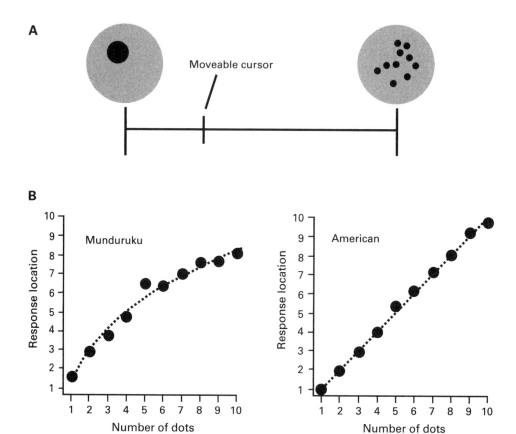

Figure 5.3
Mapping of numbers onto space in Western and Amazonian indigene cultures.
A) Participants were asked to show via a movable cursor where they thought different numbers of dots would fall on a number line with zero at one end and 10 at the other end. B) Munduruku participants map numbers onto a logarithmic scale (*left*), whereas American participants used linear mapping (*right*; from Dehaene et al., 2008).

A striking difference appeared between the two groups of people (figure 5.3B). The Americans placed the numbers at equal intervals along the line. For instance, five dots were placed right in the middle of the line. When numbers are spread out evenly on a scale in such a way, the scale is called linear. This is the number layout we are used to on meter sticks. The Munduruku, however, showed a different pattern. They placed five dots closer to the right end point of the line, where we would expect the six to be. The Munduruku thought that intervals between the numbers started large and became progressively smaller as the numbers increased. When numbers get closer in space as they get larger in values, a logarithmic number scale emerges—Fechner's law holds even in the Amazon rainforest.

This is strong evidence that we originally present numerical quantities on a logarithmic number line. Maybe this is how our brain processes numbers devoid of cultural influences. Of course, adults always live with some sort of cultural heritage. So maybe the logarithmic representation is specific for the Munduruku, and not for humans in general. This assumption, however, can be rejected based on findings in children raised in our Western culture. They show that the logarithmic mapping of numbers is not exclusive to Amazonian Indians: we are all born conceiving numbers this way.

In 2004, Robert Siegler and Julie Booth at Carnegie Mellon University presented a similar version of the number-line experiment to different groups of children.[27] First, they presented it to a group of kindergarten pupils (average age 5.8 years), then to first-graders (6.9 years), and finally to second-graders (7.8 years). The results showed in slow motion how familiarity with counting molds our number instinct (figure 5.4): The kindergarten pupil, with no formal math education, maps out numbers logarithmically. By the first year of school, when students are being introduced to number words and symbols, the curve is still logarithmic, but the estimates get more precise. And by the second year at school, the larger numbers are at last evenly laid out along the line, as expected for a linear scale. Clearly, our culture imposes the linear scale on us. Exact numbers provide us with a linear framework that contradicts our logarithmic intuition. It seems that the logarithmic scale that is also seen in animals is the evolutionary default for non-verbal numerosity representations before number symbols require a linear transformation. And as we shall see later, the logarithm as the primordial scale of the mental number line is deeply engraved in the brains of animals and humans alike.

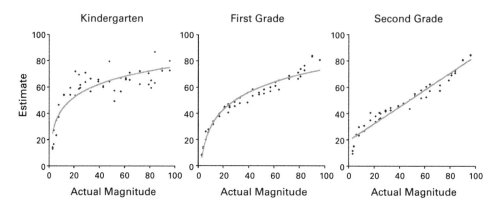

Figure 5.4
Development of spatial mapping of numbers in young children. Kindergarteners, first-graders, and second-graders were asked to show where they thought different numbers (Arabic numerals) would fall on a number line with zero at one end and 100 at the other end. Patterns of estimates progressed from logarithmic in kindergarteners, to a mixture of logarithmic and linear in first-graders, and finally a primary linear pattern in second-graders (from Siegler and Booth, 2004).

Part III Numerical Quantity in the Brain

6 Localizing Numerical Quantity Representations in the Human Brain

6.1 The Building Plan of the Cerebral Cortex

The way the brain functions, based on its nerve cells, or neurons, determines the way we experience, think, and act; in short, it determines everything we are. According to the latest cell counts done by Brazilian neuroanatomist Suzana Herculano-Houzel at Vanderbilt University, the adult male human brain, at an average of 1.5 kg, has 86 billion neurons (and about an equal number of glia cells).[1] All these neurons make connections with up to a thousand or so other neurons, resulting in a staggering number of around 86 trillion connections in the brain!

Before studying the way neurons process numerical information, it is necessary to know where mathematical functions are located in the brain. For that, a general knowledge of the brain's basic anatomy is required. Generally, a map is always helpful when entering new terrain. Just like an office building, the brain has different departments that are located in specific areas. Some areas receives messages from the senses, while others evaluate and store incoming information and yet others issue commands to generate motor movements. In order to understand how these areas function and interact, it is important to know their locations and connections to other brain regions.

Fortunately, neuroanatomists studying the structure of the nervous system provide anatomical divisions and surface maps of the brain. Like any other organ in our body, the brain must grow and develop. The different structures that emerge early in development are the foundations of the major anatomical parts of the brain. Early in our lives, in embryo, the nervous system is nothing but a fluid-filled tube. The wall of this neural tube consists of neurons that, over many months of gestation, multiply and migrate to their final destination. Later on in development, walls of the front end of the neural tube, its anterior part, bulge to become five bubbles, or vesicles. They are sequentially arranged like pearls on a string. All parts of the brain emerge from these five vesicles that we share with all vertebrates. At the most anterior pole lies the telencephalon (also called the cerebrum or endbrain), which is split along the midline into the left and right cerebral hemispheres. Next comes the *diencephalon* (with thalamus and hypothalamus), the *mesencephalon* (the midbrain), the *metencephalon* (cerebellum and pons), and finally the *myelencephalon*, which is followed by the long spinal cord reminiscent of the neural tube.

This book will exclusively focus on the endbrain (figure 6.1). From the endbrain's coating, the *pallium,* emerges the *cerebral cortex* (or *neocortex*). This area constitutes the highest integration center in our primate brain and determines "who we are." As will be demonstrated later, the cerebral cortex also determines how we deal with numbers and mathematics.

In order to prevent the endbrain from blowing up like a balloon during the course of evolution, the coating intensively folded in large primate species. As a consequence, our brains are heavily wrinkled. The grooves are called *sulci* (singular *sulcus*), or *fissures* if they are particularly deep, and it is important to remember that large parts of the cortical surface are hidden away in these sulci. One of the most important sulci, and a key landmark in the brain, is the *central sulcus*, a large notch in the middle of the lateral hemisphere running from top (dorsal) to bottom (ventral; figure 6.1). For numerical competence, another sulcus, namely the *intraparietal sulcus* (IPS) will turn out to be most important. The windings on the outer surface, on the other hand, are called *gyri* (singular *gyrus*). For example, the winding anterior to the central sulcus is termed the *precentral gyrus*, and it constitutes our primary motor cortex from which volitional movement commands are issued to make our extremities move.

Historically, many heroic scientists have charted the once-unknown territory of the cerebral cortex. One of the main questions early on was whether specific functions, such as seeing, hearing, speaking, or acting, were processed all over the brain in a holistic way, or rather in localized and dedicated places in the cerebral cortex. During the end of the nineteenth and beginning of twentieth century, it gradually became accepted that different functions are localized in different parts of the cerebral cortex. Of

Localizing Numerical Quantity Representations in the Human Brain

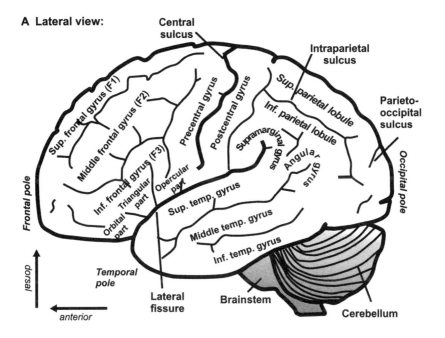

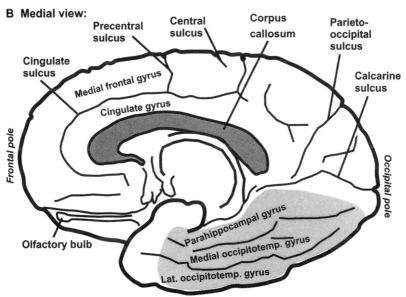

Figure 6.1
Schematic surface of the human brain with anatomical landmarks. **A)** Lateral view of the brain. White surface corresponds to the cerebral cortex. **B)** Medial view of the brain (with brainstem and cerebellum removed).

paramount importance leading to this insight were studies on patients who had lost specific mental functions after suffering from brain damage. After the patients died and their brain were examined post-mortem, lost functions could be traced back to the locations showing brain injuries.

This was the first and perhaps most important insight: the discovery that even the most complex mental faculties, such as human language, reside in specialized areas of the cerebral cortex. In 1861, a certain Monsieur Leborgne was shown to French anatomist Paul Broca (1824–1880). Leborgne, a sick man for many years, had lost his ability to speak and was able only to mumble a word that sounded like "tan." Six days after he was introduced to Broca, Leborgne died. His brain was removed to be examined. The photographs that exist from his brain show massive damage in the lower (inferior) part of the left frontal lobe, at about the height of the temple. This case study convinced Broca and the scientific community at large that language must reside in circumscribed brain regions and is cortically localized in general.[2] Since then, this key language area in the inferior frontal lobe is known as Broca's area. Subsequent studies highlighted the role of causal brain injuries in our understanding of sensory organization in the cerebral cortex.

Because brain injuries in humans occur at rather random sites, some scientists early on also used animals to study the behavioral effects of targeted brain tissue ablations. In addition, the electrical stimulation of various parts of the brain and the resulting movements of animals indicated the locations of certain functions, in particular motor outputs. The discovery of the dog cortical motor area by German physicians Eduard Hitzig (1838–1907) and Gustav Theodor Fritsch (1838–1927) nine years after Paul Broca's report must also be regarded as one of the most important experimental discoveries in the history of neuroscience.[3]

Hitzig and Fritsch used low levels of electrical stimulation applied to the surface of distinctive cortical sites in their study. They concluded that

> certainly some psychological functions, and perhaps all of them, in order to enter matter or originate from it, need circumscribed centers of the cortex.[4]

Later, British physiologist David Ferrier (1843–1928) confirmed the motor cortex experiments of Fritsch and Hitzig and produced more detailed maps of the monkey brain. Ferrier's book *The Functions of the Brain*,[5] published in 1876, opened the "modern" age of neurosurgery, because neurosurgeons could now make use of functional maps of the brain for guidance. In return, new insights in neuroscience emerged from neurosurgical treatments. These advances in cortical localization were complemented by an anatomical map of the cerebral cortex at the beginning of the twentieth century, when German neurologist Korbinian Brodmann (1868–1918) identified a total of 52 discrete cortical areas in the human brain by meticulously inspecting the arrangement of nerve cells in all areas of the cerebral cortex.[6] With this detailed map, it became possible to identify the same cortical areas in different individuals.

While the language, motor, and sensory areas of the cerebral cortex became fairly well delineated in the following years, the remaining other half of the cerebral cortex resisted functional identification. These were the so-called association areas of the brain involved in higher brain functions. The term "higher brain functions" loosely refers to our most cherished mental feats, such as consciousness or mathematics, and their supporting cognitive functions, such as working memory, attention, or decision making.

Then, in 1874, Eduard Hitzig[7] discovered that his experimental animals showed decreased intellectual capabilities after having damaged the frontal association area. He concluded that the frontal association cortex is responsible for abstract thought, an idea that still holds today. Around the same time, David Ferrier conducted experiments on monkeys and dogs and associated the prefrontal "silent" areas with inhibitory functions and attention.[8] Slightly later, the Italian neurologist Leonardo Bianchi (1848–1927), a highly dedicated student of the frontal lobes, also noted changes of "emotive states" that accompanied perceptions and actions. He came to the conclusion that lesions of the prefrontal cortex (PFC) changed the "psychical tone" of the experimental animals.[9] The ideas of Hitzig, Ferrier, and Bianchi received considerable experimental and clinical validation in the years to come.

The scientist, however, who made the most outstanding contribution in delineating the association areas based on anatomical studies was German neuroanatomist and psychiatrist Paul Emil Flechsig (1847–1929). Flechsig examined the maturation course of brains in deceased late-term fetuses and newborns. The higher cortical areas, which become mature in the developmental sequence last, he referred to as "association centers."[10,11] He chose the term "association centers" because these areas contain large numbers of association fibers, or connecting fibers between cortical areas, rather than sensory inputs from the sense organs or motor output fibers to the muscles. Flechsig's insights obtained considerable support over the following decades.[12]

Flechsig further subdivided these association centers: (1) the parietal-temporal-occipital association cortex (*posterior large association center*); (2) the temporal-limbic association cortex (*middle association center*); and (3) the prefrontal association cortex (*anterior association center*).[13] Flechsig recognized these areas' significance in higher brain functions and sometimes called them the "mental centers" (*geistige Zentren*).[14] On the quest for the neuronal foundations of numerical competence, undoubtedly a high-level brain function, these cortical association areas will be revisited often in this book.

In time, Flechsig also realized the evolutionary and phylogenetic implications of his findings. Within the zoological order Primates, the association areas grow larger the more advanced primates become. During the course of evolution, the association areas in the human brain have grown massively and become disproportionally large in humans compared to macaque monkeys.[15] A recent comparison by David Van Essen's group nicely depicts that these areas in the parietal, temporal, and frontal lobe increase by a factor of 32! They also found that the PFC is disproportionately larger in humans by a factor of 1.2 compared to chimpanzees and 1.9 compared to macaques.[16] Our intellectually privileged role in the animal kingdom is mainly due to our vastly expanded association areas in our brains and the cognitive faculties that arise from them.

Despite these evolutionary changes, the major anatomical landmarks can also be found in the macaque monkey brain, which looks like a simplified version of the human brain (figure 6.2). We will need the anatomy of the monkey brain when discussing neurons responsive to numbers in later chapters. In particular, I will often refer to the areas inside the IPS and the lateral PFC.

Due to insights gained from patients and experimental animals, most notably rhesus monkeys, the different cognitive functions can today be mapped onto the cerebral cortex, which seen from the side is parcellated into four distinct *lobes*: the occipital, temporal, parietal, and frontal lobes.[17] The *occipital lobe* is located at the posterior pole of the cerebrum and is dedicated to vision.

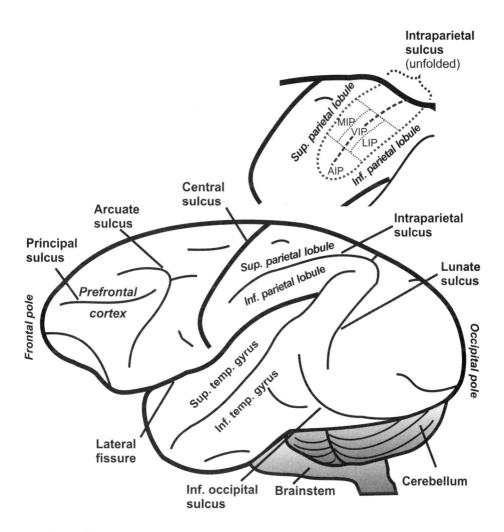

Figure 6.2
Schematic surface of the lateral monkey brain with anatomical landmarks. A) The monkey brain looks like a simplified version of the human brain. The intraparietal sulcus is unfolded in the inset at the top right to expose the lateral (LIP), ventral (VIP), anterior (AIP), and medial intraparietal areas (MIP) inside the intraparietal sulcus.

The *parietal lobe* lies in front of (anterior to) the occipital lobe and extends forward to the central sulcus. The anterior part of the parietal lobe is made up of the postcentral gyrus, which is concerned with somatosensation—that is, touch, pain, temperature, and limb position (proprioception). The posterior parietal cortex (PPC) between the postcentral gyrus and the occipital lobe is a classic association area: It integrates sensory signals from all senses and conveys this information to premotor and motor cortical areas. Within the PPC is the IPS. As a prominent and conspicuous anatomical landmark, it separates the superior parietal lobule from the inferior parietal lobule. The macaque monkey IPS has been divided into several sub-regions (such as the lateral, medial, ventral, and anterior intraparietal areas) on the basis of its histological appearance, anatomical connections, and physiological characteristics. In the context of this book and the topic at hand, the IPS—and the ventral intraparietal (VIP) area, in particular—plays a cardinal role in number processing, and its inner workings will be investigated further below. In addition, the supramarginal and angular gyri will become involved in symbolic number representations and calculation.

The *temporal lobe* is the large chunk segregated by the Sylvian fissure. It is bordered by the occipital lobe posteriorly and the parietal lobe dorsally. Two major parts of the temporal lobe can be discriminated: The first division is the dorsolateral region visible from the side. This area supports auditory functions near the fissure, but mainly cognitive functions associated with several sensory systems, as well as language function at the parieto-temporal junction. We will revisit this temporal-lobe area when discussing numeral representations. The second division is the mediotemporal lobe (MTL), which comprises the temporal areas that lie inside and at the bottom of the temporal lobe. This includes the parahippocampal gyrus, the entorhinal cortex, the hippocampus, and the amygdala. All of these areas of the MTL are multimodal and support a variety of functions, including emotion, affect, motivation, and long-term memory. As will be shown later, the MTL is the only area so far from which neurons representing numbers have been recorded in the human brain.

Finally, the cortex anterior to the central sulcus and dorsal to the lateral fissure is the *frontal lobe*. The frontal lobe is made up of three anatomically and also functionally distinct regions: the *primary motor cortex (BA 4)*, the *premotor cortex (BA 6)*, and the *PFC*. The primary motor cortex at the posterior border is located on the precentral gyrus and is responsible for voluntary movements of body parts. The *premotor cortex* on the lateral surface lies anterior to the primary motor cortex. It receives the majority of its input from the PPC and the PFC.[18] As indicated by its name, the premotor cortex projects mainly to the primary motor cortex and influences voluntary actions.

The most fascinating and largest part of the frontal lobe, and one that is very important to number representations and arithmetic, is the PFC. The PFC is located immediately in front of the premotor cortex. It is anatomically loosely defined as the cortex that receives fibers from a specific nucleus, the mediodorsal nucleus of the thalamus. The PFC is connected to the other association cortices in the parietal and temporal lobes, but also to a bewildering amount of additional brain areas. The PFC, in turn, is divided into three regions: the *lateral PFC (l-PFC)*, the *orbital PFC* just above the skull's eye sockets (orbitae), and the *medial PFC*.

The orbital PFC at the ventral surface of the frontal lobe receives input from the temporal association cortex, amygdala, and hypothalamus and is considered the highest integration center for emotional processing. The medial PFC includes the medial aspects of the PFC and the anterior cingulate cortex. Its connections are manifold and relate to long-term memory as well as to emotions processed through the limbic system. The medial PFC, along with the anterior cingulate gyrus, amygdala, insula, superior temporal sulcus, and temporoparietal junction, has been called the "social brain"[19] and allows us to interact with other people. Finally, the large l-PFC is a key area for cognitive control and will be revisited often in this book in the realm of number processing. The l-PFC extends between the longitudinal cerebral fissure that divides the two cerebral hemispheres and the lateral fissure below. This region receives processed multimodal information and has been described as a place "where past and future meet"[20] by associating memories from the past with future actions. The l-PFC is also considered a cardinal site for working memory—the capability to consciously memorize and manipulate limited amounts of information for the near future.[21] More broadly, the l-PFC is considered the brain's central executive, which integrates external and internal information, evaluates it in the light of previous experiences and current needs, and determines the means to reach often far-removed goals.[22]

As mentioned earlier, the classical association cortices in the temporal, parietal, and frontal lobe receive highly processed information from all senses, such as vision, audition, and touch. This is accomplished by anatomical pathways that project from the primary sensory cortices (primary visual, auditory, and somatosensory cortices) along a hierarchy of processing stages to the association cortices (figure 6.3). In the primate brain, information from primary sensory areas is processed via two pathways that work in parallel; one pathway via the parietal pathway, and one via the temporal association cortices. Information from both parietal and temporal pathways terminate in the prefrontal association cortex. All association cortices, and in particular the PFC, can create a multimodal representation of the outer world. This will become important when discussing number

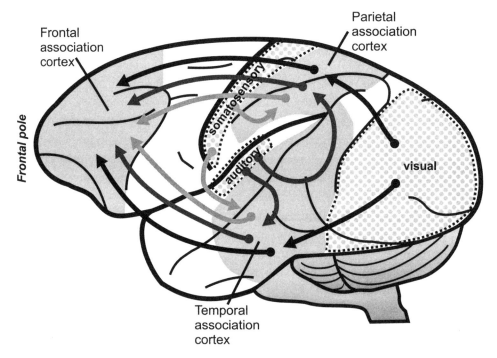

Figure 6.3
Pathways in the primate brain that convey sensory information from primary sensory cortices to the association cortices. In the primate brain, information from primary sensory areas, such as visual, auditory, and somatosensory (touch) cortices (*dotted filling*), is processed in parallel via the parietal and temporal association cortices to the prefrontal association cortex (association cortices in *light gray*). These sensory pathways in the non-human primate are equivalent to those in the human brain.

representations across different sensory modalities, such as seeing and hearing. The association cortices, in turn, project to premotor and motor cortices, primarily in the frontal lobe to elicit actions. The association areas therefore operate at the level between sensory input and motor output; they are responsible for the merging of the senses and our capacity to "think" things over before we act.

6.2 At a Loss for Numerical Quantity after Brain Injury

So far, we have seen that specific functions reside in dedicated, but of course interconnected, locations in the brain's cerebral cortex. The obvious question now is whether numbers have a special site in the cerebral cortex where they are being processed. Historically, case studies in patients with brain injuries provided the first insights into the neural roots of mathematics. At the beginning of the twentieth century, neurologists realized that brain damage could result in peculiar deficits that specifically affect calculation abilities and arithmetic. Such acquired effects in patients shed light on the location of numerical abilities in the brain.

In general, brain injuries can have different causes. For example, one injury that can be sustained is a stroke, more properly called *cerebrovascular damage*. When the blood flow to the brain, which supplies nerve cells with oxygen and nutrients, is impaired by a blockage or rupture of a cerebral artery, neurons may die due to a lack of oxygen. The affected tissue is lost permanently, because neurons in the adult central nervous system don't regrow. Sometimes neighboring brain areas take over parts of the initially lost functions, but this takes time. *Neurodegenerative diseases* are a second cause of brain damage. They are characterized by the ongoing dying away, or progressive degeneration, of neurons. Alzheimer's disease, Parkinson's disease, and Huntington's disease all are examples of this heterogeneous group of disorders.

Brain injuries can provide very direct insights about the location and parcellation of cognitive functions. If a behavior is lost after the dying away of neurons in a specific location, then this brain area was involved in producing this behavior. This gives rise to a very causal brain-mind relationship. Of course, the damaged territory is often large, sometimes even spanning across several cortical lobes, thus affecting a multitude of functions. This makes it difficult to precisely identify the region required for a particular function based on the behavior of a single neurological patient. Moreover, the site and extension of lesions vary from patient to patient, making it difficult to find the one brain location that is responsible for a specific function in a typical brain. To avoid erroneous conclusions based on single cases, more and more studies with population of patients are published. Patient populations allow to superimpose the individual lesions in an overlay plot to find the location that is commonly lesioned across patients that show the same disorder.[23]

Patients suffering from deficits in dealing with numbers have been described already at the beginning of the twentieth century. In 1908, the German neurologists Max Lewandowsky and Ernst Stadelmann[24] published the first report of a 27-year-old man who showed calculation deficits after brain damage to the left posterior, possibly parietal, cerebral cortex. This patient in particular had problems performing calculations and recognizing arithmetic symbols while language functions were preserved, Later, the German neurologist Georg Peritz investigated soldiers from World War I who returned from the battle field with gunshot wounds to the back of their heads. His studies pointed to a "confined location" for calculation, a brain center in the left posterior half of the brain that he localized near the angular gyrus at the parietotemporal junction.[25]

The Swedish neurologist Salomon Eberhard Henschen confirmed these earlier findings and proposed that these defined areas in the brain played specific roles in understanding and executing calculations.[26] Based on a review of numerous cases in previous literature and those from his own patients, he proposed anatomical substrates for arithmetic operations that differed from those required for language and music. However, rather than focusing on one specific brain center, Henschen thought that calculation incorporated a sensory and a motor function. He localized the sensory calculation center in the (left) angular gyrus (as Peritz before him) and the intraparietal fissure (presently called the IPS). The motor calculation center he postulated was in the third convolution of the frontal lobe, known as the inferior frontal gyrus, since lesions there affected continuous counting of numbers and the ability to write numerals. In 1925, Henschen also coined the term *acalculia* (Akalkulia),[27] now broadly used to describe an acquired impairment from some sort of brain damage that results in patients having difficulty performing simple mathematical tasks. In 1948, German neurologist and psychiatrist Kurt Goldstein confirmed these classical findings and proposed that acalculia typically resulted from lesions in the parietal-occipital region and occasionally the frontal lobes.[28] Moreover, the roles of the posterior parietal and frontal cortex were further confirmed in the years to follow, which will be shown in later chapters.

Understandably, all these earlier studies were primarily concerned with human-specific arithmetic skills that are part of our mathematical symbol system. These will be touched on later, but for now a more fundamental question needs to be answered: where in the brain is basic numerical quantity located? Such numerical quantity is not represented by number symbols, but rather as non-symbolic magnitudes that serve as a foundation to numerical skills. Only a few, and only relatively recent, studies address this question, but they all point to the PPC's surfaces as a key area for quantity estimation.

In one case study from 2003, a patient who was reported to have a focal lesion to the left parietal lobe demonstrated a severe slowness in estimating numerosities.[29] Also, the patient exhibited associated impairments in subitizing as well as in numerical comparison tasks, both with arrays of dots and with Arabic numerals. A similar impairment in estimating non-symbolic numbers was reported in another patient, who suffered from number deficits following cerebrovascular damage restricted to the left IPS (figure 6.4). This patient exhibited difficulties in processing the basic numerical quantity of arrays of one to nine dots.[30] More precisely, it took the patient significantly longer to compare and count the quantity of dots. Both studies suggested that the IPS is a key area of the brain's system for numerical quantities.

Similar brain locations were found to affect non-symbolic number processing in patients who suffered from neurodegenerative diseases. A patient diagnosed with posterior cortical atrophy, a syndrome of dementia, showed degeneration in the parietal lobe.[31] Besides impairments in all tasks involving dot counting, deficits appeared in tasks requiring mental number bisection. Mental number bisection requires judging the numerical midpoint between two numbers, for instance recognizing that six is the midpoint between three and nine. Numerical approximation and estimation were also impaired. In addition, an increased numerical distance effect was found in number comparisons, suggesting imprecise representations of numerical values. Numerosity deficits were also described in patients suffering from cortico-basal syndrome, another type of neurodegenerative condition that

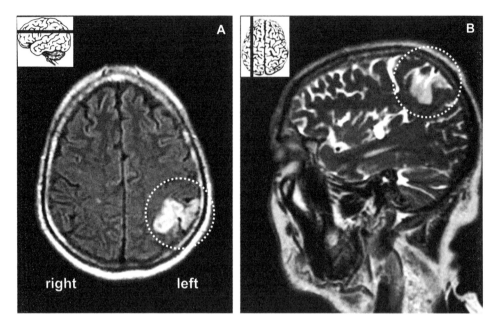

Figure 6.4
Basic numerical processing is impaired after damage restricted to the left intraparietal sulcus. Horizontal (**A**) and sagittal (**B**) MRI sections of an acalculic patient. Sections in the horizontal and sagittal plane are shown in the insets. The surrounded brain regions appearing lighter demonstrate an infarct in the left intraparietal sulcus (from Ashkenazi et al., 2008).

affects the cerebral cortex and the IPS.[32] Compared to healthy seniors, cortico-basal syndrome patients were impaired and required three times as long to judge whether a target numerosity (e.g., three dots) or a number (Arabic numerals) falls between two bounding numerosities (e.g., 1 and 5).

6.3 Mapping Numerical Quantity on the Healthy Human Brain

Brain injuries change brain functions in a radical, unpredictable, and irreversible way. Fortunately for the patients, the brain often remains plastic and sometimes can compensate for parts of lost functions by engaging neural territory adjacent to the lesion. From a scientific point of view, however, this means there is no guarantee that the locations of changed behaviors in a diseased brain correspond to normal functions of a healthy brain. For centuries, neuroscientists therefore longed for methods that would measure the locations of brain functions in healthy subjects, without invading the brain. Toward the end of the twentieth century, such methods allowing "functional imaging" became available. With the rise of methods such as positron emission tomography (PET) and functional magnetic resonance imaging (fMRI), the cerebral locations for numerical processing could be investigated in the intact human brain. Today, functional imaging is the dominant research method to study number processing in the human brain.

The first images of live, calculating brains date back to 1985 and were captured by PET. That year, Per Roland and Lars Friberg from the University of Copenhagen published a seminal article[33] with the programmatic title "Localization of cortical areas activated by thinking." One of the tasks included in their study was a mental calculation task. Human subjects received instructions to start with the number 50 and continuously subtract three from all intermediate results. This all had to happen silently and while lying totally still in the PET scanner; the participants were not allowed to say anything or to move during the procedure. During this task, the regional cerebral blood flow was measured using PET. A mildly radioactive, yet harmless, isotope was injected into the bloodstream leading toward the brain. If neurons in certain areas of the cortex are particularly active during a task, they require more oxygen and sugar, both substances which are transported via the bloodstream. Therefore, the cerebral blood flow in the blood vessels supplying these brain areas increases as a sign of higher neural activity. Because radioactive isotopes are now being transported with the bloodstream, brain areas that show a higher metabolic need and therefore higher blood flow should also be more radioactive. In other words, the high level of radioactivity in certain brain regions serves as an indirect measure of neural activity. To that aim, the head is placed inside an array of radioactivity detectors to find the spots on the cortical surface that are particularly radioactive during the mental calculation task. Of course, this method does not allow measuring fast changes in blood flow; the resulting image is basically a snapshot of activity for a given time interval during the experiment.

Roland and Friberg discovered consistent increases in blood flow during the mental calculation tasks in their subjects. In the frontal lobe, regions of the superior and inferior frontal gyrus were most active. In addition, parts of the inferior parietal lobule extending into the angular gyrus showed increased blood flow. Of course, these maps were still rather imprecise due to the technical limitations of the day. One further has to be aware that repeated mental subtractions also engage processes more general than just number representations, such as working memory and task strategy. Nonetheless, this study was a beacon for a new area in numerical cognition.

Truth be told, PET became somewhat outdated for basic research questions with the advent of functional MRI. For one thing, PET studies are rather expensive. In addition to that, being injected with radioactive substances is not very popular either. However, PET is still superior to fMRI for certain clinical issues and therefore maintains its place as an indispensable and mildly invasive imaging method.

FMRI is currently the most popular imaging technique. It was developed in the early 1990s by Seiji Ogawa[34] and Ken Kwong.[35] Just like PET, fMRI measures the differences of blood flow in the blood vessels surrounding active neurons that feed on oxygen and sugar. However, the type of signal measured is different. In contrast to PET, fMRI works with the magnetic behavior of the atoms in the body, not with radioactivity.

Functional MRI exploits the magnetic behavior of atoms, specifically in a vital molecule called hemoglobin. Hemoglobin is an important part of the red blood cells that carry oxygen to neurons in the brain. When hemoglobin is loaded with oxygen, thus becoming "oxygenated," its magnetic behavior differs substantially compared to hemoglobin that is depleted of oxygen, or "deoxygenated." Since blood oxygenation usually increases following neural activation, this difference can be used to detect brain activity indirectly. This form of MRI signal is known as the blood-oxygen–level dependent (BOLD) signal, on which fMRI relies.

In brain areas that respond to a specific stimulus (say, a picture or a sound), one would expect the BOLD signal to go up and down as the stimulus is turned on and off. These ups and downs of the BOLD signal are smeared over time, because the blood flow response lags behind neuronal activation by a few seconds. The fMRI activity in a piece of brain is defined as how closely the time course of the BOLD signal from that area matches the expected time course. This step requires quite a bit of statistical processing.[36]

Because the brain and therefore also its blood vessels are always active, simply recording the fMRI signal during a task would not show much.

FMRI activation during a test situation usually needs to be compared with a control condition that only differs in one specific stimulus component. For instance, if one wants to know if there is an area in the brain that responds specifically to speech as opposed to any sound, the BOLD signal during stimulations with non-speech sounds must be subtracted from the signal during stimulations with speech sounds. The remaining activation at a dedicated location in the brain would then indicate that this area is specifically processing speech. It is evident that the quality of the derived fMRI data depends heavily on the suitability of the control condition used during this experiment.

FMRI has become a valuable and very popular method in neuroscience due to its significant advantages. It is non-invasive and works through the intact skull; no surgical procedures are required to access the brain. And unlike PET, no radioactive substances have to be injected into the veins. Compared to other non-invasive methods, such as electroencephalography (EEG) and PET, fMRI can measure relatively small areas of brain activation to provide relatively precise activation maps. Usually the activated areas can be confined to several square millimeters, but the best machines today can provide even smaller activation areas. Because every university these days has several scanners at their disposal, and because elaborate software packages take care of the nitty-gritty details of data analysis, fMRI has become relatively easy for the experimenter to use. Or at least, so it seems.

Despite all its advantages, one needs to be aware that fMRI is not measuring the electrical activity of neurons, but merely blood flow. This cannot be emphasized enough. An fMRI image of the brain is comparable to watching a face with a thermal camera. And just as the hot-red spots in the face tell us indirectly where muscles are active and produce heat, the fMRI signal betrays the locations in the brain where lots of neurons are somehow active, but not how they transmit information. One also has to be aware that the fMRI signal results from the averaged demands of thousands of neurons, each with their own functional specializations, within a given measured brain volume. Moreover, because fMRI measures changes in blood flow and the blood system needs several seconds to increase or decrease blood flow in the vessel system, the temporal precision in detecting such changes is comparatively low. Bearing these limitations in mind, however, there is no doubt that present cognitive neuroscience is inconceivable without functional imaging.

Several fMRI imaging studies identified the PPC as a crucial area in representing the number of visual items in a collection. In a seminal study,[37] Manuela Piazza and Stanislas Dehaene localized and characterized visual numerosity representations in healthy adult subjects using fMRI. The special approach they chose to use is called *fMRI adaptation*. As the name indicates, this approach depends on adapting BOLD activity by repeating the same stimulus several times. The mechanism behind fMRI adaptation relies on what is known from neurons. Single neurons in monkeys adapt to repeatedly presented identical stimuli; neurons tuned to a specific picture progressively decreased their activity with every repetition of the preferred stimulus, an effect referred to as habituation or adaptation.[38] This effect, Piazza and Dehaene reasoned, could be exploited to explore where in the brain the BOLD signal would adapt to repeatedly presented target numerosities as a function of habituating putative number-sensitive neurons (figure 6.5A).

In this fMRI adaptation experiment, the participants laying in the MRI scanner simply had to look at sets of dots without consciously processing the numbers. This mitigated the influence of unwanted mental responses as a source of fMRI activation. Initially, visual displays of one fixed numerosity were shown on a monitor. If any region of the brain contains putative neurons responsive to that numerosity, those neurons should habituate (that is, decrease their discharge) with repeated presentations of the number. In contrast, putative neurons tuned to numerosities other than the displayed one should not be affected. After such a habituation and decrease of the fMRI response, the presentation of an occasional deviant number should then lead to a recovery of fMRI response in the same brain area, with an amount of activation inversely related to the numerical distance between the habituation and the deviant numerosity. Indeed, when the researchers presented a different number of dots after the repeated presentation of the same number, they measured a total recovery of the BOLD signal in the same brain area of the IPS. The only site of a whole-brain search that significantly habituated to numerosities was the horizontal segment of the IPS. Different populations of neurons in the human IPS are indeed responsive to the number of dots in a set. Control experiments showed that it was the number of dots, not the appearance of the dots, that caused this fMRI rebound.

My postdoc, Simon Jacob, and I repeated the very same fMRI adaptation experiment with healthy human volunteers years later and confirmed the BOLD activation in the IPS found by Piazza and Dehaene.[39] In addition to the IPS, however, we also detected numerosity-selective BOLD activity in the lateral PFC (figure 6.5B). Since then, similar prefrontal activations during fMRI adaptation have been reported.[40] This suggests that not only areas of the posterior parietal cortex, most notably the IPS[41,42,43] and regions of the superior parietal lobule,[44] but also parts of the PFC become automatically activated whenever our brain perceives numerical quantity.

Of course, the PPC is not only encoding numerosity when it is passively viewed, but also when it is actively processed in a discrimination task. Evelyn Eger and Andreas Kleinschmidt asked participants in a scanner to perform a delayed matching-to-sample task with numerosities, that is, to watch a numerosity display, memorize it during a brief delay period, and later judge whether a comparison numerosity showed the same or different number of items.[45] The authors then trained a computer program (a statistical classifier called Support Vector Machine) to analyze the systematic changes in BOLD activation patterns in the parietal lobe when the participants saw and memorized a specific number. For example, the computer program could learn that four dots elicited a specific distribution of BOLD activation spots on the cortical surface that was different from the one caused by eight dots. Based on what the computer program could learn about these specific activation patterns in the bilateral intraparietal cortex, it was surprisingly able to decipher the particular numerosity that participants had seen in novel trials. This approach also provided evidence for the graded nature of the numerical representation in the human PPC: the classification accuracy increased the more numerically distant the numerosities were, particularly in an area thought to be the human equivalent of the monkey lateral intraparietal (LIP) area.[46]

In our mind, numerical quantity is orderly—arranged from small to large, giving rise to a "mental number line." Could such a mental number line be mirrored in the brain, so that neighboring brain areas could represent neighboring set sizes, which ultimately would give rise to a map for numerosities? It is not unreasonable to suspect such a map-like arrangement. After all, orderly arrangements of external stimuli are often mirrored in the brain, so that the spatial layout of stimuli in the external world is mapped onto the brain surface. For instance, the tactile sensations on our body surface are topographically mapped onto the precentral gyrus.

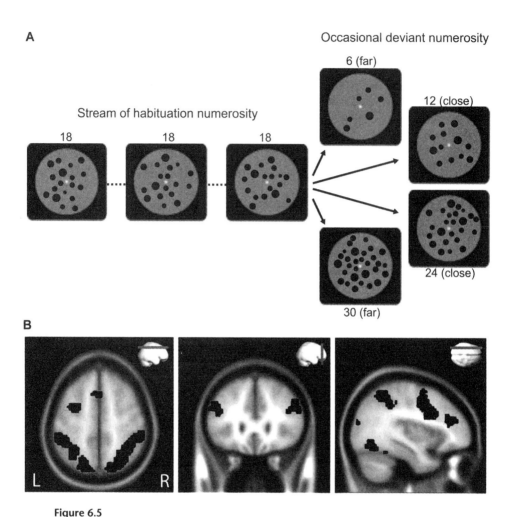

Figure 6.5
Functional MRI imaging for numerosity in the human brain. A) Experimental design for adaptation to approximate number. Human participants were scanned while they were passively presented with a rapid stream of sets of 18 dots. Unbeknownst to the participants, the number of dots occasionally changed. B) Brain areas processing number reacted to number change. Black areas show cortical regions that respond more strongly to more distant numerical deviants than to stimuli closer to the adaptation numerosity during fMRI adaptation. Sections in the horizontal (*left panel*), frontal (*middle panel*), and sagittal plane (*right panel*) are shown. The slice position is marked in each panel by the line superimposed on a schematic brain at the top right. Activation was observed in the bilateral intraparietal sulcus and precentral and prefrontal areas, as well as in the anterior cingulate and inferotemporal cortex (from Jacob and Nieder, 2009).

Adjacent cortical areas represent neighboring parts of the body: the fingers are represented next to the palm, followed by the wrist, the forearm, and so on. The entire body surface is represented as an orderly topographic map on the surface of the postcentral gyrus. Why couldn't the same logic apply to numbers then?

Serge Dumoulin and Ben Harvey from the Netherlands addressed the question of a numerosity map using high-field fMRI that measures BOLD signals with high spatial resolution.[47] They scanned adult human subjects who passively looked at dot patterns that increased and decreased in numerosity over time (spanning one to seven dots). Subsequently, they found that activation sites responsive to small numerosities in the human superior parietal lobule were organized topographically. The superior parietal cortex had been implicated in numerosity before,[48] but the new study found an orderly layout of numerical values. The activation sites formed a numerosity map that was specific for the number of dots irrespective of the exact visual layout of the set. Next to the activation site for one was the activation spot for two, and so on. FMRI activation formed a "numero-topically" organized continuum, with small numerosities represented medially and larger numerosities represented laterally in the superior parietal lobule. Along this map, the amount of cortical space devoted to a specific numerical quantity was not equal, but more cortical surface area was reserved for smaller numbers, and cortical area decreased with increasing numerosity. Interestingly enough, no spatial layout was observed for symbolic or larger non-symbolic numbers.

All fMRI studies mentioned so far only used dot arrays as numerosity stimuli. This simultaneous presentation of the items in dot arrays is of course only one possibility to present numerical quantity. In everyday life, we also enumerate items that appear one after the other across time in a sequential manner. Still, both ways to present the items of a set are instances of abstract numerical quantity. Are these two numerosity presentations encoded by different brain areas, or are certain brain areas engaged irrespective of a simultaneous or sequential presentation? The latter finding would support the idea that the brain encodes numerosity abstractly. To answer this question, activation of both presentation avenues in the same subjects must be compared.

Valérie Dormal and Mauro Pesenti from the University of Leuven in Belgium[49] used fMRI to identify the areas involved in numerosity processing while participants classified linear arrays of dots (simultaneous numerosity presentation) or flashed dot sequences (sequential numerosity presentation). As predicted by earlier studies, the processing of simultaneous numerosities induced activations bilaterally in several areas of the IPS (and the inferior temporal gyrus). However, a different activation pattern emerged for the sequential presentation; activations were restricted to the right hemispheric IPS and the inferior frontal gyrus during the processing of sequential numerosities. These findings argue that simultaneous and sequential numerosities are not recognized as identical by the brain because they engage different cortical networks. Importantly, however, these cortical networks were not completely segregated but overlapped in two locations, namely the right IPS and the right precentral gyrus extending into frontal gyrus. The involvement of the precentral gyrus, which hosts the motor cortex responsible for body movements, is somewhat unexpected. Perhaps its contribution reflects the use of hands and fingers to keep track of numerical magnitude. This might also explain why another study found the lateral premotor cortex to be most consistently activated when participants sequentially enumerated sensory items and counted motor movements.[50] Whatever the reason may be, the IPS and the right inferior frontal gyrus seem to host abstract numerical representation independent of spatial or temporal aspects.

Recently, Saudamini Damarla and Marcel Just had a more ambitious goal, namely to explore a putative cross-modal numerical representation.[51] They wondered if the fMRI activation patterns obtained for a sequential presentation of a specific numerosity in one sensory modality would be similar enough to predict the numerosity shown to the subjects in another sensory modality. They presented the numerosities one, three, and five alternating as sequential visual dots or sequential auditory sounds. After measuring BOLD signals, the authors trained a statistical classifier, similar to the one used by Evelyn Eger and Andreas Kleinschmidt mentioned previously. With this classifier, Damarla and coworkers investigated whether neural representations of quantities depicted in one modality (visual) can be identified from brain activation patterns evoked by quantities presented in the other modality (auditory). Indeed, they found that the quantities of dots were recognizable by a classifier that was trained on the neural patterns evoked by quantities of auditory tones, and vice versa. The brain regions supporting quantity classification belonged to the bilateral frontal and parietal lobes. These regions included the inferior and superior parietal lobules, the intraparietal sulci, and postcentral regions in the parietal lobes. For the frontal lobes, cross-modal quantity classification was supported by bilateral precentral, superior frontal, and inferior frontal regions. Interestingly, the representations that were common across modalities were mainly right-lateralized in frontal and parietal regions. In addition, the neural patterns in parietal cortex that represent numerosities were the same in all participants. This study showed stable neural representations of sequential numerosities across visual and auditory modalities. The findings again highlight the importance of the posterior parietal and prefrontal cortices for numerical quantity representation.

7 Number Neurons

7.1 The Language of Neurons

It is one thing to have a physical stimulus, such as the number of objects located in the outside world. It is a very different matter to have a stimulus "represented" or "encoded" in the brain. This is the job of neurons. The 86 billion neurons our human brain is made up of are biological processors that receive, evaluate, and transmit information. This process is called "neuronal signaling," and the language (or code) neurons use are tiny electrical impulses.

The body, or "soma," is the main processing part of the neuron (figure 7.1). It receives information from other neurons via multiple branched appendages called "dendrites" (from Greek "tree"), whereas the axon conducts electrical signals away from the soma to the next neuron or to other effector organs, such as muscles. The tiny gaps or junctions between the axon endings of one neuron and the dendrites or soma of the next neuron are termed "synapses." A synapse is nothing but a communication channel between neurons, and this communication can take place in two fundamentally different ways, namely electrically or chemically. In electrical synapses, currents flow from one neuron to the other across miniature membrane tubes that connect the fluid inside neurons (called "cytoplasma"). In chemical synapses, however, such bridges do not exist; rather, messenger substances called "neurotransmitters" are liberated at the nerve endings (pre-synapse) of one neuron and cross the tiny synaptic cleft to reach the (post-synapse) of the neighboring neuron. Neurotransmitters attach to receptors of the neighboring cell to influence the excitability of this neuron. Today, more than a dozen chemical substances have been

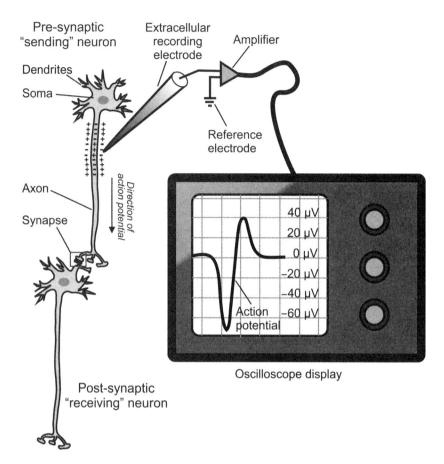

Figure 7.1
Recording electrical impulses from single neurons with microelectrodes. The left side shows a simplified presynaptic neuron that generates an electrical impulse, an action potential, which is sent along the axon toward the terminal endings. The axon's membrane segment that produces the action potential briefly depolarizes (more positive charge on the inside of the membrane). The membrane segments before and after the action potential show resting membrane potential, characterized by more negative charge inside the neuron. This excitation is conveyed via a synapse to a postsynaptic neuron. The right side schematizes extracellular recordings from the presynaptic cell. The action potential is registered at the tip of the recording electrode, amplified and filtered, and the voltage of this impulse is displayed across time on an oscilloscope. Because the action potential is recorded from outside of the neuron, it begins with a deflection in voltage amplitude due to an influx of positive sodium ions into the neuron.

identified that serve as neurotransmitters. Different neurotransmitters can influence the following (or post-synaptic) neuron in two distinct ways: either by exciting it, or by inhibiting it. These opposing synaptic effects from hundreds or even thousands of input neurons provide ample opportunity for neurons to compute information by integrating excitatory and inhibitory influences.

A sophisticated machinery located in the neuron's membrane computes the excitatory and inhibitory incoming signals and translates the resulting electrical activity into one or several electrical impulses called "actions potentials," or "spikes." An action potential is created when the difference in charge across the membrane—that is, the membrane potential which in a resting state is negative on the inside—becomes briefly flipped. Each of these impulses has an amplitude of about a tenth of a volt and lasts for about one-thousandth of a second. Because of the very short duration of a single impulse, a neuron can create tens, sometimes even hundreds of action potentials per second. The more action potentials generated per second by a neuron, the more it signals specific information. Such action potentials travel with speeds of up to a hundred meters per second down the axon until they reach terminals and are passed on via a synapse to the next neuron. In this way, neurons build electrical circuits and networks, and by doing so, they represent all our mental life.

But how can this process of neuronal signaling in tiny neurons of up to only 50 μm in diameter be monitored? This is accomplished through the method of single-cell (or single-neuron) recording. It allows neurophysiologists to directly observe the electrical activity of neurons in the living brain. Michael Gazzaniga and colleagues praise this method in their text book on cognitive neuroscience:

> The most important technological advance in neurophysiology—perhaps in all of neuroscience—was the development of methods to record the activity of single neurons in laboratory animals. With these methods, the understanding of neural activity advanced by a quantum leap.[1]

Being a neurophysiologist myself, I of course share this enthusiasm. After all, one could say that neurons are the atoms of perception, thought, and behavior. In order to understand the mind, and ultimately also its diseases, it is indispensable to understand the electrical activity of neurons that give rise to everything we experience, including numbers.

With electrophysiological recordings, the response characteristics of single neurons—the individual elements of the nervous system—can be measured. As the eminent British neurophysiologist Horace Basil Barlow once wrote: "The neuron remains the important unit of function for developing a rational account of how behavior is generated."[2]

To measure electrical action potentials, a hair-thin microelectrode is inserted into an experimental animal's brain for single-cell recording. When the electrode tip, the only electrically conductive part of the electrode, is in the vicinity of a neuron, changes in electrical activity of the neuronal membrane can be picked up. Since our very goal is to learn how neuronal activity gives rise to behavior and cognition, the best and most direct approach is to record neuronal activity while the animal is engaged in a specific, usually computer-controlled task, and produces controlled behavior. One can infer how certain neurons represent cognitive functions, such as numerical competence, by comparing behavior and neuronal responses that happen at the same time. Since the brain does not contain pain receptors and all potentially painful surgeries are done under general anesthesia, neuronal recordings can easily be achieved in performing animals. However, a valid license is obligatory to experiment on vertebrate animals, and before obtaining such a license, the details of the planned experiments and an ethical evaluation is obligatorily required and judged by an independent review board.

In chapter 2, I defined a "mental representation," a purely behavioral concept. However, the primary goal of single-cell recordings is to determine the "neuronal representation" of a sensory stimulus, a mental state, or a motor output. A "neuronal representation" can be defined as a neuronal signal with content and a function.[3] What does this mean?

The content of a neuronal signal is the message or information that the signal provides when received. This is classically determined by the changes in the responses of a neuron when certain stimuli are presented to the animal, or when the animal performs certain tasks, such as memorizing or deciding upon a stimulus. In numerical cognition, for instance, one of the first questions is whether a neuron changes its impulse rate (the number of action potentials per second) when different numbers of dots are shown to the animal. In presenting different stimuli one at a time, the responses of an individual neuron can be correlated with a specific stimulus. If a neuron is responding to a given stimulus, but not to others, this neuron's activity is selective for this given stimulus. Single neurons can be surprisingly specific to the stimulus they are encoding. For example, some neurons in the temporal lobe respond exclusively to faces, but nothing else.[4] As will be shown, the same kind of sophisticated neurons work within the number domain. One needs to be aware, however, that even such dedicated neurons don't work alone but are embedded in networks of neurons, sometimes spreading over distributed brain areas.

One way to graphically assess the content of a neuronal signal is by depicting the stimulus-response function, or input-output behavior, of a neuron in the form of tuning curves. In general, the impulse rate is plotted against varying values of a stimulus dimension. To plot a tuning curve of a neuron's selectivity to numbers, for example, means plotting the number of action potentials the neuron created per second against the range of presented numbers from small to large. Such graphs will be shown later.

The function of a neuronal signal, the second part of the definition of "neuronal representation," is its meaning for the animal's perceptions, behaviors, and internal cognitive states. The meaning of the neuronal signal can be explored with tuning curves. For instance, if the tuning curve to numbers differs in a predictable way when the animal is judging a number correctly as opposed to erring, this would indicate that the neuronal signal of this neuron is relevant for the animal's perception of number. One could also experimentally change, or perturb, the neuronal signal of a neuron by specific methods that will be explained later in order to cause predicted changes in number perception. This is a particularly elegant and telling way of learning about the function of a neuronal signal. These tests of functional relevance of neuronal signals will also be revisited in subsequent sections.

7.2 The Discovery of Number Neurons in the Monkey Brain

At the turn of the new millennium, ample evidence showed that monkeys could crunch numbers. Of course, the monkeys could only do so because their brains endowed them with such skills. Finding out how the workings of the brain would give rise to numerical competence seemed like the logical next step to me, and I was fortunate enough that I could work on this project in Earl Miller's laboratory at MIT, who taught me all

I needed to know and supported me at every step along the way in the following years.

In order to see what happens to individual neurons in a monkey's brain when a monkey thinks of a number of dots, that monkey has to actively see and discriminate numbers. To achieve this, we decided to train monkeys on a delayed match-to-sample task I mentioned earlier (figure 3.7A). We showed monkeys one set of dots on a computer monitor, and then, after a one-second interval, showed another set of dots. The monkeys were taught that, if the second set was equal to the first set, releasing a lever they were grasping would earn them a reward of juice. If the second set was not equal to the first, which randomly happened in half of the trials, then the monkeys were not allowed to respond; instead, they had to wait for a third set, which always matched the first, before releasing the bar to earn their reward. Whenever one such trial was finished, a new trial started with a different set of dots. As described earlier, we implemented all sorts of controls to ensure that the monkeys were not using purely visual features, but only the number of items to solve the task. All of this—such as the presentation of controlled images during certain trial epochs, the rewarding after correct responses, and later the recording of the neurons' action potentials at each moment during the trials—was accomplished by a properly programmed computer. And so, after hundreds of trials, we not only could find out how monkeys discriminated numerosity, but also how neurons in their brains represented these quantities.

We wanted to see what was happening in the monkeys' brains when they perceived the sample numerosity and briefly memorized it during the one-second working memory interval between the sample and test stimulus. So, after a surgery necessary to access the monkeys' dorsolateral PFC, we inserted a microelectrode a few microns in diameter into the neural tissue around the principal sulcus (figure 6.2).

When the monkeys saw and memorized numbers, we realized that certain neurons became very active.[5] On closer examination, it was obvious that some neurons responded with varying impulse rates depending on the number that the monkey was thinking of at the time of a trial. Much to our surprise, each of these neurons favored one of the shown numbers and created a burst of action potentials whenever this preferred number was shown. The neurons signaled the presence of specific numbers; they were numerosity-selective neurons—or, in short, "number neurons"! By listening to the neuronal impulses played over a speaker, one could hear how their activity gave rise to the monkey's experience of numerical quantity. It sounded like the crackling of dry logs in the heat of a bonfire. To this day, I am thrilled by listening to neurons discharging, and it reminds me of why I became a single-cell physiologist in the first place; recording from neurons is as close as one can get to experiencing the mind at work in real time.

Figure 7.2 depicts the detailed responses of one such number neuron from the PFC that is tuned to number two. The responses of this neuron in an ongoing trial are shown in several different ways (figure 7.2). In the so-called dot-raster histogram (figure 7.2A), each line represents a single trial. The action potentials the neuron generated during each trial are depicted as small dots. All even numbers (and one) up to 30 were displayed, and each number was displayed for many trial repetitions. Before the number array is presented, the monkey faces a blank screen in the fixation period during the first 500 ms (see task layout in figure 3.7A), causing only spontaneous and random action potentials by the neuron. Then, between 500 to 1,000 ms, different numbers of items are displayed. This prompts the neuron to discharge more strongly depending on the number shown, and therefore the dot pattern in the histogram becomes denser (figure 7.2A). The neuron's increased activity is delayed by about 150 ms relative to the onset of the stimulus. This time is lost until the information received in the eye reaches the PFC after a chain of neurons along the visual pathway. Number two, the preferred number of this neuron, elicits the largest number of action potentials, as illustrated by the blackening in the dot-raster histogram. For numbers more distant from two, the firing rates dropped gradually, causing neuronal tuning. When the sample number is removed from the screen during the delay phase, the neuron continues to show a selectively higher firing rate for its preferred number two. This is a neuronal correlate of working memory for number two. The same neuronal activity of this neuron is plotted as a spike-density histogram in figure 7.2B. The functions show the momentary averaged firing rate that occurred during trials, and this activity is temporally correlated with the time axis of the above dot-raster histogram and the trial sequence. It can be seen that number two excited the neuron the most. Numbers further away from two excite the neuron less and less.

The average firing rates to the different numbers during the sample phase are next used to create a numerosity tuning curve of this neuron as a

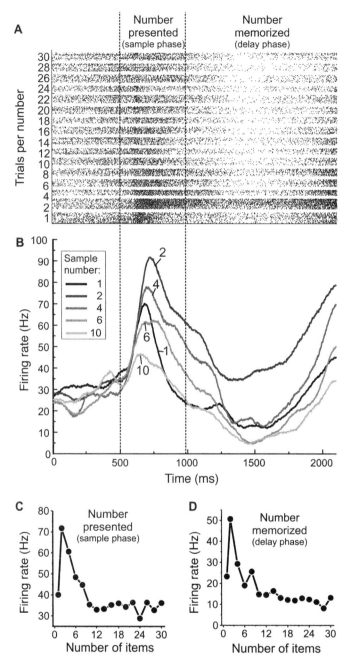

Figure 7.2

Single neurons in the monkey brain represent numbers. A) While the monkey performed the delayed match-to-numerosity task with dot arrays, single-cell activity is recorded and plotted as a dot-raster histogram. The action potentials the neuron generated during each trial are depicted as small dots. B) The same neuronal activity of the neuron in (A) plotted as a spike-density histogram. Only a selection of the numbers are shown for clarity. Different shades of grey correspond to specific tested numbers. C) The numerosity tuning curve of this neuron calculated from the activity in the sample phase. D) The same neuron was equally tuned to number two during working memory (from Nieder, 2016).

visualization of its neuronal representation of numbers (figure 7.2C). This tuning curve clearly shows that the responses to number two are strongest, and firing rates progressively decay for more remote numbers when the monkey sees numbers. This gives rise to a bell-shaped tuning curve. Finally, a tuning curve is plotted based on the neuronal responses during the delay phase, when the monkey is memorizing numbers (figure 7.2D). Clearly, this neuron shows exactly the same tuning while the monkey memorizes numbers. This neuron signals number two both during seeing number and memorizing it in working memory.

As expected of number neurons, the precise layout of the dots was totally irrelevant to them; whether the dots were large or small, densely packed or not, in all cases the neuron discharged most to its preferred numerosity. The appearance of the set did not matter to a number neuron; it was only interested in cardinality. Of course, not each and every neuron was interested in numerical information, but a significant proportion of roughly 30% turned out to be number neurons.

Since the PFC operates at the apex of the cortical hierarchy and receives highly processed visual information from the PPC and the anterior temporal cortex (figure 6.3), we reasoned that numbers could already be represented in these upstream brain areas of the parietal and temporal lobes. The PPC is considered the termination zone of the so-called "where" pathway, which processes the spatial layout of visual scenes, whereas the anterior infero-temporal cortex is the endpoint of the so-called "what" pathway, which encodes object identity information (figure 6.3). In either of these zones, visual number could have been represented. In addition, we knew that the human literature pointed toward the posterior parietal lobe as a key area in number processing.

We therefore recorded neurons from the PPC, more specifically from the superior parietal lobule and the inferior parietal lobule, as well as different areas within the IPS in our monkeys.[6] In addition, we recorded from the anterior temporal cortex in the temporal lobe. Of all these areas, the ventral intraparietal areas, or area VIP, deep in the fundus of the IPS, stood out (figure 6.2). With up to 20%, VIP contained the highest proportion of numerosity-selective neurons in the parietal lobe. The importance of area VIP and the tuning to visual numerosity was recently also confirmed by single-cell recordings in Hajime Mushiake's group in Japan.[7] In the other regions of the parietal cortex or in the temporal lobe, number neurons were not as common and were about as frequent as expected by chance. Thus, two key brain areas for visual numerosity surfaced as a consequence of our recordings: the lateral PFC in the frontal lobe, and area VIP in the IPS of the PPC (figure 7.3A).

If the primate brain contains two regions with a high proportion of number neurons, chances are that these regions are part of a larger number network. When we compared numerosity-selective neurons in VIP with those we had recorded in PFC, we saw that VIP neurons responded about 50 milliseconds earlier to the numerosity displays. Because neurons earlier in the visual processing hierarchy naturally respond faster to the incoming stimulus information, this finding suggested that the IPS is the first site of numerosity extraction in the brain. As IPS and PFC are interconnected and "talk" to each other,[8,9] that information seems to be conveyed from the IPS to the PFC, where a larger proportion of more selective neurons represent numerosity, particularly during the subsequent working memory phase, to gain control over behavior. Of course, numerosity-selective neurons do not need to exclusively encode numerical information. Evidence from single-cell recordings in monkeys suggests that single neurons have diverse coding capacities and thus can be members of different neuron ensembles that also encode other functions.

When presenting our single-cell data at neuroscience conferences, the representation of numbers in VIP often elicits some surprise. It is well known that the PFC is involved in all sorts of abstract processes from categories to concepts, but the VIP is a different matter. Neurons in VIP are generally associated with visuo-spatial processing or attention, but hardly with abstract categorization. Here are some ideas for why VIP is an ideal location in the brain to represent numerical quantity. First, number is an abstract concept that is independent from sensory modalities. The ideal regions in the brain to represent number therefore are those that receive input from all major sensory systems—visual, auditory, tactile, and the like. Unlike any other region in the parietal lobe, VIP is a multimodal association area,[10] and therefore ideally suited to represent numbers. Secondly, the number of items can be enumerated in two fundamentally different ways: across space, but also across time. Brain areas of the PPC process spatial sensory information,[11] but also temporal information.[12,13] This again makes the posterior parietal lobe an ideal candidate for number processing. Third, numerical quantity is a special type of magnitude, and the PPC is

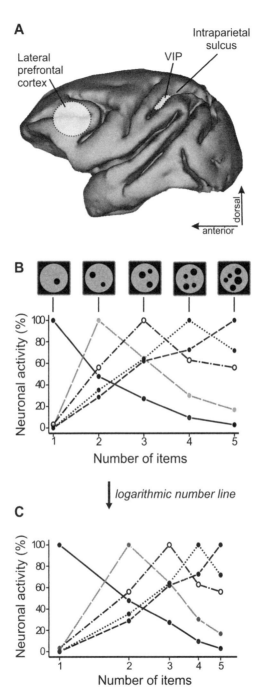

Figure 7.3
Number neurons in the monkey brain. A) A lateral view of a macaque monkey brain showing the brain areas (*light grey*) in the prefrontal cortex and ventral intraparietal area in the intraparietal sulcus that contain the highest proportion of number neurons. B) The relative level of activity of number neurons that are selectively tuned to respond to between one and five items. Notice how the tuning to smaller numbers is more precise (i.e., narrower curves). C) When plotted on a logarithmic number axis, the same tuning curves become symmetric (from Nieder, 2016).

known to encode different types of abstract magnitudes, such as object size or temporal intervals.[14] The PPC has therefore been suggested as the common processing site for general magnitudes, such as time, space, number, and other dimensions. Vincent Walsh, professor of human brain research at the University College in London, substantiated and summarized this idea in his theory of magnitude.[15,16] And, finally, numbers not only encompass the number of external sensory stimuli, but also movements that need to be enumerated. Aspects of the PPC have premotor properties and represent movement features. Finger counting is common in children when they learn to count, and parts of the parietal lobe are involved in finger counting.[17] So, by all accounts, area VIP is ideally suited to represent numbers.

In the same year in which we published how PFC neurons represented visual numerosities, a Japanese group led by Jun Tanji at Tohoku University in Sendai reported neurons in trained monkeys that kept track of the number of hand movements.[18] Tanji and coworkers trained monkeys to alternate between five arm movements of one type ("push") and five of another ("turn"). When they recorded from a specific location of the superior parietal lobule (SPL), they found neurons that preferred certain numbers of movements. The neurons were approximately tuned to the number of movements, similar to the tuning we observed toward the visual number of dots in other parietal brain areas.

While these studies of the Sendai group and our group at MIT opened the door to a new level of understanding of numerical cognition, it was not the very first time that number neurons had been reported. The earliest report of number neurons was published back in 1970, not from the behaving monkeys, but from the anesthetized cat. Richard F. Thompson and coworkers[19] recorded from the association cortex while presenting sequences of auditory or visual stimuli to unconscious cats. They found a few neurons that responded mostly to certain numbers of stimuli in the series, regardless of stimulus modality or frequency. Only five of the 500 tested neurons showed such number-selective activity. One percent of number neurons, however, can be expected by chance due to the specifications of the statistical testing. Nevertheless, this study was a heroic first attempt to decipher number processing neuronally.

7.3 The Neuronal Code for Number

Once we had discovered number neurons in the primate brain, the next question was how precisely such neurons would represent numbers based on their neuronal signal. In other words, what would be the code for numbers? Number neurons are each tuned to their individual preferred number. To clarify, a neuron that signals "three" with a maximum firing rate also responds to some extent to "two" and "four," but remains silent for "one" and "five." This highlights an important characteristic of number neurons: their tuning is approximate and relatively imprecise, because not only the preferred number but also adjacent ones elicit responses (figure 7.3B). However, this provides an important link to behavior and the approximate number system (ANS), which, as we have seen, is also imprecise and approximate.

The encoding of numbers by bell-shaped tuning functions with the peak at the preferred numerosity is termed a "labeled-line code." Just with an active telephone line, the line for a specific number is open to inform the monkey whenever a certain number neuron is active. Obviously, not only one, but many, number neurons simultaneously become active when numerosities are shown. The monkey needs to evaluate the activities from populations of neurons to decide which number it is being shown. To that aim, the monkey is listening to its neurons and what they have to say about number. These votes are taken into account just as votes are in a democratic election.

Unfortunately, the neurons' votes are sometimes ambiguous and can lead to erroneous number judgments, in particular when decisions on adjacent numbers have to be made. One reason for this ambiguity is that the neurons represent numerosity only approximately, as reflected by a bell-shaped tuning curve. Another reason is that the neuronal firing rate slightly fluctuates from one trial to the next. So when, for example, a monkey is presented with three dots, the neurons that prefer three are the most active generally, but the neurons that preferred two and four are also active, though usually less so. Sometimes it is not clear which of the neurons discharged the most. This leaves the monkey prone to mixing up neighboring numbers. This was precisely what was seen in the monkeys' behavior when they estimated set size, and it gives rise to the numerical distance effect.[20] But the numerical size effect can also be explained. Namely, the widths of the bell-shaped tuning curves systematically increase with numerical magnitudes. As a consequence, the likelihood that the monkey confuses neighboring numbers increases with increasing numbers. Both the numerical distance and size effect are characteristics of Weber's law, which we can now explain based on the responses of number neurons (figure 7.3B).

But tuned number neurons explain yet another psychophysical finding, namely Fechner's law, which states that the sensation of numerosity scales with the logarithm of numerical magnitude. As pointed out, the behavioral performance functions of monkeys are better described by a logarithmic as opposed to a linear number scale.[21] We found exactly the same relationship for neuronal tuning curves (figure 7.3C). A logarithmically compressed number line meets the requirements of symmetric tuning functions much better,[22,23] demonstrating number neurons' adherence to Fechner's law.

The logarithmic coding scheme agrees not only with Fechner's law, but has two more physiological advantages. First, it guarantees that all number neurons have the same widths of tuning for all numbers, an effect called scale invariance. Secondly, it ensures that the variability of neuronal responses is independent of number preference. Because of these benefits, symmetric bell-shaped (Gaussian) tuning functions are extensively used in neural models of information coding. This is because they show computational advantages over non-symmetric functions, and their analytical expression can be easily derived mathematically.[24]

The firing rate of tuned number neurons is undoubtedly an important clue to how numerical information is processed and transmitted in the brain. However, this is likely not the only source of information. There are reasons to believe that the temporal pattern of action potentials—how they are distributed across time when a numerosity is presented—can also provide information about set sizes.[25] Exactly how numerical information is conveyed based on the temporal structure of a neuron's firing is not yet clear and awaits further investigation.

The discovery of tuned number neurons in the monkey brain inspired cognitive neuroscientists to explore the neuronal coding of numbers in the human brain in an indirect way using functional imaging techniques. An earlier chapter mentioned the fMRI adaptation study performed by Piazza and Dehaene.[26] However, this was only half of the story. Instead of merely measuring where in the brain the BOLD signal became adapted and then recovered again, Piazza and Dehaene characterized it in interesting detail. They not only presented 32 as a deviant to the habituation stimulus 16, but also other deviant numerosities. This allowed them to systematically investigate how the BOLD signal would recover as a function of numerosities relative to the target. They found effects reminiscent of the tuning of number neurons in the non-human primate brain: the recovery of the BOLD signal in the IPS systematically increased with the numerical distance from 16, forming an inverted bell-shaped tuning curve. This was a first hint that tuned number neurons might not only exist in monkeys, but also in the human brain.

When my postdoc, Simon Jacob, and I repeated this fMRI adaptation experiment years later,[27] we found fMRI adaptation to numerosities not only in the IPS, but, consistent with my single-unit recordings in monkeys, also in the lateral PFC. Notably, the BOLD-recovery profile was significantly better described on a logarithmic number scale. This was again mirroring the scaling of monkey neurons. FMRI adaptation indirectly supported the notion that tuned number neurons might exist in the human brain that follow both Weber's and Fechner's laws.

One may think that tuning to numbers might be a special feature of our adult human brain. However, this is far from the truth. Jessica Cantlon's group[28] at the University of Rochester recently showed with fMRI adaptation that children as young as 3–4 years old already exhibit the same neural tuning to the cardinality of sets of dots in the IPS. This is an important finding, because it emphasizes that our brain can crunch non-symbolic numerical quantities even before we learn to count with symbols. This non-symbolic representation could therefore serve as the foundation for more elaborate symbolic representations later in life.

The fMRI adaptation experiments in human children and adults vis-à-vis our single-cell recordings in monkeys point to close parallels between the human and the monkey number brain. The fact that the IPS and the PFC are essential parts of the core number system in both primate species suggests that what we see in the monkey brain really is a biological precursor to our uniquely human symbolic number ability.

7.4 Number Neurons Are Necessary for Number Judgments

The results that neuronal tuning agrees with behavioral characteristics of number discrimination is evidence for the utility of number neurons to monkeys' quantity behaviors. However, this is still indirect evidence. The following are three approaches that demonstrate directly that monkeys rely on number neurons to process numbers.

One way to circumvent the previously shown issue is to explore the number neurons' responses when the monkeys make judgment errors. As pointed out above, number neurons that fire maximally to their preferred numerosity provide a signal for the monkey to make correct assignments. If this is the case, then the monkey should have a hard time discriminating numerosities if such neurons are not readily able to encode their preferences with maximum firing rates. The prediction from this scenario is that the monkey's number judgment errors are accompanied by the number neurons' decreased responses to their respective preferred numerosities. For instance, if the monkey makes an error for number three, the neuron tuned to three should exhibit less activity compared to correct trials. This is precisely what was demonstrated over the years in different monkeys and brain areas.[29] When monkeys made judgment errors, the neural activity to the preferred numerosity was significantly reduced compared to activity in correct trials. In other words, if the neurons did not encode a number properly, the monkeys were prone to mistakes.

The so-called "decoding approach," which became quite popular in neuroscience, is a computational method that can inform about behavioral relevance. The rationale here is as follows: if the neuronal signals are meaningful, then a computer mimicking a real brain also should be able to predict the number that the monkey has seen while recordings were being made. To test this prediction, a statistical classifier is presented with the recordings from many number neurons simultaneously. In the initial training phase, the classifier is informed of which target numerosity the monkey saw for any given trial. In doing so, the classifier can learn which specific number elicited which neuronal discharge. Once the classifier has learned these relationships, it is tested with a new set of recordings from the same experiment, but this time without any information about what number was shown to the monkey. The question then becomes whether the classifier can predict the correct number from neuronal impulses based on what it had learned before.

Indeed, my PhD student, Oana Tudusciuc, and I found that this type of trained classifier could accurately predict which numerosities a monkey saw based only on number neuron recordings.[30] Interestingly enough, when the classifier was tested with recordings that were collected when the monkeys made errors, it was no longer able to predict the numerosities that the monkey had seen. This suggests that the classifier can exploit neuronal signals to correctly categorize just as long as a monkey exhibits the correct number judgments.

However, the most direct way to show that number neurons are needed for numerical decisions is to turn them off for a while and see how this affects a monkey's behavior. Clearly, temporarily inactivating such neurons should cause the behaving monkeys to make discrimination errors. In 2010, Jun Tanji and colleagues provided this exact type of evidence for parietal number neurons.[31] They reported on the behavioral consequences from monkeys enumerating hand movements after the area in which they previously found neurons tuned to movement number was inactivated.

Rather than causing permanent lesions by removing brain tissue, the Japanese scientists used a more elegant, reversible method to cause a temporary inactivation of the brain area under investigation. They exploited a chemical compound called "muscimol" produced by the mushroom *Amanita muscaria* (fly agaric). Muscimol has the interesting property of activating inhibitory synapses in the brain, which in turn deactivates excitatory neurons, such as pyramidal cells. When locally injected using syringes, muscimol can therefore shut down an entire cortical area. Any function represented in that area will be lost for about half an hour until muscimol becomes inactive.

Before Tanji and colleagues applied muscimol, the monkeys were perfectly able to produce exactly five hand movements as they had been taught. But when area 5 in the superior parietal lobule was flooded with muscimol, the monkeys made significantly more errors. Most of the time they omitted hand movements and stopped at four or even three movements. At other times, they overdid it and produced six or more movements. These errors were not due to general motor processing deficits; in control tasks, the monkeys could select between different actions and produce hand movements just fine. The inactivation caused specific impairments to the number of actions. To control for the experimental procedure of injection, Sawamura and Tanji likewise injected plain saline water containing no drugs in control sessions; here, the monkeys were performing just fine, indicating that the specific inactivation of area 5 was responsible for the monkeys' number problems. In combination, these experiments clearly indicate a crucial role of number neurons in parietal cortex for selecting actions on the basis of numerical information.

7.5 Number Neurons Represent Different Presentation Types and Modalities

In 2004, I left Earl Miller's laboratory and moved back to Germany to accept a position equivalent to an assistant professorship at the University of Tübingen. At that time, I had already recorded from several areas of the frontal, parietal, and temporal lobes, but only using dot arrays as number stimuli. An unresolved question at the time was whether number neurons might encode numerosities irrespective of presentation type and independent of sensory modalities. It was time to explore how abstract number neurons really were.

In order to compare how neurons would react to these two types of presentations, we first needed monkeys to perform simultaneous and sequential numerosity discriminations. The first task in my own laboratory therefore was to train two monkeys to perform the well-established delayed matching-to-sample task, but this time not only with dot arrays, but also dot sequences as sample stimuli (figure 7.4A). The controls for dot arrays were already implemented based on the previous studies. The sequential format, however, required additional controls to eliminate temporal factors that may co-vary with increasing numbers of sequential items. For example, if the presentation time for the dots and pauses were kept constant, then showing four sequential dots would take twice as long compared to showing two sequential dots. In this case, the monkeys could learn to discriminate the differences in duration rather than in number. We had to make sure that temporal factors were unreliable to solve the task. And so, we alternatingly showed sample stimuli to the monkeys in which the total duration of the sample period, the duration of individual items and pauses, the total visual intensity, and the regularity, or rhythm, of the item sequence were controlled. The investigation of their performance showed us that the monkeys indeed extracted the number of items irrespective of the control displays. In addition, and as shown in previous experiments, the monkeys also proficiently discriminated numerosity in trials in which the simultaneous format was shown. The stage was set to see what neurons had to say about these two number-presentation methods.

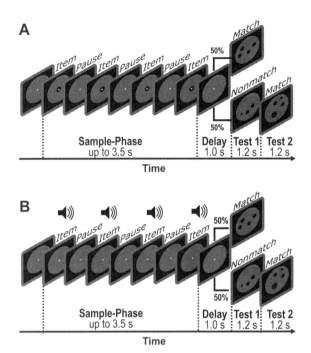
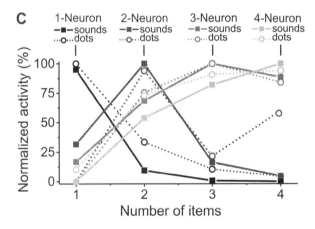

Figure 7.4
Number neurons in the prefrontal cortex represent the number of dots and sounds. A) To assess how neurons encode the number of items presented over time, a sequential visual enumeration protocol was used. The monkey had to enumerate items one by one and respond if the sample period and test display showed the same number of visual items, respectively, and withhold response if they did not (probability 50%). Sample numerosities one to four were cued randomized by sequentially presented dots separated by blank displays. B) To find out if number neurons generalized across sensory modalities, different numbers of sounds (indicated by speaker icons) had to be enumerated by monkeys and matched to the number of visual items in multidot test displays. This sequential auditory enumeration task was presented alternately with the sequential visual protocol in a given session. C) Using this protocol, neuronal responses of single neurons to both visual and auditory numbers could be tested. Tuning functions of four example supramodal prefrontal neurons with preferred numerosities one, two, three, and four are shown.

Together with my PhD students, Ilka Diester and Oana Tudusciuc, I recorded single-cell activity in area VIP of the IPS while the monkeys performed the delayed matching-to-sample task in which sample numerosity was alternatingly shown either by single dots appearing one by one, or by multiple-dot patterns.[32] The first good news was that we again found our previously described neurons were selective to numerosity in multiple-dot patterns. Much to our delight, we also found roughly 25% of VIP neurons that encoded sequentially presented numerical quantity. When we analyzed and compared the neuronal selectivity in more detail, we found that the neurons that signaled numerosity during the sample period in simultaneous and sequential presentations were actually different. Neurons encoding the number of sequential items were not tuned to numerosities in multiple-item displays, and vice versa. The sensory encoding of the two number formats was obviously represented by distinct populations of neurons.

But this was only half of the story. After the enumeration process in the sample period was completed, the animals had all the information about the cardinality of a set. Once the enumeration process was completed and the monkeys had to memorize information throughout the following delay period, a third population of neurons emerged that coded numerosities both in the sequential and simultaneous protocol. In fact, about 20% of the VIP cells were tuned to both types of number presentations in the working memory delay.

These neuronal findings argue for two different neuronal mechanisms. Initially, when the dots are actually seen during the encoding stage, the processing of simultaneous and sequential dot presentations is segregated. However, when this initial step is finished, the previously segregated number information converges onto number neurons that represent numerosity abstractly and irrespective of the presentation ways. The means that the initial representation of visual numerosity and the memorization of the numerosity are accomplished by different neuron populations. Therefore, the sensory registration of numbers is presentation-specific, whereas the maintenance of the derived set size information is abstract and presentation-independent.

So far, we had learned how neurons represented visual numerosities. But number is an abstract concept that pertains to all kinds of sensory experiences. Three light flashes or three calls are both instances of "three." Given its abstract nature, number representations need to be independent of sensory modalities; that is, encoding must occur "supramodally." For the brain, however, this is not a trivial task. After all, we process sensory information separately in specific organs and brain areas. The energy of light is processed via the eye and the downstream visual cortex, whereas the energy of sound is received by the ears and the auditory cortex. For an abstract and modality-independent representation of number, the information in these segregated sensory pathways needed to converge.

The only way to concretely discover whether neurons encoded the number of items irrespective of sensory modality was to record from the same neuron while the monkey performed numerosity discriminations across different sensory modalities. Fortunately, we had monkeys available that knew how to enumerate the number of sequential visual dots in a delayed match-to-sample task (figure 7.4A). All we had to do was to train them to assess the number of sounds in addition to visual sequences (figure 7.4B). In one trial, the sample numerosities were cued by visual dots, while the next one was cued by sound bursts. The monkeys learned this task and switched from visual numerosities to auditory numerosities from one trial to the next. After the monkeys were proficient in these tasks, we recorded from the brain, only this time simultaneously from both the PFC in the frontal lobe and VIP in the IPS.[33] This turned out to be beneficial, because interesting differences between these brain areas surfaced. The first exciting finding was that neurons in both brain areas were tuned to either the number of auditory pulses, visual items, or to both of these. This meant that these neurons in the frontal and parietal association areas indeed had access to visual and auditory numerical information.

However, responsiveness to sequences of dots and sounds is not the same as supramodal coding in the brain. This turned out to be a crucial difference between brain areas because, as a second result, we found that number neurons only in PFC, but not in VIP, responded abstractly and across modalities. For supramodal coding to be present, neurons have to fulfill two criteria. First, they have to respond to both visual and auditory numbers. Second, they also have to be tuned to the same preferred numerosity in both sensory modalities. In the VIP, only a very few neurons matched both criteria, and then only for a numerosity of one. If VIP neurons were tuned to visual and auditory numbers, they usually were tuned to different preferred numbers. PFC neurons, on the other hand, responded much more abstractly. Here, supramodal tuning to each of the four tested numerosities was indeed present. Figure 7.4C displays the activity of four PFC neurons that were tuned supramodally to a numerosity of one, two, three, or four. The tuning functions to the number of visual items and the number of sounds were identical for the respective neurons. Here it was finally—the truly abstract and sensory-independent representation of numerosity by single neurons!

Indirect evidence for supramodal coding has also been collected using fMRI in humans. Modality-independent BOLD activation during numerosity estimation has likewise been observed in a right fronto-parietal network.[34] The same study identified a similar network during the verbal counting of visual and auditory items, with additional activation in prefrontal, parietal, and premotor areas. The IPS also seems to encode number symbols in an abstract, supramodal way. Evelyn Eger and Andreas Kleinschmidt, from the French Institute of Health and Medical Research at the NeuroSpin in France, conducted fMRI studies while subjects were asked to merely detect numerals, letters, or colors in visual sequences or acoustic streams.[35] To avoid confounds by response selection and associated cognitive states (such as attention), the authors analyzed the presentation of non-target numerals (numerals that were not required to be detected) and compared it to non-target letters or colors. The IPS was the only region that exhibited higher activation for numerals, both visually and acoustically. Therefore, numerical activation in the IPS seems to be automatic (task-independent), modality-independent (visual and auditory), and notation-independent (irrespective of whether numerals are spoken or written).

The presence of a supramodal number code is not mandatory from a theoretical point of view, but it might provide a computational advantage: once number neurons respond supramodally, they can easily be linked to visual shapes or auditory sounds when learning to establish symbolic representations of numbers, such as numerals and number words. Also, models of number processing assume such numerical representations. For instance, the influential triple-code model (see figure 10.5) by Stanislas Dehaene and Laurent Cohen postulates that numerical cognition initially involves a lower step of modality-specific analysis, followed by a higher processing stage where these representations reach an abstract level.[36] Supramodal number neurons could well constitute the neurophysiological basis of such an modality-independent representation.

7.6 Convergent Evolution of Number Neurons: Lessons from Crows

As shown in previous chapters, different levels of numerical competence are widespread throughout the animal kingdom, ranging from insects to fish, birds, and mammals. This may be surprising, given that some of these animal groups had a last common ancestor several hundred million years ago. Since then, these respective lineages have evolved in parallel. During this parallel evolution, all their organs experienced dramatic changes, causing, most notably, extremely different organized endbrains. How is it possible that numerical capacities are ubiquitous in the animal kingdom despite a wide range of brain types?

From an evolutionary point of view, a comparison of neuronal mechanisms between insects (as protostomes) and primates (as deuterostomes) would be most informative to answer this question. After all, their last common ancestor lived more than 600 million years ago. Unfortunately, the neurophysiological basis of number competence is still unknown in insects, so this comparison is impossible. However, a comparison is possible with other remotely related groups of animals, namely between birds and primates. Their reptilian-like last common ancestor lived about 320 million years ago, so fundamental differences in their brain design can be expected.[37,38] In fact, striking differences do exist in their endbrains, which happened to be a rather insignificant part of the brain in prehistoric reptiles, but turned into the dominant intelligence center in birds and mammals.

What would any brain require to process abstract categories such as numbers? One requirement would be high-level brain areas that can integrate sensory information across time, space, and senses before planning motor commands. In primates, sophisticated circuitries in the six-layered neocortex of the parietal and frontal lobes fulfill these requirements. These classical association areas receive highly processed information from all sensory modalities and project to premotor structures (see figure 6.3). As we have seen, a dedicated number network for processing numerical information indeed exists in the parietal and frontal association cortices[39] (figure 7.3).

Birds, on the other hand, have an independently evolved endbrain design. After birds and mammals split from the last common ancestor 320 million years ago (figure 3.1), birds expanded different parts of the endbrain to become intelligence centers. Notably, the brains of corvid songbirds (crows, jays, and ravens) contain twice as many neurons as a primate brain of equal mass.[40] In both mammals and birds, the bulk of the endbrain stems from the ontogenetic mantle, the pallium, and thus shares common ancestry.[41,42] However, the overall architecture of the corvid endbrain is very different and has emerged based on "convergent evolution."[43,44] Convergent evolution is the independent evolution of similar features and functions in species of different lineages. For example, the wings of birds and insects evolved independently from each other and built on different structures (from the forelimbs in birds, but from the body wall of the thorax in insects), but still both are used to perform the function of flying. In the case of birds and mammals, convergent evolution gave rise to distinct endbrains from which intelligent behavior can originate.

As a consequence of this convergent evolution, a six-layered neocortex with a PFC that endows primates with the highest levels of cognition is absent in birds and all other non-mammalian vertebrates (figure 7.5A). Instead of layers, the avian endbrain consists of nuclear organized circuits. They originate from a different part of the telencephalic pallium.[45,46,47] The so-called "dorsal ventricular ridge" is one of the main components of the sauropsid (i.e., reptilian and avian) pallium that gives rise to these associative circuits. A particularly integrative region originating from the dorsal ventricular ridge is the "nidopallium", which also contains the nidopallium caudolaterale (NCL; figure 7.5A). The NCL is a high-level cognitive structure in birds and considered to be the functional equivalent of the mammalian PFC.[48,49] Just like the PFC, the NCL integrates highly processed sensory information from all senses and projects to premotor structures, is modulated by the neuromodulator dopamine, and interacts with structures concerned with affect and memory. Recent single-unit recordings in behaving crows confirm the resemblance of the NCL to the PFC.[50] In short, primates and corvids provide a most interesting case of the convergent evolution of numerical competence because of the independent evolution of the associative endbrain areas in both animal groups.

Now, how is numerosity represented in monkeys and crows? In general, the neurophysiological solutions to a common behavioral problem may take different paths when species evolve in parallel. Sound localization in birds and mammals provides an illustrative example for some of the physiological differences that result from parallel evolution. How location

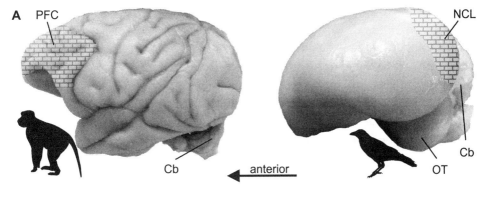

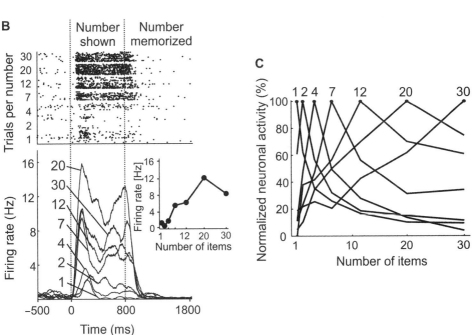

Figure 7.5

Neurons responsive to numbers in the convergently evolved crow endbrain. A) The brains of macaques and crows. *Left:* Lateral view of a macaque brain highlighting the prefrontal cortex (PFC). *Right:* Lateral view of a crow brain with the nidopallium caudolaterale located in the telencephalon. Cb: cerebellum; OT: optic tectum. **B)** The crow nidopallium caudolaterale contains neurons that are tuned to number. The top panel depicts a dot-raster histogram of the firing of a single neuron for multiple presentations of different numbers of items. The bottom panel shows the neuron's activity as a spike-density histogram. The inset on the right represents the numerosity tuning curve of this neuron with the preferred numerosity 20. **C)** The relative level of activity of number neurons that are selectively tuned to respond to between one and 30 items. Notice how the tuning to larger numbers is less precise (i.e., broader curves; from Ditz and Nieder, 2016).

is computed from time differences of sounds arriving at the two ears and passed on to homologous auditory brainstem nuclei is fundamentally different in birds and mammals.[51] Different solutions to the same problem—sound localization—have been implemented during the course of evolution. Could a similar scenario have happened for numerical representations?

My laboratory is interested in such comparative neurobiological questions that help to address evolutionary trajectories. To that aim, we conduct behavioral and neurophysiological experiments not only in monkeys, but also in crows. To explore where and based on what neuronal code numerosity is represented in avian brains, my PhD student, Helen M. Ditz, and I trained crows in a delayed matching-to-sample task to discriminate the visual number of items in controlled displays.[52] At the same time, single-cell activity was recorded from their NCL.[53] We found that the behavioral and neuronal data showed an impressive correspondence with data described earlier from the primate brain. First, NCL neurons in crows were tuned to individual preferred numerosities, and an analysis of error trials showed that the neuronal discharges of NCL number neurons proved to be relevant for the crows' correct performance (figure 7.5B). Second, NCL neurons were approximately tuned to number and showed a neuronal distance and size effect that followed Weber's law (figure 7.5C). Finally, both the neuronal and the behavioral tuning functions were best described on a logarithmic number line. This argues for a non-linearly compressed coding of numerical information, as predicted by the psychophysical Weber and Fecher laws. All of these characteristics are reminiscent of findings in primates.

Based on the data, we conclude that crows show exactly the same code for numbers as primates do, despite having very differently organized endbrain. This suggests that this way of coding numerical information has evolved based on convergent evolution because it exhibits a superior solution to a common computational problem. I think findings like this emphasizes that a comparative approach is indispensable for deciphering evolutionary stable neuronal mechanisms and codes, and not only in the realm of numerical competence, but for all neurobiological questions.

7.7 Number Neurons in the Human Brain

For almost two decades, we had intensively explored number neurons in the neocortex of non-human primates. That we humans also possess number neurons in our brains was suggested by the imaging literature. However, the final proof—number neurons recorded from the human brain—was still missing. This is because single-cell recordings from human brains are rare and exceptional. But sometimes the treatment of brain diseases may provide the unique opportunity to record from neurons inside the brain. Such opportunities may emerge when surgical procedures are required to treat cases of epilepsy. In this case, intracranial electrodes may be inserted as part of the procedure to localize the brain abnormality in preparation for surgical removal. Of course, single-cell recordings during experimental tasks are always a byproduct of a therapeutic intervention, and are only done with the consent of the patient.

I was fortunate enough that Florian Mormann, a professor of cognitive and clinical neurophysiology at the department of epileptology at the University of Bonn, was willing to collaborate with me on this quest for human number neurons. Together with our PhD student, Esther Kutter, we could solve this long-standing question of number neurons in the human brain once and for all. Florian Mormann routinely implants electrodes in patients who undergo treatment for debilitating and pharmacologically intractable epilepsy at the University Hospital in Bonn, Germany (figure 7.6). After the electrodes are implanted, the patients rest for a few days while their neuronal activity is monitored via electrodes. This procedure is necessary to unequivocally identify brain tissue that causes epileptic seizures. Once the continuous recordings pinpoint the location of epileptic activity, that exact pathological tissue (and the electrodes, of course) can later be removed by a neurosurgeon in a second surgery. While the patients wait for this second surgery, they may participate in scientific studies.

Nine participants who underwent such treatment gave their written consent and participated in our number study.[54] They were implanted with chronic depth electrodes in the medial temporal lobe (MTL), which includes the hippocampus, parahippocampal cortex, entorhinal cortex, and amygdala. As outlined before, parts of the parietal and prefrontal cortices are the core number system in primates, and, therefore, these regions would have been the most obvious choice for recordings. However, this was not possible, because the electrodes were inserted for purely medical reasons. Nonetheless, the MTL is still a suitable brain region to learn about numerical representations because it is incorporated in the wider cortical number network.[55] The MTL comprises highly associative brain areas that

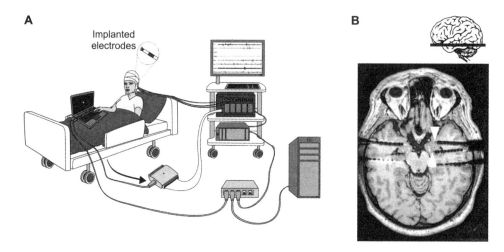

Figure 7.6
The treatment of brain diseases can provide the unique opportunity to record single-cell activity in humans. A) After the electrodes are implanted, the patients rest for a few days. During this period, the patients perform tasks on a laptop while their neuronal activity is monitored via the electrodes. The picked-up neuronal signals are processed and sent together with behavioral performance information from the notebook to a data acquisition system that is connected with a recording PC. All these data are stored and can later be analyzed (modified from Knieling et al., 2017). B) Magnetic resonance image (horizontal section) showing the implanted depth electrodes targeting areas of the medial temporal lobes (up is the face region with the eyes). White arrow heads point to the tips of the microelectrodes that protrude from the clinical electrode (1.25 mm diameter). The clinical electrodes appear enlarged due to MRI artifact (MR image by courtesy of Florian Mormann).

are directly and reciprocally connected to the frontal number network.[56] Most importantly, functional imaging studies in humans showed that the hippocampal system is involved in learning to count and arithmetic skill acquisition, specifically during childhood.[57,58,59] We had a fair chance of detecting number neurons in the MTL, if they existed.

With the electrodes inserted, the participants performed simple sequential addition and subtraction tasks using a computer display brought to their beds (figure 7.7A). We chose this calculation task to ensure the numbers shown were actively processed by the participants. The numerical values of the operands ranged from one to five. In half of the shuffled trials, the numerical values were presented non-symbolically as the numbers of

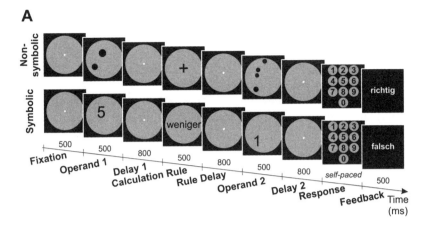

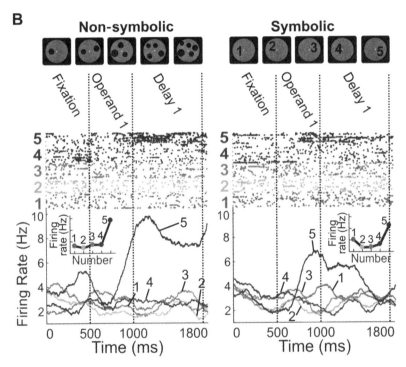

Figure 7.7
Number neurons in the human brain. A) Participants performed simple sequential addition and subtraction tasks in a notebook display. In half of the shuffled trials, the numerical values were presented non-symbolically as dot arrays (*top*). In the other half, Arabic numerals were shown as symbolic number representations (*bottom*). B) When the participants calculated, a proportion of the tested neurons from the medial temporal lobe showed activity that varied exclusively with the number of the first operand. The responses of the same single neuron to numerosities (non-symbolic format; *left side*) and Arabic numerals (symbolic format; *right side*) are displayed as dot-raster histogram (*top*) and spike-density histogram (*bottom*). This particular neuron was tuned to quantity five in both non-symbolic and symbolic formats (from Kutter et al., 2018).

randomly placed dots in an array (numerosity). In the other half, Arabic numerals were shown as symbolic number representations. As always, the non-symbolic and symbolic numbers were shown in standard and control displays in order to control for low-level visual features. The results of our study presented in this chapter will focus on the non-symbolic dot numerosity presentations.

We recorded from hundreds of single neurons in the medial temporal lobes while the participants performed the calculation tasks.[60] We were primarily interested to see how human neurons would represent numerical values during the sample and the first delay period, but before any calculation was required. Indeed, we found neurons responding to the number of items, and the results confirmed everything we had predicted based on recordings in non-human primates. When the dot displays were presented, a substantial 16% of the randomly selected neurons exclusively responded to the number of items, irrespective of the dot array layout. In addition, each selective cell was tuned to a preferred numerosity (figure 7.7B). This was exactly the type of approximate number code we had previously seen in monkeys and crows. This same code was found to be present in our own brain!

With 29% of the neurons being selective to the number of dots, the parahippocampal cortex (PHC) showed the highest proportions of number neurons among the four tested MTL areas. This may not be a chance accident. The PHC is part of a large network that connects regions of the temporal, parietal, and frontal cortices and has been associated with many cognitive processes.[61,62,63,64] Interestingly, the PHC has prominent connections with association areas, including the parietal lobule.[65] We therefore suspect that representations about numerical quantity most likely do not originate within the PHC (or other areas of the MTL), but are reflections from processing in other anatomically connected areas of the parieto-frontal core number system.

Of course, we were interested to see if the neuronal responses could explain human number judgments. So we performed the decoding analysis mentioned earlier by training and testing a statistical classifier. It turned out that numerosities one to five could reliably be decoded and predicted from MTL neurons. This held true for the populations of number neurons, but irrespective of response selectivity also for the entire population of recorded neurons. In addition, the bell-shaped number tuning curves also explained the rather large numerical distance effect found for non-symbolic numerosity comparisons even in humans.[66] Both findings provide evidence for these neurons as the physiological correlate of number representations.

Our data, for the first time, showed how non-symbolic numerical quantity is encoded by neurons of the human brain. The data supported the hypothesis that high-level human numerical abilities are rooted in biologically determined mechanisms. I will come back to MTL neurons' representation of symbolic number during the same task in a later chapter of the book.

7.8 An Innate Number Instinct

For a long time, number neurons were investigated in monkeys that had been trained to discriminate set sizes. Training animals usually takes many months, and during this time, the connections in the brain can change. Could it be that number neurons are just a byproduct of intense experience rather than a preexisting, or innate, faculty? After all, it is well known that neuronal responses in the association cortex can be heavily modified by learning.[67,68]

My PhD student, Pooja Viswanathan, and I therefore decided to explore if number neurons existed spontaneously—that is, in monkeys that had never been trained to discriminate numerosity.[69] To test this, we of course needed to present different numerosities, but we did not want the monkeys to pay attention to the number of the dots in the displays. Therefore, we presented one to five dots in different colors and trained two new monkeys to only discriminate the color of the dots in the usual delayed matching-to-sample task. Discriminating color was an easy task for the monkeys. When we occasionally removed the colors and showed all black dots, they were completely lost because they really ignored the number of dots in the display.

Next, we recorded single-cell activity from those brain areas of the numerically naïve monkeys where we previously had found a high proportion of number neurons in trained monkeys, namely the VIP and PFC. To our delight, we found that about 10% of the randomly sampled neurons selectively responded to number in both areas in these monkeys that had not been trained to judge numbers. Just as in monkeys before that were trained to discriminate number, the number neurons in numerically naïve monkeys were spontaneously tuned to preferred numerosities. Therefore, we reached the conclusion that number neurons are an inherent part of the primate brain.

We also found that numerosity was again encoded earlier in the VIP, suggesting that numerical information is automatically extracted in the parietal cortex, and then conveyed to the frontal lobe. Indeed, when the same initially numerically naive monkeys were retrained to actively discriminate visual numerosities, we found contrasting neuronal effects on PFC and VIP neurons as a result of training[70]: although PFC neurons became more responsive and more selective during active discrimination, none of these effects were observed for VIP neurons. This indicates that PFC neurons become more engaged when the task requires it, whereas VIP neurons continue to encode numerosity as a visual stimulus regardless of behavioral relevance. Complementing these findings from non-human primates, imaging evidence also suggests that direct and automatic extraction of numerosity occurs in the adult human brain.[71,72]

But there is more to come. Number neurons are not only present by nature in the cerebral cortex of primates, but also in the differently organized endbrain of crows. When my PhD student Lysann Wagener recorded single-cell activity from the NCL of numerically naïve crows,[73] we again found that a proportion of NCL neurons were spontaneously responsive to numerosity and tuned to the number of items, even though the crows were never trained to assess numerical quantity. This shows that number-selective neuronal responses are naturally present in distinct endbrains of different vertebrate taxa.

Collectively, these findings suggest that the brain is innately endowed with the capability to extract numbers. This sounds very reminiscent of the notion of a "number instinct" that I alluded to in chapter 1—the idea that we and animals have an intuitive understanding of what numerical quantity is. How else would it be possible that newborn or freshly hatched chicks can discriminate numerosity so early on? In fact, there is much evidence that shows that we don't have to learn how to represent numbers; instead, we are something like natural-born reckoners.

But there is another connotation to the idea of a "number instinct," namely that numbers are sensed in a perceptual-like way, just like the direction of a moving object or the frequency of a sound is sensed directly by the visual and auditory system, respectively. Strong support for such a perceptual-like assessment resulted from psychophysical experiments performed by David C. Burr, professor of physiological psychology at the University of Florence in Italy. Starting with the programmatic paper "The visual sense of number" in 2008,[74] Burr's laboratory since then has deciphered the psychophysical signatures of number intuition in a series of influential papers. They showed that visual numerosity assessments are subject to adaptation, like other sensory perceptions. After a period of observing dense or sparse dot clouds (for approximately 30 seconds), the apparent numerosity of subsequently viewed dot clouds changes considerably. More precisely, set sizes are underestimated after adaptation to a high numerosity.[75] For instance, after adapting to 400 dots, subjects judge 100 dots in a subsequent display as being equal to 30 dots. The finding that brief periods of adaptation are sufficient to elicit large changes in apparent numerosity could be traced back to the IPS in an fMRI study.[76] This adaptation effect is even present for sequentially presented items, across sensory modalities, and across spatial arrays and temporal sequences.[77] Even adapting to actions affects number perception. Burr and coworkers had subjects tapping their fingers in mid-air either rapidly or slowly, then judging the numerosity of sequences of flashes, or of arrays of dots. Surprisingly, adapting to slow tapping caused overestimation and adapting to fast sequences caused underestimation.[78] Does this mean that numbers are a property that is perceived with a special sense? This assumption would be wrong. First, numbers comprise an abstract category that applies to all seen, heard, or otherwise sensed items. To elaborate, let's say I see three pencils and hear three phone rings. From a sensory point of view, these objects are very different, but from a numerical stance they are all the same cardinality of three. Numbers rely on all of these sensory channels, but they do not comprise a true sensory system. Second, numbers not only are susceptible to adaptation in the visual domain, but also in the auditory and even across domains. This indicates that number adaptation may, at least partly, happen at a stage that follows purely sensory processing, and therefore outside of these separate sensory systems. It may well be that adaptation effects are not restricted to the classical senses, but also apply to more abstract features, such as space or time.

There is every reason to believe that we have an intuitive and possibly innate instinct for numbers. This is what Tobias Dantzig had in mind when he first used the term "number sense."[79] Ample evidence from single-cell recordings, psychophysics, and brain imaging studies suggests that humans and animals perceive numbers innately. This requires dedicated brain mechanisms, and number neurons are most likely the neurobiological foundations of this ability. The next section on neural networks will also show how this number instinct can emerge spontaneoulsy from the processing of cortical systems.

7.9 Number Models and Networks

One of the ultiamtive goals of neuroscience is to have a quantitative explanation of how the brain works by actually being able to compute the processing of information by neurons with numbers. Such models can help us to develop new ideas on how the brain may actually compute information, which in turn allows us to test such predictions in subsequent experiments. Several models of neural computation have been proposed over the years that try to elucidate the abstract principles that underlie number processing. The architecture of these models ranges from very artificial and remote from the brain to biologically inspired and mimicking major components of brain processing. Of course, the quality of such models depends on whether or not they correspond with what is currently known about number processing in real brains, and in particular the code for number.

Traditionally, two main models have been proposed to account for the cortical extraction of quantity from sensory inputs, and they assume different number codes. The first is called the "mode-control model" and was suggested in 1983 by Warren H. Meck and Russell M. Church, then at Brown University, to explain number and time estimation.[80] The mode-control model was designed for serial presentation of item sets and suggests that each item is encoded by an impulse from a pacemaker, which is added to an accumulator. The magnitude in the accumulator at the end of the count is then read into memory, forming a representation of the number of a set. Thus, it is assumed that quantity is encoded by "summation coding"—that is, the monotonically increasing and decreasing response functions of artificial neurons in response to summed up items.

The second type of traditional model, the "numerosity detector model" by Stanislas Dehaene and Jean-Pierre Changeux from the Collège de France in Paris, is geared towards representing numerosity from item arrays.[81] This model also assumes that activity from the number of items is initially pooled and encoded by summation units. However, the summation units finally project to tuned numerosity detectors as the output stage that resemble tuned number neurons in the brain. A similar architecture was proposed by Tom Verguts and Wim Fias from Ghent University in Belgium using a back-propagation network.[82] Also in their network, summation units developed spontaneously in the second processing stage (the "hidden units") after tuned numerosity detectors were determined as the output stage—and conversely, cells tuned to numerosity developed at the output level in an unsupervised model with summation units at the input level. Despite fundamental differences in design, both the mode-control model and the numerosity detector model require summation units that accumulate number in a graded fashion, either as final output, or at intermediate processing stages. One of the controversial issues therefore is whether or not these summation units are required or even present at all in real brains.

Among dozens of single-cell studies reporting tuned numerosity neurons, only one study reports neuronal responses that resemble summation units. Jamie D. Roitman and Michael Platt, then at Duke University, recorded neurons in the monkey LIP in IPS (see figure 6.2 for anatomical location) whose discharge rates primarily increased or decreased monotonically with a change in the numerosity of the stimulus set that was presented in the neurons' response field.[83] The authors suggested that the two classes of number-selective neurons—monotonic versus tuned cells—may be the physiological cognates of the summation units and numerosity units proposed in neural network models of numerical representation. According to this logic, monotonic magnitude coding by LIP neurons may provide input to the tuned neurons in VIP and PFC.

While this scenario is plausible, it still remains an unresolved question whether it is realized in the brain. First, the neurons recorded by Roitman and coworkers cannot be the end stage of number coding, as these neurons only encoded the numerosity of sets placed within their spatially restricted response fields. Extraction of abstract numerical information requires integration across all visual space as well as time and modality for some tasks. In contrast to area LIP, area VIP and the PFC integrate multimodal input and their neurons exhibit global cognitive processing properties that are no longer spatially restricted. Second, in the Roitman study, the monkeys were not discriminating numerosity but were actively engaged in a delayed saccade task in which numerosity served to predict the amount of reward that the animal would get after a correct saccade. The responses of the neurons therefore may have been correlated with a coding of reward magnitude rather than number. By contrast, all studies that required the monkeys to use cardinal numerical information explicitly found a labeled-line code, irrespective of stimulus modality, presentation format, and recording site.[84] Importantly, this code was not imposed onto neurons by intensive training, because the same code is present in numerically naïve monkeys[85] and crows.[86]

Currently, the most successful neural networks in computational neuroscience are so-called "deep neural networks." These are artificial neural networks with multiple hidden layers between the input and output layers. The "deep" refers to the number of layers through which the data is transformed; the more layers, the "deeper" the network. From the input layer to the output layer, the data flows through a cascade of multiple layers of non-linear processing units for feature extraction and transformation. "Non-linear" means that the output is not a linear function of its input. Each successive layer uses the output from the previous layer as input. Each level learns to transform its input data into a slightly more abstract and composite representation, thus mimicking the processing of the brain's visual stream. Each layer consists of a collection of connected units called "artificial neurons." The response, or state, of artificial neurons can be analyzed and compared to real neurons.

Ivilin Stoianov and Marco Zorzi from the University of Padua used a deep network that had one "visible" layer encoding the sensory data and two hierarchically organized "hidden" layers.[87] They trained the network with more than 50,000 numerosity (i.e., dot-pattern) stimuli. Crucially, information about number was not provided during the learning phase. They found distinct populations of artificial neurons in the second hidden layer that estimated numerosity by monotonic (summation) coding. Stoianov and Zorzi concluded that the emergent monotonic encoding is consistent with the Roitman-study reporting single cells in the monkey LIP by a summation code. However, and as emphasized earlier, this model cannot explain the tuned numerosity detectors that are usually found in electrophysiological studies.

My student Khaled Nasr and I therefore implemented a more biologically inspired deep neural network[88] that neither relied on hard-coded connections, nor training on non-naturalistic dot pattern stimuli. In fact, our model was merely trained to classify natural images in a task that was unrelated to numerosity. Much to our delight, about 10% of the network units exhibited tuned numerosity selectivity, just like number neurons in the brains of numerically naïve monkeys and crows.[89] Moreover, and in agreement with real number neurons, the network units exhibited approximate tuning that decreased in precision with increasing numbers, and were best described on a logarithmically compressed number line. The activity of the network's numerosity-selective units' obeyed the Weber–Fechner law known to be followed by neurons in the human, monkey and crow brains. Neither the output layer, nor the preceding intermediate layers contained a meaningful proportion of summation units, and their activity was not relevant for the network's number discrimination performance. Our network results therefore indicate that summation units are not relevant for the number instinct. The most important finding, however, was that numerosity selectivity can emerge simply as a by-product of exposure to natural visual stimuli, without requiring any explicit training for numerosity estimation. The basic sense for numbers may capitalize on already existing cortical networks. This could explain why newborns and animals in the wild are innately endowed with numerical capabilities.

7.10 Numerical Working Memory

So far, this chapter has only covered the perception of numbers. But what happens after a numerical stimulus goes away and can no longer be seen or heard? Somehow, a memory of the stimulus needs to be kept in the mind to evaluate and compare numerical information later on. Numerical cognition would therefore be impossible without the capacity to briefly store and manipulate numbers in the mind.

The capacity to consciously memorize something is called "working memory"; it is our mental sketchpad. Working memory allows us to "think about" quantitative information, and to weigh and measure it in relation to previous experiences and current goals as a hallmark of cognitive control. Consider the simple calculation "2 + 3"; it relies on the ability to mentally maintain "2" before it can be incremented by "3" to end up with the result, "5." Without working memory, we would be entirely lost.

Neurons represent working memory by so-called persistent, or sustained, activity. This can be seen in delayed response tasks that include a short gap in time (seconds) between a sensory cue and the opportunity to act based on that cue. Neurons show elevated levels of neural activity over a delay without sensory stimulation, as if they are bridging the gap by sustaining their firing to the cue. This was first reported in 1971 by Joaquin M. Fuster, professor of cognitive neuroscience at the University of California at Los Angeles. He trained monkeys in a delayed matching-to-sample task to match colors. When recording neurons from the PFC of the behaving monkeys, some neurons selectively increased their firing rate to specific colors throughout the 10-second delay.[90] Through sustained activity, neurons actively buffer and process information to bridge the gap until a response is required. Later, Fuster could indeed show that temporarily inactivating these neurons by cooling the PFC caused the monkeys to forget the stimulus. Because the short-term buffering of information in an active "online" state is a keystone of working memory, sustained delay activity has become virtually synonymous with "working memory" among neuroscientists.[91] Michael Shadlen from Columbia University properly says that persistent activity during delay periods provides us with "freedom from immediacy"[92]; we don't have to immediately act on everything that affects us, we can take the liberty to think about it for a while and then come up with the most suited solution to a problem.

Neurons in the reciprocally connected cortical association areas are particularly well suited to maintain information across time to exert cognitive control. The posterior parietal and lateral prefrontal cortices are connected not only to one another, but also to up to over a dozen other widely distributed cortical areas,[93] which are interconnected by a common thalamic input.[94] The areas in this network display so-called "non-canonical" circuit properties.[95] Non-canonical circuit properties contrast "canonical" circuits of the sensory and motor cortices. That means that information in the sensory and motor cortices is transmitted directionally from bottom to top (feed-forward) or from top to bottom (feed-back) along the processing hierarchy. However, non-canonical circuits lack clear feedforward and feedback projection patterns. Instead, the non-canonical network of the prefrontal and posterior parietal association cortices seems to be designed for parallel and repetitive processing, so that information is processed along multiple paths and kept over time. This creates a global neuronal workspace that is thought to enable access to the conscious processing of mental representations.[96]

The delayed match-to-numerosity task introduced earlier (figure 3.7A) allows the investigation of persistent activity of single neurons during numerical tasks. This is because it requires animals to remember a number of items over a delay period to subserve a future selection process. As a physical correlate of working memory for number, a significant proportion of neurons (20–30% in the PFC and 10–20% in the PPC) show persistent activity during delay periods when monkeys memorize the numerosity they have just perceived. Like neurons that are active during sensory stimulation, delay-selective number neurons are tuned to the remembered numerosity. These cells are even tuned to the same numerosity in both the sample and the delay periods; the example neuron in figure 7.2 shows such responses. In addition, many delay-selective neurons integrate across spatial and temporal presentation formats[97] and visual–auditory modalities.[98] Therefore, numerical delay selectivity can be abstract and found irrespective of the exact spatial or temporal appearance of the memorized sets.

But memorizing events is not as easy as it sounds. Anyone who has tried to remember long telephone numbers knows that working memory has a severely limited capacity. We can only keep short sets of numbers active in the mind. Even worse, different numbers compete for limited capacity in working memory. As a consequence, such number representations are vulnerable to distractions. Consider a simple example: you look up your friend's phone number to call, but while you pick up the phone, someone utters an arbitrary number. Only with considerable effort will you be able to ignore the distraction and still manage to correctly call your friend's number. Distractors potentially replace important representations, so that the relevant content may be lost entirely. The process of restraining or overriding irrelevant representations to keep hanging on to relevant information is termed "cognitive inhibition."[99] Where and how number representations compete to gain access to the working memory workspace then becomes a crucial question.

My postdoctoral researcher, Simon N. Jacob, and I confronted monkeys with a situation mirroring the aforementioned phone call example.[100] For that purpose, a behaviorally irrelevant number display was inserted as a distractor in the delay period of a delayed match-to-number task. Because cognitive inhibition cannot entirely eradicate distracting influences, a worsening of performance reflects increased distraction. Indeed, the monkeys managed to resist the distractor numbers most of the time, but made slightly more errors than when no distractors were present.

While the monkeys performed this distractor task, we recorded single-unit activity simultaneously from the PFC and the VIP. After sample presentations, PFC neurons encoded the relevant sample number during the first delay period and exhibited the well-known tuning curves and sustained activity. However, when the distractor number was shown subsequently, the PFC neurons did not resist interference. Rather, sample information was overwritten, and the neurons now responded strongly and in a tuned fashion to the distractor numbers. Surprisingly, however, target number representations were not permanently lost but regenerated in the subsequent delay after the distraction, just in time to solve the task. Neurons in the VIP, by contrast, were largely unaffected by the distractors and continued to retain the working memory representation of the sample numbers, even if a distractor had been presented.

These results point toward the PFC as a control or selection stage in the number network. Activity in the PFC might be better understood not as a signature of memory storage,[101] but rather as a top-down signal that influences other parts of the number workspace,[102] such as the VIP, where the actual working memory representations seem to be maintained. To adopt this position of a selection stage, the PFC needs representations of both relevant and not immediately relevant information in order to filter distractors and flexibly guide stimulus selection for the upcoming response later in the task. This could explain why the ability to resist interfering stimuli is compromised in monkeys[103] and humans with lateral PFC lesions,[104] not because stored information is lost, but because the control and selection processes in the global workspace are dysfunctional.

Functional connectivity studies in humans suggest that the PPC is connected with three parietal-frontal working memory-related circuits.[105] These three distinct subdivisions of the inferior parietal cortex are the IPS, the supramarginal gyrus, and the angular gyrus (see figure 6.1 for anatomical locations) The IPS is part of an intrinsically connected parietal-frontal system that includes the frontal eye fields, supplementary motor area, anterior insula, and ventrolateral PFC.[106] In contrast, the supramarginal gyrus is more tightly linked to the dorsolateral PFC, together with which it forms the canonical parietal-frontal central executive network.[107] The angular gyrus is strongly connected with ventromedial PFC and posterior cingulate regions comprising the default mode network,[108] a system active when individuals are not involved in a task and therefore with no direct involvement in working memory. Thus, leaving the angular gyrus aside, the IPS and supramarginal gyrus form distinct parietal frontal working-memory–related circuits.

The parietal-frontal working memory systems emerge most prominently in developmental studies. Children have immature problem-solving abilities which require them to break down numerical problems into more basic components that require greater reliance on working memory. For example, children rely more on counting strategies during simple arithmetic problem solving and need to access multiple working memory components, including short-term storage, rule-based manipulation, and updating of the stored contents. Consistent with these behavioral findings, Susan Rivera and Vinod Menon from Stanford University found that, relative to adults, children tend to engage the PPC less, and the PFC more, when solving arithmetic problems.[109] This likely reflects the increased role of visuospatial processing and the concurrent decrease in demands on cognitive control.

Traditionally, the neuroscience of numbers hovers around the question of which brain areas represent numerical quantity, and how abstractly these areas can bring about numerical competence. However, this core number system needs the support from other, more general cognitive systems. A comprehensive systems neuroscience account of the neural basis of mathematical learning and knowledge acquisition in both children and adults will therefore need to branch out to connected general brain systems of cognition, such as working memory, with the core number network.

Part IV Number Symbols

8 Signs for Numbers

8.1 Evolution Pushed *Homo sapiens* toward Symbolic Thinking

We encounter numbers daily in many different forms and notations. I read numbers from my watch and press the keys on my keyboard to insert numbers into my writings. In fact, we are so used to number symbols and notations that we take them for granted. We tend to forget that these symbols didn't even exist for a long time in human history.

What might have pushed our *Homo* ancestors toward symbolic thinking? One key event certainly was the development of the large and capable brain of modern humans. The evolutionary driving forces for the emergence of these big-brained modern humans are disputed, but it probably arose as a result of local demographic,[1] geographic, and climatic[2] pressures. Whatever drove the massive increase in brain size and in the association cortex, in particular—whether it was the climate, the use of tools, enhanced sociality, or high-quality nutrition in the form of cooked meat—these changes laid the foundation for what archaeologists call culture in early modern humans, including the appearance of symbols.

Unfortunately, we have no direct evidence of how the brains of our ancestors may have looked, because brains do not fossilize. However, paleoanthropologists can study the internal casts of the bony braincase to get an idea about the evolution of overall human brain anatomy. These internal casts of the hollow skulls, known as endocasts, can be measured by computed tomography (CT) scans. Endocasts approximate outer brain morphology because the brain surface and the cranial bones interact in an integrated and highly coordinated way during early development. As a consequence, the inside surface of the skull is a rough imprint of the outer surface of the brain. Based on endocast measurements, paleoanthropologists can explore brain shapes and even the sizes of endbrain lobes of fossilized skulls and can retrace how brains evolved throughout our evolutionary lineage.

Recent explorations of endocasts in combination with craniofacial (skull and face) morphology and dental development revealed fascinating new insights. The archeological record suggests that our modern human brain shape was not yet established at the origin of our species more than 300,000 years ago. The brain of *Homo sapiens* experienced visible changes during the hundreds of thousand years our species has existed. Our brains differ significantly from the brains of the first modern humans, despite belonging to the same species. This, of course, can only be caused by genetic changes that accumulated over time.

What is even more fascinating is that these brain changes correlate with behavioral changes of our species. For the longest time of its existence, our species still remained rather unobtrusive. Present-day human brain shape was reached between about 100,000 and 35,000 years ago.[3] As a consequence, dramatic changes occurred with the advent of the Upper Paleolithic Age (the youngest Old Stone Age) about 50,000 years ago, when these brain changes triggered the "human revolution."[4] All of a sudden, behavioral modernity could be seen in the archeological record of *Homo sapiens*, such as worked bone, ornaments, pigments, stone-age technology, art, and, most importantly, the usage of signs and symbols. All of these features are clear indications of abstract and self-reflective thinking in our species. Consequently, the Upper Paleolithic was a period of great transition in the world. The Neanderthals in Europe became edged out and disappeared 30,000 years ago, probably due to the doings of our species, and modern humans began to have the world to themselves. Next, around 12,000 years ago, the Neolithic (Young Stone Age) Revolution brought about major cultural changes. This period is characterized by the transition from a lifestyle of hunting and gathering to one of agriculture and settlement, the domestication of plants and animals, the first cities, and the rise of the first high cultures with number notation systems in Mesopotamia around 4,000 BCE. Finally, the Bronze Age dawned as the third phase in the development of material cultures around 3,000 BCE. At that point, mankind had metal at its disposal.

Cognitive archaeologists attempt to reconstruct the ways in which the human mind operated in the past and to elucidate how it has changed over the course of human history.[5] I guess many would think that our ability to use language endowed us to construct number symbols; after all, we first learn to speak and then to count as children. Recently, however, a fresh and provocative account of how numerical symbols may have started has been put forward by Lambros Malafouris from the University of Oxford in his "material engagement theory."[6] He argues that before linguistic quantifiers or other symbols were available in human culture to express numerical quantities, the ability to conceptualize discrete numbers was acquired through engagement with material artifacts, such as bones or clay tokens. Malafouris claims that only material artifacts, not language, have the necessary *representational stability* to acquire larger numerical concepts. According to him, "the ability that enables humans to conceptualize the quantity 10 in the absence of language or symbol does not refer to a process of learning but to a process of "enactive discovery and signification."[7] To Malafouris, the visible and tangible properties of these tokens "brought forth" numbers by inducing subsequent brain changes. As Karenleigh Overmann from the University of Colorado summarizes Malafouris's position: "Inchoate concepts of quantity, once externalized in material form, become tangible; in becoming tangible, they also become more explicit, more manipulable and more readily shared between individuals and generations."[8]

Indeed, it is interesting to see how "things," such as fingers, bones and knots that serve as numerical "signs," helped us humans to begin to understand number theory. At a certain point in human history—and also during our development as children—we started to replace the elements of a collection by other objects. And finally, someone had to start writing down numbers in a smart way to create a permanent record of numerical quantity, a number notation system. This move beyond a non-symbolic number instinct implies a mental leap of which no other animal is capable. The following sections retrace the cognitive and neuronal stages of this process.

8.2 Number Signs: Icons, Indices, and Symbols

The term "sign" is an umbrella term for all possible associations between a object or event and the entity it stands for. In general, a sign denotes something else, namely the signified. Signs can take the form of sounds, images, or even actions. However, signs themselves have no intrinsic meaning; they become signs only when they are associated with meaning. Signs come in many forms, from simple to complex, and it is helpful to classify them according to their meaning, or semantic complexity. The different classes of signs tell us a great deal about how a symbolic understanding of numerical quantity arose in evolution, development, and human history.

The American philosopher Charles Sanders Peirce[9] distinguished three kinds of signs: *icons*, *indices*, and *symbols*. These sign categories are ordered according to increasing complexity of the relationship between the sign and the signified. Since we possess two powerful symbols systems, natural language and number faculty, I will illustrate signs in both realms before focusing on numbers again.

The simplest sign, an icon, is characterized by similarity between sign and object (reference based on similarity). For example, the sign ☎ is an icon for a telephone. In the numerical domain, ••• could be an icon for three objects. The first number signs used in human history were icons. The iconic origin of number symbols can easily be seen in many ancient notation systems, such as the as the number of lines that represent the Roman numerals I, II, and III.

An index, on the other hand, is more abstract. An indexical sign does not look or sound like what it denotes. Rather, it appears together with what it denotes and then is associated with it. Sometimes the signified and the sign are causally related. For example, increasing temperature causes an expansion of mercury, so the mercury position in a thermometer is an index for temperature. Similarly, a good mood causes smiling, so a smile is an index for happiness. Often, however, random signs are simply being associated with the signified based on learning. For example, we all learn that the horn of an ambulance is an indexical sign for an emergency situation that prompts us to clear the street. In the number domain, the cardinality of sets can be associated with arbitrary shapes or sounds to give rise to a numerical index. Of course, we can invent many signs that apply for different situations in the same realm. Nautical flags, for instance, carry different meanings for communication on water, but this meaning is constant and fixed for each individual flag. Importantly, indices are signs that work in isolation from other indices; they are not part of a sign system.

This is a crucial difference to the third, and most complex kind of sign, the symbol. Symbols are also arbitrary links between the signifier and signified based on convention. When we learn language, we all agree the sounds the letters "c-o-w" produce refer to a large mammal that has horns, feeds on grass, and gives milk. Nothing in those sounds bears any similarity to the actual animal. This, however, is the same for indices. As emphasized by Terrence W. Deacon from the University of California at Berkeley, the defining difference is that symbols are part of an entire system of signs in which the meaning of symbols is dependent on how they are combined into longer expressions.[10] A symbol is part of a combinatorial sign system. The rules we use to structure and order symbols in expressions, called syntax, determine the meaning of an expression, such as a sentence or a mathematical formula. A very simple example can show this important aspect. Let's take three symbols—in this example, words, namely "child," "cat," and "bite." Now we arrange them in two different ways. One arrangement reads "child bites cat," the other is "cat bites child." It is immediately clear that the two expressions, even though they consist of the same symbols, have very different meanings. Therefore, reference shifts from the sign to the system.[11] Symbolic reference is crucially a link between sign-sign relations, not between individual sign-object relations.[12]

What accounts for words, the symbols of natural language, is equally true for number symbols in mathematical expressions. Take the simple calculation $2 + 5 \times 3$. Whether we first perform the addition and then the multiplication, or the other way around, influences the result. The result is 21 in the first case, but 17 in the latter. Nonetheless, the value of the individual number symbols remains identical. Symbols, as understood according to the above definition, are human-specific.

This taxonomy of signs can now be applied to numbers. We will see a progression from the simple to the complex, from icons to indices, and finally to symbols. This suggests that each more advanced sign builds on the previous one. Such a progression of signs can be witnessed across different time scales: a cultural progression in human history, an ontogenetic progression during development from infant to adult, and to some extent even an evolutionary progression in the animal kingdom.

8.3 Invention of Number Symbols in Human History

Before indices and symbols were invented, mankind first used simple icons, such as fingers, to signify numbers. Fingers are the oldest tools for recording numbers known to man; it is said that all people able to count have an early stage of finger or body matching.[13] Fingers are clearly used as icons: every finger represents one element of a set, so that the number of fingers and the to-be-represented set elements bear clear perceptual similarity. Moreover, fingers are ideal icons as a prelude to counting, as I will show later.

Despite their pragmatic advantages, the scope of finger usage is limited. Holding up fingers leaves no permanent trace, and at a certain point in human history, this became unbearable. As modern humans matured from nomadic hunter-gatherers to settled farmers, temporary tallying methods became inadequate for keeping track of livestock, possessions, and other everyday purposes. Mankind therefore had to resort to more permanent methods.

The first stage in the emergence of written numerals and later number notation systems was the carving of tally marks. As with fingers, every notch on a tally stick corresponds to one element of a set. Tally sticks therefore are permanent representations of numerical quantity. During the Upper Paleolithic, humans started to carve notches on whatever was most available and suitable, such as sticks of wood and bones of dead animals.

A number of anthropological artifacts have been conjectured to be tally sticks. The earliest have been found in South Africa. The Lebombo bone seems to be the oldest known tally tool and is around 44,000 years old according to radiocarbon age estimates.[14] It was discovered in the Border Cave located in the Lebombo Mountains near the border between South Africa and Swaziland. The Lebombo bone itself is a tally bone made from a fibula, one of the two lower leg bones, of a baboon. It presents an incomplete sequence of 29 notches. The surface of the object also is heavily polished, suggesting long-term usage.

Another prehistoric tally tool is the so-called Wolf bone, a prehistoric artifact discovered in 1937 in the Czech Republic during excavations at Vestonice. The wolf bone is approximately 30,000 years old, is about 18 cm long, and carries 55 marks. Because a sculpted ivory head of a woman was excavated nearby, we know that the people who made this tally bone also had developed some form of art.

The Ishango bone made from the fibula of a baboon is another fascinating bone tool. It was found in Zaire in Central Africa and is estimated to be approximately 20,000 years old. It carries groups of different numbers of

Signs for Numbers

notches and has been the subject of all sorts of wild speculations as to what kind of arithmetic they could have been used for.

Earlier mentions of fingers as counting methods credited them as the first iconic stage of number signs. But what if the number of elements in a set is larger than 10 and the counter runs out of fingers? One way out would be to continue with the toes as icons and thereby increase the number space to 20. But because this is quite cumbersome and also not a large enough number, many indigenous cultures resort to using body parts. The switch from fingers to body parts nicely illustrates the transition from an iconic sign to an indexical sign: while fingers represent single elements as icons, other body parts already represent the total number of elements of a set, its cardinality or numerosity, as indices.

The work conducted by Geoffrey B. Saxe from the University of California at Berkeley provides a nice example of body parts as numerical indices.[15] Saxe studied the Oksapmin, a remote community in Papua New Guinea. Traditionally, the Oksapmin used a counting system that makes use of 27 parts of the body (figure 8.1). After the five fingers on the right hand are used up, they switch over to the wrist, denoting cardinality 6, and use the

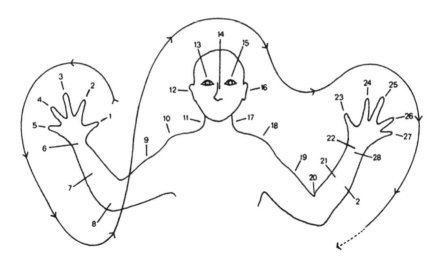

Figure 8.1
Indigenous cultures resort to using body parts to keep track of numbers. The Oksapmin in Papua New Guinea employed a counting system that makes use of 27 parts of the body (from Saxe, 1981).

underarm to signify numerosity 7, the elbow to represent numerosity 8, and so on.[16] How would we know which number the wrist, the underarm, and the elbow represent? Only by convention can we understand these signs, a clear characteristic of indices. Another feature of body parts as indices is that they denote isolated cardinalities, and traditional uses of the system did not involve arithmetical functions.

The usage of knots is another example of numerical indices.[17] Knots mark the transition away from body parts toward human artifacts for creating permanent records of quantity. Using knots has been a fairly widespread practice throughout the American, European, and Asian continents, even though it never evolved into a more elaborate spoken or written notation system. Probably the most developed system of knot numbers we know of is that of the Incas of Peru from the twelfth to the sixteenth century. Knots were tied on twisted woolen cords, and such a knotted cord is known as a *quipu*. Three different types of knots represented different kinds of numbers. Because the meaning of these three different types of knots had to be learned and therefore is a convention, they represent indices. Figure-eight knots represented units of one, slip knots with the appropriate number of loops signified units from two to nine and, finally, the appropriate number of single knots represented tens and hundreds. For instance, the number 235 would be displayed on a single cord with two knots near the top of the cord (for the two hundreds), followed by three knots for the three tens, and finally a slip knot with five loops for the five units. Remnants of knot systems can be found in different cultures all across the world. Examples include the knotted woolen cords of Tibetan prayer strings, or the threaded rosary beads of the Roman Catholic Church.

As pointed out, fingers are used as customary number icons. However, the position of fingers at the hand can also be used as conventional and indexical signs for cardinality that are culturally passed on from one group to the other. The most famous reference to this sort of indexical finger counting in Europe is that of Venerable Bede. After his death in 735, Bede left the details in an introduction to a work dealing with the computation of the date of Easter, an important calculation of that time. Interestingly, the first three finger signs are iconic, because for every element of a set, one finger is put down. However, the transition from icon to index is clearly visible starting with the sign for four: the thumb, index finger, and little finger are held up, while the middle and ring finger are put down. Similar complex hand postures are used for the larger numbers. As a result, people can only know the cardinal value of this hand posture by convention, which is again characteristic for indices.

Besides artifacts as counting devices, counting words also exhibit indexical qualities. The number words of the Australian Chaap Wurrong tribe in Victoria, for example, display this trait[18]: four is the verbal expression "two and two," five is "one hand," six is "one finger, hand," and 10 is "two hands." Similarly, the Russian word *sorok*, or "40," goes back to Old Nordic *sekr*, the word for furs that were traded in bundles of 40.[19] Here, links between words and cardinalities are established, thus forming indexical reference.

True number symbols first appeared with the rise of the first ancient civilizations in Mesopotamia and Egypt. In these ancient times, thousands of people started to gather in cities, and the invention of bargaining and the tax system required some sort of a number system suitable for calculation. As a result of this demand, numbers became part of a symbol system. Numerical notation was first developed in Mesopotamia around 3,500 BCE, slightly earlier than the Egyptian hieroglyphic numerals that were used as early as 3,250 BCE.[20] The Mesopotamian numerical notation systems were cuneiform systems written down using wedge-shaped marks. Numerals were written almost exclusively on tablets using a stylus to impress signs onto wet clay.

Many of the first visual number symbols, or numerals, actually developed out of notches and tallies. In these systems, symbols for smaller units are either short lines carved into rock, indentations on clay tablets, or written strokes that are repeated as often as necessary. Using former icons as number symbols comes in handy, because the ability to enumerate very small collections at one glance is found in all explored human cultures.[21] For instance, the signs || and ||| represent two and three, respectively, both in Egyptian engraved hieroglyphic texts and the familiar Roman numerals system that are still displayed on clock faces. Entirely independent from the Western numerals, the Maya used dots for small numbers, so that •• and ••• also represented the numbers two and three. The fact that these cultures could not possibly have had contact at the time of the invention of these systems strongly suggests that iconic tally signs were a universal blueprint for the later developed numeral systems.

In ancient Egypt, Greece, or Rome, specific number values were denoted by dedicated signs. In the Roman numeral system, for instance, units up to three were denoted by I, five was represented by V, 10 by X, 50 by L, and so on. Such signs were simply added to represent numbers. The number 71 in Roman notation therefore was written as LXXI. However, as a result of such a primitive and inflexible non-positional system, mathematics couldn't be performed properly.

It took humanity quite a while to invent the most powerful and universal type of number notation system that we still use today: the "positional notation system," also called the "place value system." In a positional notation system, the same numeral adopts different numerical values according to its position in a record. In our decimal position system, from right to left, the positions correspond to units, tens, hundreds, and so on. For instance, the "3" in the number 302 stands for three hundred, whereas in the number 203, it denotes three. We owe this positional notation system to the Indians. Mediated by the Arabs, this superior notation system gradually became introduced to the West after the twelfth century. As a result, we still refer to these symbols as "Arabic numerals."

Of course, the units don't have to go to 10, as in our base-10 notation system; the base is an arbitrary convention. The units could also go up to 20, as in the base-20 system, or only adopt two values, as in the binary base-2 system. However, any effective place value system requires a sign to denote the absence of a positive digit, and for this purpose, zero was first used.[22] The last chapter of this book is dedicated to this fascinating number zero.

8.4 How Children Learn to Deal with Number Signs

Just like in human history, evidence for iconic stages can be found in young children's acquisition of numbers.[23] The clearest one is again the usage of fingers to assess the number of items, a habit that many of us maintain even into adulthood. At least in our culture, hardly any child will have learned to count without using fingers. In fact, finger counting is so deeply rooted in us that the skilled use of fingers has even been found to be a predictor of math achievement.[24]

The intimate relationship between the usage of fingers and counting skills is also shown in the brain. Finger representations and mental arithmetic share common parietal substrates.[25] Damage to the inferior parietal lobule, particularly around the angular gyrus, of the language-dominant hemisphere often results in a neuropsychological disorder called Gerstmann's syndrome.[26] This syndrome is characterized by four primary symptoms: agraphia (deficiency in the ability to write), left-right disorientation, acalculia (difficulty in comprehending numbers), and finger agnosia, an inability to distinguish between fingers on one's hand. The presence of Gerstmann's syndrome has been taken as evidence of a tight association between finger recognition and numerical ability.[27]

However, children also use verbal icons. When taught to count verbally, young children at around age two start producing one word for each element of a set. For example, when presented with four candies and asked to report how many candies lay on the table, they repeatedly recite the sequence "*one-two-three-four.*" Interestingly, they initially don't understand that the last word in the sequence represents the cardinality of the entire set.[28] Up to this point, the individual words are simply used as spoken icons, just as each finger is used in counting. The meaning of the words is not important at all; they could be any words as long as they are unique and come in a stable order. Indeed, even non-human primates have the capacity to learn arbitrary (visual) lists,[29] and so it is likely part of an innate capability. Of course, parents teach little children to use number words correctly early on, but these words are initially only placeholders until they come to have their eventual numerical meanings.

The indexical stage surfaces when children learn to associate cardinality with visual shapes—for instance, three items with the arbitrary visual shape "3." At this stage, the shape "3" is an index for cardinality three, but not yet a symbol. This is because each visual shape individually represents a specific cardinality, and these signs are not yet understood as symbols in a system.

To arrive at a symbolic number concept, we must master the transition from indices into a symbolic system, a stage at which only humans arrive during childhood. At first, children painstakingly have to learn to associate a cardinal value with a word that, at this stage, still constitutes a sign. But this is only necessary for numbers up to four. Once children learn to count symbolically, they suddenly begin to generalize the counting procedure to numbers larger than four with no specific training and with no evident upper bound.[30,31] They understand that they are dealing with an ordinal list of numerals in which every successive numeral represents a number value incremented by one. This is referred to as the "successor relation."[32] During this period, these signs are becoming part of a symbol system. Next, children acquire a set of elementary arithmetic facts and calculation procedures to perform arithmetic operations on all the numbers they can count. Finally, they extend number knowledge beyond the limits of their counting procedures, using arithmetic operations to represent fractions, zero, and negative numbers. These and many other developments distinguish numerical skills in human children from the brightest animals. It marks a profound discontinuity between human and non-human minds that will be elaborated later on.

Once numerical symbols are acquired, they can be regarded as mathematical objects that stand for ordered cardinality. They come in visual shapes, such as Arabic numerals or as number words that we read or hear. Of course, the way these symbols look or sound is arbitrary and only defined by our culture. Some researchers therefore posit that number symbols are categorically different from any non-symbolic representation of numerical quantity; they assert that there is no link between non-symbolic and symbolic quantity representations.[33] But in fact, there is ample evidence pointing to a fundamental relationship between these two types of representations. More precisely, number symbols seem to be grounded in non-symbolic quantity representations, and therefore share fundamental similarities. The remnants of the non-symbolic number system in symbolic notations are in fact, revealed by behavioral, neuroimaging, and neurophysiological commonalities, as we will see in chapter 12.

8.5 Teaching Number Signs to Animals

Interestingly enough, the usage of iconic signs is not uncommon among animals, and some animals even spontaneously use them for communication. One of the most famous cases is the production of alarm calls by vervet monkeys, close relatives of macaques. As Richard Seyfarth and Doris Cheney from the University of Pennsylvania discovered, wild vervet monkeys use three acoustically distinct alarm calls to denote the presence of their enemies, whether in the sky ("eagle call"), running on ground ("leopard call"), or crawling ("snake call").[34] For quantities, however, the spontaneous usage of indices is largely absent, perhaps with the exception of the previously mentioned chickadees that use the number of "dee" notes to indicate the dangerousness of predators.[35]

In the laboratory, several avian and mammalian species have been trained to map visual shapes to cardinality. Superficially, such behavior looks a lot like symbolic competence, particularly when Arabic numerals are chosen as signs. However, below the surface lie fundamental differences. For example, the African grey parrot by the name of "Alex" was a particularly fascinating animal prodigy. Parrots have a reputation of imitating human speech sounds, and Alex, who died in 2007, responded not by pecking at response keys, but instead by talking to his trainer, Irene Pepperberg from Brandeis University. In behavioral tests, Alex responded in spoken English and began to name different objects, shapes, colors, and numbers.[36,37] `In response to certain questions, he could vocalize both the quantity and the name of the type of object, such as "four keys." Interestingly, he could also report the number of objects in a subset within a heterogeneous array. For example, when asked "How many corks?" he could answer "two" when shown a random mixture of two corks and three keys. As a sign of conceptual understanding, he accurately transferred labels to other objects for which he knew the names but that had not been used in numerical training. Alex's performance was a very convincing impersonation of a human using language to respond. Hardly anyone escapes this fascination when listening to records of Alex giving responses in flawless English. But besides his mastery of some speech, his behavior is fundamentally similar to the performance of other, less vocal animals. The key difference is the different kind of motor output Alex used as a response. Instead of pecking with his beak at conditioned signs, Alex used a restricted set of human speech sounds to denote numbers and other categories. He never rearranged the learned labels or expressed sentences that would indicate that the vocal signals would have been more than indices associated with categories. Training a parrot to discriminate abstract categories and respond with human speech sounds is heroic and fascinating in its own right. It demonstrates the wide range of iconic signs. However, it is still far from symbolic number understanding.

To explore the usage of number signs, most animals are trained on some variation of an association task. In this task, an animal learns by trial and error to link a numerosity to an arbitrary visual shape and to store this association in its long-term memory. Pigeons, for example, have been trained successfully to associate the number of dots from one to four with the letter signs A, B, F, and G.[38] These pigeons first had to peck each item of a dot array. The numerosity display was then replaced by an array showing the five different signs, and the pigeons had to peck the sign that was associated with the number of items it had pecked before. Sometimes the pigeons erred, but the errors were not random. In error trials, the pigeons tended to choose the signs that carried a numerical value similar to the correct sign, thus showing a numerical distance effect. As a result, this behavior indicates that they learned to associate signs with approximate number representations.

In my own laboratory, we wanted to see if discriminations based on numerical signs would allow rhesus monkeys to gain more precision in judging numerical values.[39] After all, the usage of number signs coincides with the development of an exact counting capability in children. Therefore, we trained monkeys in a delayed association task (see figure 9.3B) to associate visual shapes to numerosities in dot displays. For simplicity, we chose the well-known Arabic numerals as shapes; Arabic numeral 1 was associated with one dot, numeral 2 with two dots, and so on. The monkeys' performance showed that they had learned to associate numeral shape with number, albeit again only in an approximate way, as witnessed by the presence of a numerical distance effect that was similar to the one observed for dot patters. Obviously, they first had to translate the meaning of the numerals into an approximate number code before they could correctly match the shapes to numbers. And this process caused the Weber's law characteristics that we had seen in the data. This suggests two findings. First, the monkeys were indeed judging the approximate numerical values that were mapped onto the numeral shapes. And second, the numerical signs did not improve discrimination performance. The latter finding argues that the shapes were simply placeholders for approximate numerical values, instead of symbols enabling precise number representations.

To find out if the monkeys truly associated numerical values with the trained visual shapes, we had them match shapes to shapes. In other words, a given numeral had to be matched to an identical one. Obviously, the monkeys could have done this entirely without thinking about numbers, simply by matching shapes. But this is not what they did. Instead, the performance of the monkeys still showed a numerical distance and size effect. They mixed up signs that were associated with similar numerical values, such as two and three, more often than signs that carried remote values, such as one and four. The monkeys obviously continued to judge these shapes as numerical signs, as they had been taught to do for many months. As a result, the visual shapes became automatically linked to approximate cardinal values.

Instead of perceiving numbers, some animals have also been trained to produce a certain number of movements cued by a numerical sign. Pigeons, for example, learned to reproduce a numerical value signified by a sign by producing a specific number of pecks.[40] Similarly, chimpanzees have been trained to produce a certain number of movements based on a learned sign.[41] In this study, two chimpanzees, using a joystick to control a cursor on a computer monitor, were trained to produce a specific number of cursor movements (from one to six) in each trial. To begin a trial, the chimpanzees moved the cursor until it made contact with a target numeral that was associated with a specific numerical value. When the cursor made contact, the computer presented an array of up to 10 randomly placed dots in the lower half of the monitor. The chimpanzees then had to touch individual dots with the cursor. Every time this happened, a brief auditory cue was given, and the dot disappeared from the lower portion of the screen and reappeared on the top portion to indicate the number of items that had already been "collected." For a subject to perform a trial correctly, the dots at the bottom of the screen had to be collected, one at a time, until the collection was equal in value to the previously given numerosity. To complete the task, the chimpanzees had to move the cursor back into contact with the target numeral. As expected for numerosity estimation (ANS), the chimpanzees' performance decreased systematically with successively larger target values, and the performance functions broadened with increasing numerical values, all of which are familiar characteristics of Weber's law.

Overall, different animal species show this important first step toward symbolic understanding. Such data in animals are therefore instrumental to understand the evolutionary basis of our own counting capabilities. If animals can learn to map visual signs onto representations of numerical quantity, does this mean that animals can be taught number symbols, as several studies have claimed? Unfortunately, such claims are premature; with a closer look, animals actually lack symbolic understanding.

Claims for symbolic number competence have been made for Alex, the African grey parrot,[42] and for a chimpanzee.[43] Let's focus on the study involving our closest relative in the animal kingdom, the chimpanzee. A female chimpanzee named Ai belongs to the most famous of these "counting" animals. Ai lives at the Primate Research Institute of Kyoto University, where Tetsuro Matsuzawa has been working with her for almost 40 years. Using a touchscreen to respond, Ai is able to map dot numerosities onto Arabic numerals from one to nine. She also knows the ordinal relationships among these numerals and can touch a list of numerals in ascending numerical order. The most well-known paper from 1985 reports on Ai's ability to map visual shapes onto the number of objects.[44] This learning process is quite revealing and shows an important difference to the way children learn to count. Ai was first taught to associate the visual shape "1" with sets of one item and "2" with sets of two items. After she mastered this mapping, she was introduced to "3" and sets with three items. She retained her high performance on "1" but was discriminating "2" and "3" at chance values. This suggests that she had first associated "2" with all values of two and higher. After thousands of additional trials, Ai mastered all three numbers. Subsequently, whenever a higher numeral was added to the set, Ai would display the same pattern of performance as mentioned before. Obviously, Ai never understood that each number sign added to the set was to be associated with a precise numerosity. As each new sign was introduced, up to "9," she was always at chance between the last item learned "n" and the newly introduced item "n + 1." This suggests that she had previously learned that "n" was to be associated with all numerosities equal to or higher than "n." This is in stark contrast to how children learn to count. Children show this learning pattern only up to "four," but then they understand that every new numeral denotes "n + 1." I therefore follow the conclusion reached by Susan Carey from Harvard University for all animal studies that attempt to show an understanding of symbolic numbers:

> It is likely that the process underlying this achievement is nothing like the process children go through. Years of operant condition was required in each case, and there is no evidence that any animal has induced how the numeral list works.[45]

Ai's failure might be explained by two aspects. First, she obviously never grasped that she was dealing with an ordinal list of numerals, in which every successive numeral represents a number value incremented by one. This is called the successor relation. For every new sign, she painstakingly had to learn to associate a cardinal value with a visual sign, indicating that these numerical signs were not part of a symbol system, but were actually isolated indices. A second possible reason why Ai failed could be the lack of her understanding of recursion. Recursion is the determination of later parts of an expression that refer to, or depend on, preceding parts of an expression. Take the sentence "Hillary's lawyer's secretary." In order to understand which secretary one is referring to, I need to understand that it is the secretary of the lawyer, which in turn is the lawyer of Hillary. The reference returns from the secretary to Hillary. Similarly, recursion is a principle present in counting: if 2 is (1 + 1), then 3 is ((1 + 1) + 1), and so on. The ability to understand recursion has long been suggested to separate animals from humans.[46] It seems that symbolic competence in numerical cognition is, after all, human-specific and beyond the reach of animals.

9 Neural Foundation of Counting and Number Symbols

9.1 The Patient Who Lost All Numbers beyond Four

As mentioned earlier, the study of patients suffering from brain damage was historically the first source of information confirming that numbers had a special place in the brain. The most conspicuous deficits that emerged in such patients were counting and calculation impairments—all of which, of course, relied on the symbolic understanding of numbers.

The case of an Italian woman who suffered a stroke and was examined at a neurological clinic in Padua, Italy, by Lisa Cipolotti and coworkers exemplifies how striking the loss of numerical functions can be.[1] To maintain patient confidentiality, the patient is only known by her initials, C.G. This woman was well educated and attended school for a total of 13 years. Before her illness, she dealt with the administrative side of her family hotel. In October 1987, at the age of 56, C.G. suffered from a stroke affecting the left middle cerebral artery. The left middle cerebral artery, one of the three major cerebral arteries, serves the majority of the lateral parts of the frontal, parietal, and temporal cortices. As a result of this injury, C.G. immediately developed a right hemiparesis (weakness of the right side of the body) and pronounced language disturbances (global aphasia). A magnetic resonance scan showed brain damage in large parts of the left fronto-parietal region (see figure 9.1). Fortunately, C.G. improved over the following days and months. Ultimately, her motor impairment disappeared completely. Her initially massive aphasia showed rapid and almost complete recovery, and she could eventually speak again. However, she complained of severe difficulties in dealing with numbers in her everyday life. The report says that

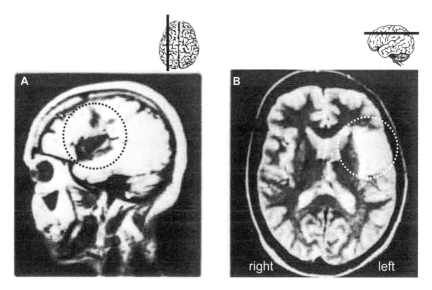

Figure 9.1
The patient who—after brain injury—lost all numbers beyond four. MRI images of sagittal (**A**) and left-hemispheric horizontal (**B**) sections of the brain of patient C.G. suffering from brain damage involving large parts of the left fronto-parietal region (encircled). For orientation, sections in the sagittal (left) and horizontal plane are shown in the insets (from Cipolotti et al., 1991).

she was no longer able to read the time, dial a telephone number, recognize money or check change. She could not produce her own age, clothes size, telephone number, the number of her house, the ages of her children and husband, etc.[2]

When she underwent neuropsychological examinations, she scored normal verbal and performance intelligence. However, she exhibited a significant and selective impairment on numerical tasks.

The most striking feature of her innumeracy was her apparently total inability to deal with numbers above four. She could not recognize, perceive, or hear numbers beyond four, and could not give any answer for numbers above it or for zero. C.G. could also not produce numbers beyond four herself. She was able to count one by one just up to four, and after that, she said "La mia matematica finisce qui" (my mathematics stops here). Moreover, her calculation ability was gone, except for some additions and subtractions with operands and solutions below five. Surprisingly, however, her ability to deal with numbers up to four was largely preserved:

she could recognize their names; she could count and recite them, and she could judge these numbers and dot patterns, though with some errors. As the authors of the study point out,[3] these deficits with numbers were not due to a more widespread degeneration of memory: when asked to sort pictures according to categories, such as animals, fruits, musical instruments, vehicles, and so on, she could do it. She was also able to name these categories when shown pictures, and she could also define concrete and abstract words, or find synonyms for words. Moreover, C.G. did not suffer an impairment in logical reasoning. She simply wasn't able to deal with numbers above four anymore.

Of course, patient C.G. was not the first patient with an acquired and selective deficit in dealing with numbers and calculations. As discussed earlier in chapter 6, such patients have been reported on since the beginning of the twentieth century. In 1925, the Swedish neurologist Salomon Eberhard Henschen coined the term *"acalculia"* (akalkulia),[4] a term still used today to describe an acquired impairment in processing numbers and arithmetic.

The German neurologist Hans Berger, who later developed EEG,[5] proposed a separation of two categories of acalculia after reviewing several cases from his patients' files who suffered from some type of calculation disorder[6]: "primary acalculia," which develops independent of other impairments, and "secondary acalculia," which is present in conjunction with disturbances of memory, attention, or language. Just as Henschen had, Berger rejected the idea of a single calculation center in the brain; rather, he argued, different localized components work together in order to orchestrate diverse mathematical abilities.[7]

The roles of the posterior parietal and frontal cortices were further confirmed in the years to follow. In 1948, Kurt Goldstein proposed that acalculia typically followed lesions in the parietal-occipital region and occasionally the frontal lobes.[8] Including only patients with lesions in the occipital and parietal lobes, Henry Hécaen and colleagues[9] similarly noted in 1961 that the majority of acalculia patients showed bilateral lesions primarily in parietal-occipital regions. Lesion studies in patients confirmed the association cortices as locations for number symbols and arithmetic. In addition, the often-observed number-specific deficits in acalculic patients strongly argued that the number symbol system and the language system are separate in the brain, a finding I will later discuss in more detail.

9.2 Imaging in the Human Brain During Symbolic Numerical Tasks

In agreement with findings in acalculic patients, functional imaging studies have consistently observed activations in human parietal and frontal cortices in a variety of symbolic numerical tasks.[10,11,12] Several studies have reported evidence of numerical tuning in the IPS not only for set representations, but also for numerals, counting words,[13,14,15] and even fractions.[16]

Earlier in this book, I mentioned the fMRI adaptation approach that Manuela Piazza and Stanislas Dehaene applied to explore representations of dot numerosities. A few years later, the authors used the same protocol again, but this time to study representations of number symbols and how they relate to representations of numerosities.[17] In this experiment, participants were presented with either arrays of dots (non-symbolic number) or Arabic numerals (symbolic number) of a constant numerical quantity. This repetition of a constant number led to a reduced response in the bilateral IPS regardless of the presentation format. When numerical deviants of either format were infrequently presented, they led to a recovery in prefrontal and posterior parietal activation that was proportional to the numerical distance between the habituated and the deviant number. The authors focused on the IPS and found that this distance-dependent recovery occurred for non-symbolic and symbolic numbers when deviants were presented with their respective formats. Interestingly, this distance-dependent recovery applied across formats as well: it was present when deviant Arabic numerals were presented among dots and vice versa. This suggests that non-symbolic and symbolic numbers are represented (at least partly) irrespective of their format. However, a notation-independent representation has also been challenged for symbolic notations of Arabic numerals and number words.[18] In addition, an interesting difference between the hemispheres surfaced: greater precision of coding was observed for symbolic number representations in the left IPS than in the right IPS. It seems that acquisition of number symbols sharpens and enhances the IPS on the left hemisphere, which usually is also the language-dominant hemisphere.

In a comprehensive review from 2011, Marie Arsalidou and Margot Taylor gathered and combined the substantial body of fMRI data derived from 52 brain imaging studies to provide a quantitative picture of which brain areas process numbers.[19] They explored brain areas in which fMRI activation related to number tasks overlapped across different studies. In doing so, they generated fMRI maps for numerical processing in a standard human brain. One of their goals was to see whether numerical magnitude representations might be affected by the format—that is, whether the presentation of sets or numerals made a difference. When they compared activations in studies in which the participants had to evaluate non-symbolic and symbolic numbers, they found significant overlap mainly in the posterior parietal lobe (SPL, IPS, and IPL), but also in the superior, medial, and inferior frontal gyri, the precentral gyrus, the cingulate gyrus, the insula, and the left fusiform gyrus (figure 9.2). Outside cortical areas, regions of the cerebellum and the basal ganglia became activated.

In the IPS, such number representations also seem to be largely independent of whether we hear number words or see numerals: when numbers, letters, and colors were presented visually or acoustically in the absence of a number task, stronger responses to numerals than letters and colors (present in both modalities) were observed in a bilateral region in the horizontal IPS.[20] This finding supports the idea of a modality-independent, or supramodal, number representation that is automatically accessed.

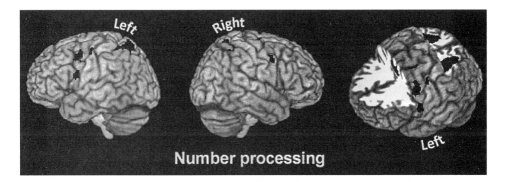

Figure 9.2
Our brain hosts a network dedicated to the processing of numbers. Brain regions consistently activated in fMRI studies during non-symbolic and symbolic number tasks are marked in black color. The left image shows the left hemisphere of a human brain (left is anterior), the middle panel shows the right hemisphere (right is anterior). In the brain on the right side (left is anterior), parts of the frontal and parietal lobes are graphically removed to show the activations within the intraparietal sulcus and the prefrontal sulci as well as the medial frontal lobes (from Arsalidou and Taylor, 2011).

To summarize, the fMRI studies discussed argue for a shared representation of what numerical quantity means in the IPS and parts of the PFC. This representation of quantity in parietal and frontal lobes is largely independent of the representation format. Of course, other aspects of numbers, such as their notations as words or numerals, may additionally address different brain areas. But mostly, the IPS and parts of the PFC encode what a quantity is irrespective of how it is denoted.

9.3 Numerical Association Neurons in Monkeys

Nonetheless, the question remains as to how single neurons in the brain represent numerical signs. Recordings in non-human primates fortunately can help answer this. One might think this is a contradiction, because I have argued previously that animals cannot grasp symbols. So what can we learn from them about number symbols? Although we cannot address genuine number symbols in any animal species, there is a lot to be learned about how number symbols develop from more primitive associations of signs with numerical quantity. The previous chapter showed that animals can learn numerical indices, which also is an important step in children who learn to count. At some early point in our lives, we, too, had to establish long-term associations between numerical categories and initially meaningless shapes that later become numerals. This semantic association is a prerequisite for the use of signs as numerical symbols.[21] By no means is it sufficient, but it is a necessary step toward the utilization of number symbols in humans. And the mechanisms for this indexical step can be distilled down to the level of single neurons in monkeys.

To investigate how single neurons associate the quantitative meaning of a set to an arbitrary visual shape, my PhD student, Ilka Diester, and I recorded from monkeys that had learned to associate visual shapes with the numerosity in multiple-dot displays.[22] I outlined the behavioral protocol we used, known as a delayed association task, and the performance data in the previous chapter. Briefly, the monkey sees a numeral displayed on a computer screen, memorizes it over a brief delay period, and releases a bar only when the next display shows the same numerical value of a set of dots (figure 9.3B). As always, only correct responses are rewarded. In addition to this sign-to-quantity association task, the monkeys also performed our standard delayed matching-to-sample task with numerosities (figure 9.3A)

Neural Foundation of Counting and Number Symbols

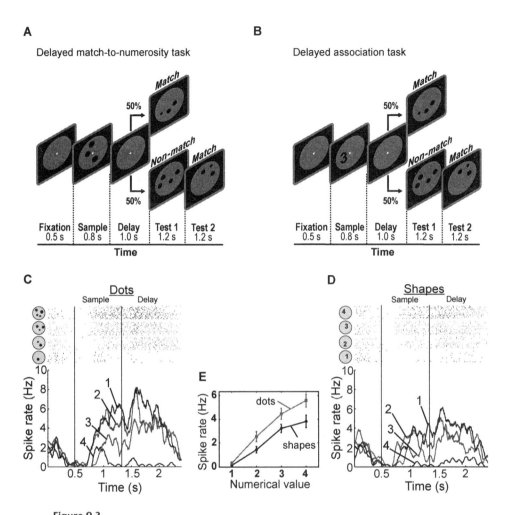

Figure 9.3
Neurons in prefrontal cortex of monkeys learn to associate the quantitative meaning of a set to an arbitrary visual shape. Monkeys were trained to match the number of dots (A) and also to associate the shapes of Arabic numerals 1–4 with the number of dots (B) in a delayed response task. C and D) A single prefrontal cortex neuron is encoding the same numerical values of both numerosities (C) and associated numeral shapes (D). Neuronal responses are shown as dot-raster histograms (*top*) and spike density histograms (*bottom*). The first 500 ms indicate the fixation period. Black vertical lines mark sample-onset (500 ms) and offset (1,300 ms). E) The neuron showed similar tuning functions to dots and shapes (after Diester and Nieder, 2007).

in half of the trials, in which they simply matched the numerosity of a sample dot display to the numerosity of a test dot display. Combining the matching task with the association task in one recording session allowed us to compare the responses of individual neurons to both dot numerosities and associated signs during the sample and working memory phase.

After the monkeys were proficiently performing the tasks, we simultaneously recorded single-cell activity from the lateral PFC and VIP. Much to our delight, a relatively large proportion (24%) of randomly recorded PFC neurons encoded numerical values, irrespective of presentation style. They had the same preferred numerosities for dot stimuli and for the associated numerical values, in addition to showing surprisingly similar tuning to both formats. For example, the neuron depicted in figures 9.3C and 9.3D responded strongest to numerical value four in the sample and delay period. Note the similarity in this association neuron's temporal discharge pattern and number tuning in response to the multiple-dot displays and the shape of Arabic numerals. We called such neurons "association neurons." The activity of association neurons interestingly predicted the monkeys' judgments. When we compared the firing rates of association neurons to their preferred numerical values for correct and error trials, we found them to be systematically lower during error trials. These findings argue that PFC association neurons are a neuronal substrate for the mapping processes between numerical signs and categories.

It dawned on us that the responses of PFC association neurons were indeed special when we analyzed the responses of VIP neurons from the recordings. As you recall, we recorded from both areas simultaneously; therefore, everything was identical for VIP and PFC recordings. Despite the monkeys being in exactly the same state of mind, the proportion of association neurons in VIP was disappointingly low. Hardly any of the recorded IPS neurons associated signs with numerosities, even though we could confirm our earlier findings that a significant proportion of VIP neurons respond to dot numerosities, but to numerosities only. These results revealed that even though monkeys use the PFC and IPS for non-symbolic quantity representations, only the prefrontal part of this network is engaged in the mapping of the numerical meaning of shapes in relation to set size. This may change with years of association training, but we have not attempted this experiment yet. Whatever the case, it is clear that the PFC has a key role in acquiring numerical signs.

At first glance, our findings for IPS neurons seem at odds with the dominant role of the posterior parietal lobe for numeracy in adult humans. In fact, many lines of evidence show that number symbols are represented in the human IPS. This, however, is primarily true only for numerate adults; it is different for children. Children that are still in the process of learning how to deal with number symbols engage the PFC more than they do the parietal lobe. Six- and seven-year-old children recruit the inferior frontal gyrus for notation-independent numerical processing to a much greater degree than adults do.[23,24] Similarly, a greater engagement of frontal brain regions during numeral judgments[25] and symbolic arithmetic tasks[26] has been described in children.

The prefrontal region, both in human and in other primates, is strategically situated to establish semantic associations[27]; it receives input from both the anterior infero-temporal cortex, which encodes visual shape information,[28] as well as from the PPC, which contains number neurons.[29] Neurons in the PFC encode learned associations between two sensory stimuli without intrinsic meanings (e.g., the association of a certain color with a specific sound, or pairs of pictures).[30,31] In addition, the PFC is important for the conscious retrieval of associative representations.[32] Our findings demonstrate that neurons in the PFC represent semantic long-term associations not only between pairs of pictures, but also between arbitrary shapes and systematically arranged categories with inherent meanings (i.e., the numerosities of sets). In that respect, the primate PFC may not only control the retrieval of long-term associations, but may in fact constitute a crucial processing stage for abstract semantic associations. Evolutionarily speaking, once such associative networks were established in the primate brain, they could easily have been used for symbolic associations in humans.

These results highlight the PFC as a cardinal structure in acquiring a symbolic number concept during human development. However, this may only be half the story. At the transition from childhood to adulthood, this activation in PFC seems to shift to the parietal lobe once children grow older and become more proficient with numbers and arithmetic.[33] Perhaps the indexical association processes in the phylogenetically older brain of non-human primates mirror those present early in human life. The human IPS may become specialized only later in life with years of learning and experience with symbolic number representations.

9.4 Symbolic Number Neurons in the Human Brain

Numerical association neurons in the monkey brain provide the first hint to how number signs may be processed in our brain. However, due to the lack of symbol systems in monkeys and other animals, monkey studies cannot inform us about how number symbols are represented in the brain. The only brain that can deal with true number symbols is our own.

In order to learn about how single neurons represent symbolic numbers, single-cell recordings in humans are required, which are exceptionally rare and only performed for medical reasons. Luckily, my cooperation with epileptologist Florian Mormann at the University of Bonn provided us with the opportunity to do just this. As mentioned earlier, we presented simple calculation tasks to epileptic patients who were implanted with chronic depth electrodes in their medial temporal lobes (MTLs; see figure 7.6).[34] In this experiment, we not only presented the numerical values as nonsymbolic dot numerosities, but also as symbolic Arabic numerals in half of the trials. This allowed us to investigate whether single neurons also encoded number symbols, and whether different presentations of numbers were represented by the same neurons.

When participants were presented with numerals, a small but significant proportion of neurons (i.e., 3%) responded selectively to those displays. These numeral-selective neurons were tuned to preferred values, just like number neurons. This suggest a labeled-line code, not only for nonsymbolic, but also for symbolic numbers. Some of the numeral-selective neurons were also tuned to non-symbolic numbers, and a few of those had identically preferred numerical values for both formats. Figure 7.7B shows one of these number neurons tuned to number five in both non-symbolic and symbolic formats.

The numeral-selective neurons had preferred values ranging from one to five, and their normalized tuning functions only mildly overlapped. All of these neurons together carried enough information to allow a statistical classifier to significantly above chance predict the numerical values the patients had seen based on recorded firing rates. The decline of activity from the preferred to the non-preferred numerals was brisk and categorical, with only a mild progressive decrease with numerical distance. This means that tuning to number symbols was sharper than tuning to non-symbolic numbers and barely showed a numerical distance effect.

This finding is in agreement with behavioral studies[35] and neural modeling,[36] which show that the distance effect is substantial for the comparison of non-symbolic numerosities, but minute for judgments on exact number symbols. Consequently, the accuracy of number discrimination based on the neuronal discharges exhibited large distance effects for the populations of broadly tuned numerosity neurons, but small distance effects for sharply tuned symbolic number neurons. This provides further evidence for these number neurons being the physiological correlates of mental number representations.

The distance effect for number symbols is thought to be inherited from more basic non-symbolic number representations. Its presence in human number neurons therefore argues that high-level human numerical abilities are actually rooted in biologically determined mechanisms. Number symbols seem to acquire their numerical meanings by becoming linked to evolutionarily conserved set size representations during cognitive development.[37] Symbolic number cognition thus appears to be grounded in neuronal circuits devoted to deriving precise values from approximate representations.

Whether segregated populations of "format-dependent" number neurons that process either non-symbolic or symbolic numerical formats is a special feature of the MTL or the general way of how our brain encodes different formats of numbers remains to be explored. If it is a general feature, the representation of differently formatted number information by two distinct populations of number neurons should also be seen in the fronto-parietal core number system; this would indicate that the MTL simply inherits format-dependent number neurons. Alternatively, neurons in the fronto-parietal core number system could potentially encode number more abstractly and be format-independent. After all, neurons in the PFC of monkeys responded abstractly to numerical. Only more recordings in different brain areas will settle this question.

9.5 A Brain Area Dedicated to Numerals

The role of the PFC and PPC for quantity processing is well established. But how does the human brain come to associate numerals as visual symbols of numerical information? Given that numerals are so important in our daily lives, one could assume that they are processed in a privileged way compared to other complex visual stimuli. However, what is the neurobiological evidence that such a brain module for the visual properties of number symbols does exist?

Functional imaging provides a first hint for such a "number form area." The ventral occipito-temporal cortex (VOT) is regularly activated for numerical stimuli compared to other complex visual shape stimuli.[38] It is a rather fitting brain region for visual number signs, because different parts of the VOT are associated with representations of visual object properties, such as shapes,[39] object categories,[40] and even written words.[41] Moreover, patient studies suggest that lesions within the region may cause dyslexia, a specific learning disability in reading, but not for numbers.[42] Finally, electrical stimulation of the VOT can impair the reading of numerals differently than the reading of letters.[43]

The best and most direct evidence for the existence of a "number form area" stems from direct electrical activity recordings from the cortical surface of patients, a method called *electrocorticography,* or ECoG. In this method, flat electrodes are placed directly on the exposed surface of a brain. In order to access the surface of the brain, a surgeon must first perform a craniotomy by removing part of the skull. This procedure is of course performed on patients under anesthesia, and is performed with the aim to treat patients with severe epilepsy by precisely localizing, and later removing, the region of origin that causes seizures. Once the patients have recovered from the surgery and have agreed to participate in the studies, they are given a behavioral task while the electrical activity is monitored. Unfortunately, ECoG electrodes are not fine enough to record action potentials of single neurons. Instead, they measure local field potentials, which are the combined potential changes at the synapses of thousands of neurons that are active simultaneously. Parameters, such as the intensity, or power, of the potential changes, in addition to the frequencies of the voltage changes of this signal, are taken as measures of the responsiveness of neuronal populations to certain stimuli.

Using electrocorticography, the laboratory of Josef Parvizi at Stanford University identified the location of the "number form area" within the inferior temporal gyrus (ITG) close to the occipital lobe.[44] In this location, neuronal populations preferentially responded to visual numerals compared to other types of symbols, such as letters or words (figure 9.4).

The identification of this region is an interesting finding in relation to the question of why the human brain is able to represent visual symbols in

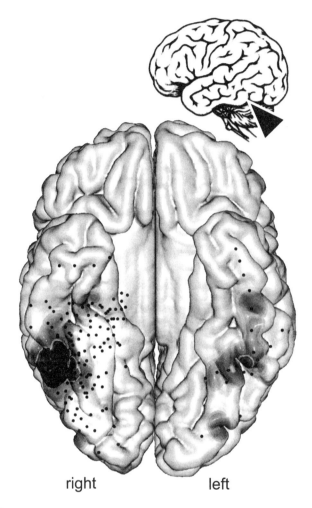

Figure 9.4
Arabic numerals engage a special brain site in the temporal lobe known as the number form area. The image shows the underside (ventral side) of a human brain, with dark areas on the temporal lobe indicating regions that show more electrical activity to numerals compared with other types of symbols (dots represent electrode locations). The location of the number form area in the ventral temporal lobe is indicated by the arrow pointing to a lateral view of a human brain (from Shum et al., 2013).

the first place. Relative to the pace of brain evolution in our ancestors, the invention of numeral symbols is a very recent event, dating back around 5,500 years. This short interval of time, however, is not sufficient for brain regions to become heritably specialized for symbol recognition through the process of natural selection. Therefore, we all have to acquire number symbols during our development as children and learn their meanings. And, somehow, these symbols become engraved onto a part of the ITG so that it becomes dedicated to representing numerals as a cultural and extremely important visual category.

Why precisely this part of the cortex becomes responsive compared to others is a fascinating question. One hypothesis is that the ITG is evolutionarily primed for such representations. The region is already a high-level visual area, in which neurons represent complex visual shapes in both humans and non-human primates. In monkeys, for instance, some of the inferior temporal areas are dedicated to complex visual categories, such as faces[45] and body parts.[46] Visual numerals are similarly complex visual shapes with a specific, numerical meaning. However, for the numerals to acquire their meaning, this area needs connections to the brain region that hosts quantity information, namely the PPC. How else would the ITG be able to learn the numerical meanings of these visual shapes?

How this preference for number symbols emerged, and whether it actually depended on visual processing, was addressed in an fMRI study with blind subjects performed by Sami Abboud and Amir Amedi at the Hebrew University of Jerusalem in Israel.[47] The investigated participants were blind from birth and therefore had no experience whatsoever with visual numerals. So how would the researchers present number symbols? They used an electronic device that translated the shape of visual objects into music which the participants could hear. As a result, every number sounded differently and distinctly. Of course, the participants were first familiarized with the new meaning of these sounds. Would the number form area still become activated even if numbers were presented acoustically? Indeed, greater activation was observed in the right ITG when participants processed acoustic number symbols compared to control tasks with symbols that had no numerical meaning. These findings suggest that numerical specificity in the temporal cortex can emerge independently from visual modality and visual experience!

But why would a brain area geared toward visual processing suddenly encode numbers played as acoustic stimuli? This probably results from the connectivity of the ITG to other number-related areas. The evidence for this claim comes from the same study,[48] because it was found that the visual number area in the right ITG was activated together with the right IPS in the parietal cortex, a sulcus that has been repeatedly implicated with the representation of quantities. Therefore, the function of the visual number area seems not to depend on sensory modality, but rather on the connections to associative cortices and their respective function.

The finding of a "number form area" in the high-level visual cortex is in agreement with the idea of *cultural recycling* of brain areas.[49] This hypothesis suggests that the acquisition of novel cultural inventions is only feasible if these inventions capitalize on existing anatomical structures. Thus artificial creations, such as numerals, invade preexisting brain networks capable of performing a function sufficiently similar to what is needed. This is not only seen for numerals, but also for other symbols, such as words: a temporal lobe area known as the "visual word form area" shows preference for reading visual words. This area is active in subjects blind from birth during Braille reading, which is a way of reading via tactile stimulation.[50] The acquisition of literacy induces a reorientation of cortical maps toward letters at the expense of other categories, such as faces.

The idea of a cultural recycling of brain areas, however, suggests one unfavorable consequence. If numerals invade this area in the ITG during development and cause it to change its function, the functions originally residing within this or in nearby cortical areas of the ITC have to retreat. It is hard to believe, but this is exactly what an imaging study with mathematicians suggests. FMRI responses to faces are significantly reduced in mathematicians relative to control subjects in the right-hemispheric inferior-temporal cortex.[51] Similar findings were reported for the visual word form area: in children learning to read, written words invade a sector of the visual cortex that was initially weakly specialized and slightly responsive to pictures of tools.[52] It is tempting to speculate that the acquisition of expert numeral processing shifts the responses of this area toward numerals. According to this, we seem to lose some high-order visual representations in order to gain symbolic number functions. So don't be surprised if mathematicians don't recognize your face—they simply may have better, that is mathematical, objects to represent in their brains!

From an evolutionary point of view, one might wonder whether nonhuman primates already show the humble beginnings of learning-induced brain changes when they are trained to represent artificial, but meaningful, visual signs. Given the long time scale necessary for such brain areas to evolve, precursors of the number form area or the visual word form area might be expected. Consistent with this idea, fMRI studies show that juvenile rhesus macaques trained to recognize Helvetica characters develop a specialized form area for these letters at a reproducible location in the inferior temporal cortex.[53] Even more, if the monkeys are trained with cartoon faces, Tetris blocks, and Helvetica characters, areas selective for each of these visual stimuli appear by the end of training in the inferior temporal cortex.[54] Whether these temporal lobe areas in monkeys are homologous to the human form areas is difficult to say, because functional areas migrated quite a bit during the evolutionary expansion of the human temporal lobe. But it is clear that some sort of pre-adaptation exists in macaque monkeys, with which we shared a last common ancestor some 30 million years ago. Once again, the importance of our evolutionary heritage cannot be overestimated.

10 The Calculating Brain

10.1 Non-Symbolic Calculation in Indigenous People, Infants, and Animals

After numerical quantities have been extracted from the sense organs and maintained in memory, numbers need to be processed to arrive at outcomes that guide decisions. Numbers need to be compared, summed up, and so on. This brings to mind solving equations. However, calculation operations do not necessarily require symbolic numbers and formal mathematics. Additions and subtractions can actually be performed with approximate set sizes. Innumerate people, infants, and animals show us how this works.

Earlier in this book, I mentioned the fascinating work by Pierre Pica in the Amazon rainforest of Brazil, where he studied the numerical skills of the indigenous Munduruku. You may remember that these people did not develop a counting system. Rather, they possess only five number words for "one," "two," "three-ish," "four-ish," and "five-ish." When Pica visited the Munduruku, he also brought calculation exercises with him that he showed the Munduruku as videos on his laptop.[1] He found that the Munduruku can perform arithmetic operations, such as additions and subtractions, albeit only in an approximate way.

In one test, he asked them to add large numbers. The volunteers were shown a movie of two sets of dots, 15 dots in one set and 15 dots in the second set, falling into a can (figure 10.1A). They were then asked to indicate if these two sets added together in the can—and no longer visible for comparison—amounted to more than a third set of 60 dots which followed on the screen. Indeed, the Munduruku participants had no difficulty in the task. When literate French adults were tested with the same task for comparison, the Munduruku showed a precision identical to that of the French controls. Interestingly, although the Munduruku lack words for numbers

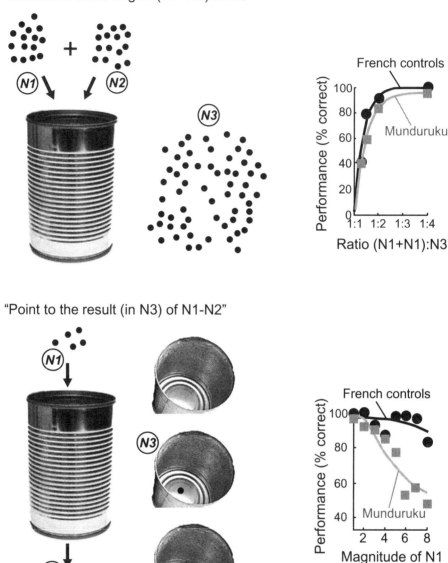

Figure 10.1
Arithmetic in an innumerate Amazonian indigene group, the Munduruku. A) In the approximate addition and comparison task, the participants judged whether the magnitude of two sets of dots (N1 and N2) that drop inside a can is larger than the set size N3. The graph on the right side shows that the Munduruku were equally successful as numerate French adults. B) In the exact subtraction task, the participants judged which dot number shown inside the can (from three possible choices N3) is equal to the number of dots that remains after dots fell inside the can (N1) and others dropped out of the can (N2). For set sizes larger than four, the Munduruku were significantly worse in this task compared to numerate French adults (from Pica et al., 2004).

beyond five, they are able to compare and add large approximate numbers that are far beyond their naming range. They can do this based on the ancient approximate number system our brain is equipped with from birth.

However, differences between the Munduruku and numerate Westerners emerged for small number calculations, in this case intuitive subtractions. Numerate adults immediately form precise and symbolic number representations that are inaccessible to the Munduruku. For example, Pica played a movie showing five dots falling into a can (figure 10.1B). The can was upright, so that the dots were no longer invisible after having fallen into the can. In the following sequence, four objects dropped out of the can. How many objects are still in the can? The Munduruku were shown the inside of three different cans showing zero, one, or two dots as possible outcomes and had to choose the outcome they thought was correct. In most cases, the Munduruku chose the can with one object. So with nothing but an approximate number estimation system, they were, on average, able to arrive at the correct result. This time, however, the Munduruku were vastly surpassed by the French controls that hardly made any errors. This means that the Munduruku still deployed approximate representations, subject to Weber's law, in a task that French controls easily resolved by exact calculation based on number symbols.

Might a similar calculation capacity be present in infants that are not yet able to speak? Human infants have earlier been demonstrated to distinguish numbers. As if that were not impressive enough, developmental psychologists investigated further whether infants could also transform numerosity representations and determine the correct result of basic calculations. This was the question Karen Wynn, now at Yale University, addressed in a groundbreaking study in 1992 with five-month-old infants.[2] Wynn used a "violation of expectation" protocol, in which infants witnessed sets of objects disappearing and reappearing behind screens as perceptual manifestations of additions and subtractions. The rationale of this protocol is that if infants were able to calculate numerosities then they should expect a certain numerical outcome after witnessing an addition or subtraction. If the outcome conforms to their internally calculated expectation, they should not be surprised and only look moderately long at the outcome of a seen transformation. However, if the outcome deviates from the expected, they should stare for a longer time since, in a way, they can't believe their eyes. Again, the looking time in the two conditions is an indirect measure of the infants' expectations based on their internal number calculations.

The infants were divided into two equal groups and placed in front of a puppet show while their looking time was recorded (figure 10.2). The first group of infants saw a Mickey Mouse doll placed on the stage, after which a screen rotated up and hid the doll from view. In the next scene, the infants could see that a human hand brought a second Mickey Mouse doll onto the stage in clear view. The hand then placed the second doll behind the screen and out of the infant's sight. The infants therefore could clearly see the arithmetic process, namely that a second doll was added to the first. However, they could not see the outcome of this process, because the screen hid the objects.

Finally, the screen dropped to reveal the outcome. To see if the infants had formed a calculated expectation, the experiment was terminated in two ways (figure 10.2A). The first group of infants saw the correct, expected outcome, namely two dolls standing on the stage. With this, the infants looked moderately long on the stage. For the second group, however, an invisible experimenter removed one doll behind the screen. When the screen came down, only one Mickey Mouse was standing on the stage. And indeed, the infants stared significantly longer at this impossible result. This was interpreted as the infants being surprised by an apparently impossible result. As shown from this, infants at the age of five months can clearly add one and one and expect that the outcome is two.

This experiment was modified and adapted to also investigate subtraction (figure 10.2B). This time, two dolls stood on the stage, the screen came up, and a hand visible to the infants removed one object from behind the screen. When the screen came down, once again infants looked significantly longer at the sight of two dolls (the impossible outcome) compared to the expected result of one doll (possible outcome). As a result from this study, infants at the age of five months can also subtract one from two and expect that the outcome is one.

Initially, it was believed that infants' ability to engage in arithmetical reasoning is limited to small sets of one or two items. However, a subsequent study by Koleen McCrink and Karen Wynn showed that by at least nine months, infants can also calculate the approximate result of additions and subtractions over large sets.[3] In this study, infants viewed animated displays depicting either a 5 + 5 event, or a 10 − 5 event. In the 5 + 5 event, five geometric objects moved onto the display and after several seconds were hidden behind a screen. Then five more objects appeared and moved

The Calculating Brain

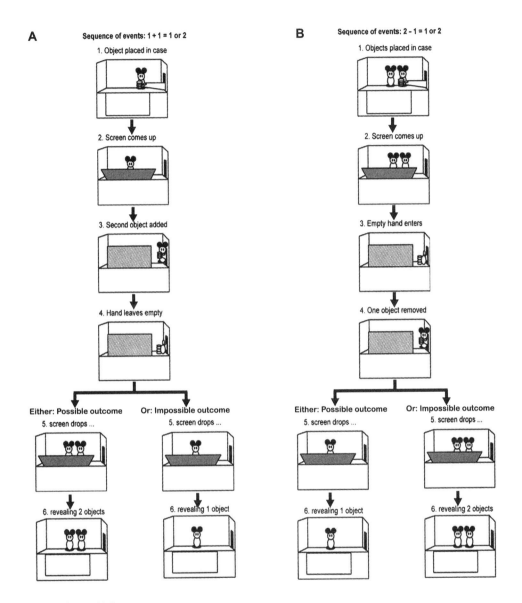

Figure 10.2
Baby arithmetic in 5 months old infants. Sequence of events shown to infants in front of a puppet show. **A)** Addition test; **B)** subtraction test (from Wynn, 1992).

behind the screen. Conversely, in the 10 – 5 event, the opposite occurred. For both groups of infants, the screen was lowered to reveal five objects on some trials, and 10 objects on other trials. Infants who had seen addition looked longer when five objects were revealed, whereas infants who had seen the subtraction looked longer when 10 objects were revealed. This suggests that infants had computed the approximate result of the calculation with large numerosities, and again looked longer when the result deviated from the expected outcome. Importantly, non-numerical cues were controlled; if the infants responded, it was only to number. This and other findings suggest that infants' inherent sense of numerical quantities underlies not only their ability to differentiate between sets based on numerosity, but also allows them to engage in simple arithmetic computations. Thus, even before the advent of language, the human brain is already equipped to represent numbers and to perform rudimentary arithmetic with non-symbolic numbers.

If preverbal infants can perform approximate calculations, maybe non-human species can too. It turns out that, in addition to the representation of approximate numerosities, animals can also perform elementary arithmetic computations. In one study with free-ranging rhesus monkeys in the Puerto Rican island of Cayo Santiago, the monkeys demonstrated that they can spontaneously add two quantities of food.[4] To test wild monkeys, the violation-of-expectation design and looking time measures were applied, just like in the infant studies. This time, the experiment was performed with four lemons disappearing behind a screen, instead of dolls. When the screen was removed, the monkeys either saw the possible outcome, eight in this case, or the impossible outcome of four lemons. Like infants, the monkeys looked longer at the impossible outcome relative to the expected one, suggesting that they added the two sets of lemons. A few years later, Jessica Cantlon and Elizabeth Brannon trained rhesus monkeys to approximately add two sets of dots that were shown on a computer screen, and to choose a subsequent display that showed the correct sum of the two sets.[5] The monkeys were as proficient as college students who did not have enough time to count the dots and had to rely on estimation. While these are impressive results, these studies only tested one fixed calculation rule, either addition or subtraction. Can monkeys also flexibly switch between such abstract rules?

The capability to switch between operations was demonstrated in trained Japanese macaques (*Macaca fuscata*) by Sumito Okuyama and Hajime

Mushiake from Tohoku University in Japan.[6] In this elegant study, the monkeys first saw the target numerosity, a set of dots on a computer screen, of a calculation they subsequently had to perform. After a brief delay, another set of dots was shown which constituted the first operand. The monkeys knew that they had to do something to match the target numerosity. If the number of dots shown as the first part in this "equation," that is the first operand, was smaller than the target numerosity, the monkeys had to add dots, and vice versa remove dots if the number of dots were greater than the target. In the experimental setup, the monkeys could actively add dots by turning a switch on the left side of the screen. Alternatively, they could remove dots from the first operand by turning a switch on the right side of the screen. Once they reached the correct target numerosity, they had to stop using the switches and wait for a while to receive a reward. Here is an example of an addition task for illustration: the monkeys first sees three dots on the screen as target numerosity. Next, one dot is shown as first operand. The monkeys then turns the switch on the left side two times, thereby adding two more dots to the single dot to reach the target. In other trials, however, a subtraction was necessary, and thus the monkey acted accordingly by using the right switch to remove dots.

All in all, this is an extremely difficult task for a monkey. But despite the complexity, the monkeys matched the target numerosities (zero to four) with astonishing precision. Lacking a symbol system, they relied of course on their estimation skills and, as a result, sometimes mixed up the target numerosity with adjacent higher or lower numbers. For example, when the target numerosity of one was presented, the monkeys erroneously chose numerosities two and zero more often than other numerosities. However, this is to be expected for operations based on the approximate number system. But the results showed more: not only were the monkeys able to add or subtract dots until they reached the target numerosity, they also understood which calculation rule—addition or subtraction—to apply on a trial-by-trial basis to reach the target. This is goal-directed numerical behavior at its best.

In 2016, Jessica Cantlon and Elizabeth Brannon used a less training-intensive, more intuitive calculation task to assess the characteristics of rhesus monkey arithmetic.[7] Specifically, they asked whether three effects that are commonly observed in children's math performance in school would be present in monkeys. These three effects are the *problem size effect,* the *tie effect,* and the *practice effect*. The problem size effect represents a systematic decline in accuracy and response time as the magnitude of the operands in an arithmetic problem increases (e.g., 5 + 7 is more difficult than 3 + 4). The tie effect represents better performance for addition problems in which the two operands are identical (e.g., 2 + 2 is easier than 1 + 3). Finally, practice effects are simply improvements in performance with repeated exposure to a given problem.

In their experiments with monkeys, the arithmetic problems were presented as movies in which two sets of moving dots constituted the first and second operands of the calculations, respectively. In these experiments as well, the monkeys were able to solve the addition and subtraction tasks. When tested with novel sets of dots they had never been trained on, they could flexibly adapt their arithmetic knowledge. Moreover, monkeys exhibited the problem size and tie effects, just like humans do. Unlike humans, however, monkeys did not exhibit a practice effect. These findings provide new evidence for a cognitive relation between non-symbolic calculation in non-human primates and symbolic arithmetic that remains reserved to humans.

10.2 Single-Neuron Arithmetic

We have seen already that non-human primates can perform non-symbolic arithmetic tasks on par with college students. This suggests not only a shared system for non-symbolic number representations, but also an evolutionarily primitive system for non-symbolic mathematical competence shared among all primates. So far, the neuronal correlates of addition and subtraction have not been studied in monkeys. But what has been investigated is how single neurons respond to more basic mathematical operations, namely "greater than" (>) and "less than" (<) operations. Such comparative relationships are some of the first mathematical rules we learn in school. In its simplest form, pupils learn to judge whether a greater than or less than judgment is true. For instance, five apples is greater than three apples holds true. Such numerical relationships are formalized later in education to read "5 > 3" (or "3 < 5," respectively) with the symbolic relational operators (>, <) that govern how set sizes are to be evaluated.

However, such operators can also be elaborated to become behavioral rules. In formal terms, this would read "if x > y then a" versus "if x < y then b." Rules can be understood as conditional "if-then" statements that determine the logic of a goal-directed behavioral task. In contrast to concrete behavioral rules ("stop at red—go at green"), arithmetic rules are special in that they operate on abstract categories, namely numbers. Such instructions based on a mathematical rules are highly relevant not only for mathematicians, but also for animals.

Imagine chimpanzees in the jungle of central Africa and replace x in the statement above with "number of friends," y with "number of foes," a with "attack" and b with "flee." Then the above expression reads, "If the number of friends is greater than the number of foes, then attack," whereas the second expression translates to "If the number of friends is less than the number of foes, then flee." As shown earlier, these are exactly the rules that chimpanzees in the wild follow, and it saves lives.[8]

We implement such instructions based on mathematical rules in the laboratory, but of course in a much less violent variation.[9] My PhD student, Sylvia Bongard, and I trained rhesus monkeys performing a sort of computer game with sets of dots to flexibly switch between "greater-than" and "less-than" rules (figure 10.3A). In each trial, a number of dots (the sample stimulus) indicated the reference numerosity that the monkeys had to remember over a brief time interval (the memory delay). Next, a colored circle that served as a rule cue instructed the monkeys to follow either a "greater-than" or a "less-than" rule. A red circle signified the greater-than rule, whereas a blue circle indicated the opposite. (In addition, a second set of rule cues was thought to the monkeys that later allowed to disentangle the rule from the sensory appearance of the rule cue.) After a second delay (the "rule delay"), the monkey had to respond according to the currently valid rule and pick either more or fewer numbers of dots, respectively, than it had previously seen in the sample display. For instance, if the sample numerosity was five, and the red circle indicating the greater-than rule was in effect, the monkey had to release a lever when eight dots were shown, but not when three dots appeared. In contrast, the opposite was true if the blue circle showed that the less-than rule was in effect.

The monkeys performed this task with different numbers of items and generalized to novel numerosities, indicating that they had learned an abstract numerical principle. While the monkeys performed this task, we recorded from single neurons in the PFC. We found that about 20% of PFC neurons represented the relational rules during the delay period following

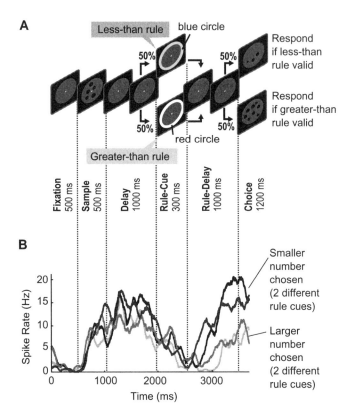

Figure 10.3
Neurons encode numerical rules. A) How primates and their neurons process numerical rules has been investigated using a rule-switching task. Here, monkeys had to chose more or less dots than presented in a sample display. B) Prefrontal neurons recorded in monkeys that were engaged in this numerical rule-switching task selectively responded to one or the other of these numerical rules. The activity averaged over all trials (spike density histogram) of an example neuron was systematically higher during the rule delay when the monkey followed the "smaller than" rule, irrespective of the sensory features of the rule cue. The plot is temporally correlated to the task layout that is shown in (A) (from Eiselt and Nieder, 2013).

the rule cue presentation (figure 10.3B). This means that a neuron selective to the less-than rule increased its discharge rate whenever the monkey was informed that, at the end of the trial, it had to pick a numerosity that was smaller than the reference. Other rule-selective neurons preferred the greater-than rule and fired strongest when the monkey was prompted to pick a number greater than the reference numerosity. About one-half of the rule-selective neurons preferred the greater-than rule, whereas the other half preferred the less-than rule. Therefore, each rule was represented by about equal numbers of selective rule neurons.

It was fascinating to see how such neurons encoded the abstract rules and did not care about absolute numerical values. Whether the monkey was required to choose a numerosity smaller than five or any other number was irrelevant to a less-than neuron; the neuron always responded greatly no matter the reference numerosity. The same was true for neurons preferring the greater-than rule. Moreover, rule-selective neurons did not depend on the sensory appearance of the rule cues. For a less-than rule neuron, it was irrelevant whether the rule cue was indicated by a blue circle or any other sensory cue; it always responded with the same high number of impulses. This was also the case with greater-than neurons. The responses of rule-selective neurons are not for the preparation of an action, but really for the concepts of abstract rules. This is because we saw rule-elective activity starting to emerge while the comparison number was still missing in the rule delay and the monkey therefore did not know how to respond.

A very convincing argument for why such rule-selective neurons are really needed by monkeys to mentally apply the rule is that their responses predicted the monkeys' behaviors: if the monkeys mixed up the cued mathematical rules and made wrong decisions in certain trials, the neurons' responses to the preferred rule were markedly reduced. In other words, if neurons preferring the less-than rule did not respond with maximum discharge rates in less-than trials, the monkey was prone to apply the greater-than rule and erroneously chose more numbers of dots. This observation indicates that there is a direct relationship between the neurons' rule selectivity and the monkey's task performance.

In a follow-up study, my PhD student, Daniela Vallentin, and I recorded not only from neurons in the PFC, but also cells in other frontal lobe areas, as well as in the VIP area of the IPS.[10] The VIP also showed cells selective to mathematical rules, but the proportion and selectivity of rule-cued VIP neurons was nowhere near to that of frontal lobe neurons. This suggests that the frontal association areas are more important when it comes to these mathematical rules. The same is true for humans, as will be shown in the next section.

It is obvious that relational rules not only can be applied to numbers, but also to any magnitude, such as length, size, or value. Could rule neurons represent these relations in an overarching way simultaneously to different types of quantities instead of just to numbers? My PhD student, Anne-Kathrin Eiselt, and I tested this by training monkeys to apply these rules not only to numerosities, but also to the lengths of lines.[11] When we again recorded from neurons of the PFC, we found that most rule-selective neurons responded only to quantitative rules applied to one specific magnitude type; for example, to the longer lines, but not to more items. We termed such rule neurons "rule specialists."

But this was not all we found. A smaller, but significantly larger group of neurons than expected by chance indeed generalized the magnitude principle. Such "rule generalists" encoded the overarching concept of magnitude rules by equally representing the quantitative rules related to both numerosities and line lengths. The presence of both rule-specialist and rule-generalist neurons indicates that the primate brain uses a mixture of selective rule neurons, perhaps embedded in functional hierarchies in which rule-generalist neurons are built based on rule-specialist neurons. The rule-generalist neurons would provide a computational advantage over rule specialists and could operate at a higher functional hierarchy[12]; indeed, they may allow the generalization and adaptation of quantitative rules to new circumstances. Overall, our findings suggest that the rule neurons we described in the monkey brain may serve as an evolutionary precursor for formal mathematical rule coding in humans. Symbolic mathematical operations may co-opt or recycle prefrontal circuits to dramatically enrich and enhance our symbolic mathematical skills.

When studying rule processing, or any other cognitive function for that matter, we pretend that neurons are very stable and static processing units. While this might be true as an approximation, neurons in the brain are constantly under the heavy influence by endemic drugs that modify their functioning, often with noticeable behavioral consequences. One group of such drugs are the so-called neuromodulators. These are chemical substances produced by a dedicated group of neurons in the brain. Neuromodulators do not directly excite or inhibit neurons, but alter impulse transmissions. Such neuromodulator systems in the brain are usually targeted by neuropharmacological substances to either increase or decrease their influence dependent on the psychiatric disease to treat.

One neuromodulator that is particularly prominent in the PFC is dopamine. Dopaminergic neurons in the midbrain eject the molecule into the surrounding fluid of PFC neurons. This then causes dopamine to affect the responses of neurons by activating dopamine receptors on the neurons' cell membranes. The neuromodulator dopamine affects the basic coding properties of PFC neurons via the two major receptor families, the D1- and D2-receptor families.[13,14] In addition to its roles in movement and reward circuitry, dopamine plays an important role in cognitive control functions, such as working memory and decision making.[15] We therefore hypothesized that dopamine might also influence the processing of mathematical rules.

My PhD student, Torben Ott, and I tested the influence of different dopamine receptor families on rule-selective neurons. To that aim, we used a method called "micro-iontophoresis" in combination with single-cell recordings.[16] With this method, small amounts of drugs that activate or block dopamine receptors can be ejected right at the tip of the recording electrode. This enables us to test how the neurons recorded at the electrode tip react to different levels of dopamine that would normally occur in the brain. In addition, we used drugs that separately influenced these two major dopamine receptor families.

We found that activating D1- and D2-dopamine receptor families enhanced rule coding in PFC neurons. In other words, the neurons differentiated the preferred and non-preferred rules more clearly by a starker contrast in firing rates. However, the mechanism of this enhancement in rule coding differed between the two receptor families: while an activation of D1-receptors increased the firing rate to the preferred rule (but did not alter the discharges to the non-preferred rule), the activation of D2-receptors suppressed the firing rate to the non-preferred rule (while leaving the firing rates to the preferred rule unaffected). In both scenarios, the result is a clearer differentiation between the rules, causing higher selectivity. Clearly, dopamine plays a role in processing numerical information, a finding that needs to be investigated in more detail in the future. At the same time, exploring numerical operations can be instrumental in deciphering the influence of neuromodulation in high-level cognitive processing.

I am particularly fascinated by abstract task rules, whether with numbers or in other domains, because they are an irreducible component of "executive functioning," sometimes also called "cognitive control." Executive functions are a set of cognitive processes that are necessary for synthesizing information from external (sense organs) and internal sources (body status), evaluating them in the light of previous experiences and current needs, and providing the means to achieve often far-removed goals. These functions include attention, cognitive inhibition, working memory, and also rule following.

The neural substrates for executive control therefore need to have access to a wide range of external and internal information. This no doubt depends on many different brain areas, and cortical association areas in particular. However, one cortical region is especially necessary: the PFC. If the brain were a business organization, the PFC would be the CEO. It is this exact cortical area that reaches the greatest relative expansion in the human brain and is thus thought to be the neural location of the mental feats we deem "intelligent." The PFC is anatomically well situated to play the role as the brain's CEO.[17] It receives information from, and sends projections to, endbrain systems that process information about the external world, motor system structures that produce voluntary movement, systems that consolidate long-term memories, and even systems that process information about affect and motivational state.[18] This anatomy has long suggested that the PFC may be important for synthesizing the external and internal information needed to produce complex behavior in general,[19] and mathematical thought in particular.

10.3 Cortical Location of Calculation

Consistent with the general role of the PFC in executive functions, deficits with numbers and calculation after frontal lesions in humans are complex matters to puzzle through. Alexander R. Luria in his classic book on higher cortical brain functions[20] suggested that the difficulty patients with frontal lobe lesions experience in carrying out arithmetical operations stems from a disturbance of intellectual activity, a general problem of solving a complex problem, rather than primary acalculia. Quantitative problem solving is necessary for cognitive estimation, and cognitive estimation deficits in various quantitative domains (size, weight, numerosity, and time) have been frequently reported for patients with frontal lesions.[21,22,23,24] Since semantic number representations tend to be conserved in such patients, estimation deficits of frontal lobe patients are instead attributed to executive deficits that disrupt the translation of number representations to structured output. As a most peculiar deficit that affects rule switching in arthmetic, "task-switching acalculia" was reported in a stroke patient with brain lesions in the left ventral and dorsolateral frontal lobe.[25] Interestingly enough, the patient's calculation ability *per se* was not affected, but he showed a specific deficit in switching between different operations in simple calculations; for example, from multiplication to addition or subtraction, and vice versa. This "task-switching acalculia" is consistent with the idea that frontal lesions lead to weak top-down control and an inability to switch between mathematical rules.

In the field of functional brain imaging, symbolic calculations were the first aspects of numerical cognition that were investigated as soon as the new technique was available in the 1980s. Neuroscientists initially did not bother with "dull" number representations, but jumped right into investigating arithmetic operations and mental calculation. Only later did they realize that calculation processes, in fact, address a plethora of cognitive functions in the brain that are necessary, but by no means specific for mathematics. Over the years, one of the important research agendas therefore became the anatomical parcellation of calculation-specific functions from more general processes.

In 1985, one of the very first functional imaging studies by Per Roland and Lars Friberg from the University of Copenhagen used PET to explore arithmetic "thinking."[26] I briefly mentioned this study already in chapter 6. In one of the tasks, a repetitive mental subtraction task, human participants started with the number 50 and then mentally and continuously subtracted three from the result: 50–3, 47–3, and so on. During this mental calculation, brain activation was measured based on the flow of radioactively labeled blood, which was found in larger quantities in the superior PFC and quite selectively bilaterally in the angular gyrus. Of course, the technical ability to pin down specific brain areas was humble in these early days of functional imaging. However, the important cortical brain areas for calculation surfaced and were repeatedly found in later and more specific imaging studies. A few years later, several studies using PET[27,28] and the increasingly popular fMRI[29,30,31] described a distributed network that subserves the performance of calculation tasks, including prefrontal, premotor, and parietal cortices.

In order to learn more about the specific involvement of various brain areas in calculation, subsequent studies tried to differentiate different types of calculations. For example, in 1999 Stanislas Dehaene and coworkers[32] compared exact and approximate calculation tasks. In the exact addition condition, subjects selected the correct sum from two numerically close numbers, such as 4 + 5 = 9. In the approximate addition condition, they were asked to estimate the result and select the closest number; for instance, the sum of 4 + 5 is closer to 8 than to 3. Relative differences in brain activation produced by exact and approximate calculations were found in the prefrontal-premotor-parietal calculation network. During exact calculation, the left inferior frontal lobule, but also the angular gyrus and the bilateral middle temporal gyrus, were more active. In contrast, the areas in and around the IPS were more active during approximation, accompanied by higher activation in the left dorsolateral PFC and left superior frontal gyrus.

Of course, arithmetic tasks also access general cognitive processes that are required in mathematical as well as in other tasks during calculation. We cannot just calculate without a working memory for numbers or an understanding of the calculation rules. In addition, we immediately try to label numbers with symbols. In a study from 2001, Oliver Gruber and Andreas Kleinschmidt[33] tried to remove this sort of general cognitive activation from their study. To that aim, they measured fMRI activity during one type of exact calculation, either subtraction or multiplication. They subtracted any BOLD activation that was evoked by more general cognitive operations that were necessary, but not specific for arithmetic. One such non-mathematical control task was a "letter substitution task." After a letter pair was shown to the participants in the scanner, they were required to substitute either the first or second letter of the pair according to an instruction cue, memorize the resulting letter pair, and repeat the letter substitution with different instructions four more times. Obviously, this operation is only possible with working memory and the application of a transformation rule, but it has nothing to do with calculation. In fact, the researchers found differences between calculations and the non-mathematical tasks in parietal sub-regions. Calculations predominantly activated the left angular gyrus and the medial parietal cortices. However, more complex calculation tasks with larger numbers involving the application of calculation rules increased activity in left inferior frontal areas. The left inferior frontal cortex is known to subserve symbolic processing, working memory, and executive functions. This study confirmed that a bilateral prefrontal-premotor-parietal network subserves mental calculation. However, to a large extent, these cortical areas are not exclusively involved in arithmetic procedures. Instead, they are also engaged in other cognitive tasks relying on similar cognitive components, such as working memory or processing symbolic information. Either calculation recruits these more general brain networks, or, alternatively, cells specifically involved in calculation are, in fact, intermingled with neurons serving other cognitive processes within the same brain network. Unfortunately, fMRI does not have the spatial resolution yet to disentangle these alternatives.

In their meta-analysis from 2011, Arsalidou and Taylor[34] also analyzed the sites activated by arithmetic tasks, such as subtraction and multiplication (figure 10.4). When they compared activations in studies in which the participants performed symbolic calculations (addition, subtraction,

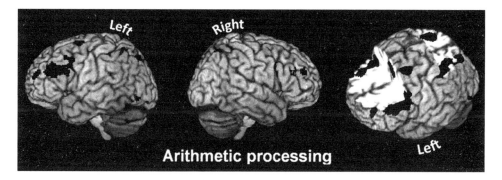

Figure 10.4
The brain's network for arithmetic. Black areas correspond to brain regions that show consistent fMRI activation during basic calculation tasks. The left image shows the left hemisphere of a human brain (left is anterior), the middle panel shows the right hemisphere (right is anterior). In the brain on the right side (left is anterior), parts of the frontal and parietal lobes are graphically removed to show the activations within the intraparietal sulcus and the prefrontal sulci as well as the medial frontal lobes (from Arsalidou and Taylor, 2011).

and multiplication), they found a large overlap among areas activated during number tasks and calculation tasks, particularly in the parietal regions. However, in contrast to pure number tasks, more prefrontal regions were active for calculation tasks, such as the middle and superior frontal gyri. This is consistent with the well-established role of the PFC representing numbers, but also in working memory, rule following, and outcome decision. After all, these are all higher brain functions that are needed when solving arithmetic problems. They also registered cortical areas that were specific to calculations and not active during simple number comparisons, such as the right angular gyrus, the bilateral middle frontal gyrus, and the left superior frontal gyrus. In addition, they detected areas outside of the cortex, such as the right caudate body and the right thalamus.

Arsalidou and Taylor's meta-analyses of several imaging studies largely confirm the currently most influential brain model on calculation processes, the "triple-code model" proposed by Stanislas Dehaene and Laurent Cohen.[35] The triple-code model was originally established based on neuropsychological findings in patients and suggests anatomical locations of certain sub-components required for number processing during calculations (figure 10.5).

The triple codes refer to three components: First, an "analogue quantity" code that describes what a number quantitatively means (semantic magnitude representation). It stores numerical information as logarithmically compressed approximate magnitudes, independent of specific number formats and notations (e.g., dot arrays, numerals, and number names). While logarithmically compressed approximate quantity representations are ubiquitous for non-symbolic representations, evidence suggests that semantic number representations can be culturally modified to follow our base-10 place-value system with separate representations for units, tens, hundreds, and so on.[36]

The second code is a "verbal word" code by which numbers and arithmetic facts are represented as verbal words. It is used to comprehend and produce spoken number words and is also a repository for learned arithmetical facts and tables (e.g., "three times three is nine"). Closely connected to language functions, the verbal code is assumed to be based only in the left angular gyrus of the inferior parietal lobule of the PPC.[37]

The third and final code is the "visual Arabic form" code for the visual representation of numerals. It also acts as a "workbench" for conducting

The Calculating Brain

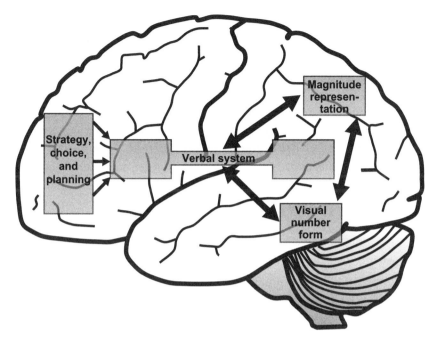

Figure 10.5
The triple-code model by Dehaene and Cohen. The model suggests three components for number processing during calculation and their anatomical locations: the semantic magnitude representation is located in the intraparietal sulcus, the verbal system comprises the angular gyrus and the inferior frontal lobe, and the visual representation for numerals resides in the inferior temporal gyrus. The verbal system, which is only present in the left hemisphere, receives input from the prefrontal cortex to guide decisions during calculation (after Dehaene & Cohen, 1995).

multidigit operations (e.g., 234 + 123). It is supposed to lie bilaterally in the fusiform gyrus at the transition between the occipital and temporal lobes (bilateral ventral occipito-temporal cortex). As described earlier, the ECoG studies by Josef Parvizi's group lend support to the location of a *"number form area"* within the ITG close to the occipital lobe.[38]

The triple-code model assumes the existence of format-specific and culturally dependent codes, namely the "verbal word" code and the "visual Arabic form" code, for representing numbers (written "8" versus spoken or written "eight") in different brain regions. Indeed, patients have been reported who could still read and write numbers, but not letters or words,[39] and vice versa.[40] Since these separate codes are different instantiations of numbers, they then need to be transcoded to be generally accessible. Transcoding is the means by which one symbol is translated into another of a different type. It encompasses processes such as reading (written symbol to verbal ones), writing (verbal labels to written symbols), and others (e.g., written symbol to a hand gesture). Consistent with the model's assumption, a number of neuropsychological studies have provided evidence for a direct route between numeral recognition and verbal output that bypasses a representation of number meaning; that is, a potential "analogue quantity" code. This explains impairments in patients who can perform calculations with high accuracy, but make errors on reading or writing Arabic numerals.

In addition to these three core codes for number processing, the triple-code model also assumes a prefrontal module necessary for strategic choice and planning during calculations, but the function of this module is not further specified. The fMRI meta-analysis[41] enabled Arsalidou and Taylor to suggest a refined picture of prefrontal functions according to the difficulty in solving arithmetic tasks. The inferior frontal regions are involved in processing simple numerical tasks that have few storage or procedural requirements. In contrast, the middle frontal regions are involved if the task requires several cognitive procedural steps (e.g., carrying a number in two-digit addition) or increased effort. Last, superior frontal regions are involved in generating strategies for solving multistep problems. For instance, more activity should be elicited in this region by ($6 \times 12 + 8$) compared to ($72 + 8$). Further extensions of the triple-code model concern brain regions that have been repeatedly reported to be involved in numerical tasks, but thus far have not been studied systematically in numerical cognition research, namely the cerebellum (suggested to come into play whenever calculation activities require goal-directed visual-motor sequencing) and the insula (thought to switch between task-relevant behavior and resting state default mode).

10.4 Dissociation of Calculation Types: Procedure versus Facts

Calculating with numbers usually means applying one of the four basic arithmetic operations: addition, subtraction, multiplication, or division. Conceptually, these arithmetic operations go together; they are the basic calculation procedures we learn early on. But surprisingly, we seem to solve these operations each with different strategies.[42,43] Self-reports of participants indicate that we usually solve subtraction (e.g., 23 − 16 = 7) and division problems by really "doing the math" and perform arithmetic transformations with numerical magnitudes. This genuine calculation method is called the "procedural strategy." In contrast, we learn the results of recurring addition and multiplication facts by heart and subsequently use memory strategies to retrieve the correct solution (e.g., 5 × 6 = 30). This approach is called the "fact-retrieval strategy." Why this difference emerges is not known; perhaps it is a consequence of our school education. After all, addition and multiplication are taught prior to subtraction and division in most school curricula, and we are encouraged to memorize at least multiplication tables by heart.

Based on self-reports from healthy subjects about their strategies to solve calculations, one might expect that the above two calculation strategies work independently of each other and therefore also have their own distinct brain sites. Coincidentally, scientists are able to determine whether mental processes are operating independently from each other, and whether they are specialized to certain areas of the brain, by establishing "double dissociation." Double dissociation is based on the idea that one of two related mental processes is affected by brain damage, whereas the other is fully functioning, and vice versa.

Indeed, several case studies found such a double dissociation and show that the procedural (subtraction and division) and the fact-retrieval strategies (addition and multiplication) function largely independently of each other. In 1933, a patient was already reported with selective impairments to subtraction and division, but not to addition and multiplication.[44] Many cases have followed since then that describe patients who were either better at solving subtractions compared to additions,[45] or showed selective preservation of one calculation strategy but not the other.[46,47] All of these studies were in agreement with the idea of a distinction between a procedural (subtraction and division) and a fact-retrieval strategy (addition and multiplication) in calculations. However, some neuropsychological studies disagree with the grouping of subtraction and division against addition and multiplication.[48,49,50] Nevertheless, there is consensus among cognitive scientists that the behavioral observation of a dissociation of arithmetic operations is grounded in different neuroanatomical substrates for these skills. But which brain areas precisely account for the respective calculation strategies?

In order to solve this question and to be able to causally associate calculation strategies with distinct brain regions, changes in calculations need to be correlated with lesion sites in the brain. The few studies available indeed point toward distinct neural regions for different calculations. In one report, a 56-year-old woman who suffered from an intracranial hemorrhage at the left parieto-temporal junction had distinct calculation problems.[51] The lesion appeared to include the angular gyrus and the supramarginal gyrus, but left the IPS intact. Neuropsychological evaluation several days after onset found her to be within normal limits of attention, language, memory, and visuospatial functions. However, her arithmetic performance was abnormal depending on the operation type; while multiplication was unreliable, subtraction was spared. However, the patient showed a peculiar error of consistently omitting the minus sign when the result was negative. In other words, the patient could only come up with the correct absolute value, instead of negative numbers during subtractions. Selective preservation of subtraction but severe problems with addition and multiplication were also reported in another case study in a patient with lesions in the angular and supramarginal gyrus.[52] In contrast, a patient with a more superior parietal lesion showed an opposite pattern, namely more impairments with subtraction rather than with multiplication.[53]

While all of the studies mentioned so far were only single cases, Juliana Baldo and Nina Dronkers from the University of California tested a large group of 68 patients with left hemisphere strokes with arithmetic and language comprehension tasks.[54] They used a new method, called "voxel-based lesion-symptom mapping," to analyze with high spatial precision the relationship between the location of brain damage and behavior. With this method, the detailed sites of a standardized human brain that cause specific impairments across all (or most) patients in the population can be identified. They found that subtraction and division were specifically associated with a number of significant foci, including the superior parietal cortex, precentral gyrus, and inferior frontal gyrus for subtraction, and the inferior parietal cortex and inferior frontal cortex for division. In contrast, addition and multiplication could not be associated with specific brain sites, because these types of calculations were represented in the same brain areas that are also required for language comprehension, namely the middle and superior temporal gyri.

Overall, this study in a large group of patients supports the notion that types of calculations are anatomically segregated and can be assigned to specific brain regions. It also supports the earlier findings that subtraction and division, both requiring more quantitative processing, are represented around the IPS in the parietal lobe, and around the inferior frontal gyrus in the frontal lobe. In contrast, addition and multiplication, which both rely more on verbal fact knowledge, are processed in areas that are equally engaged in language comprehension. The IPS is thought to be responsible for subtraction (procedural strategy), whereas the inferior parietal lobule, along with the angular and the supramarginal gyri are implicated in multiplication (fact-retrieval strategy).

Besides the study of permanent lesions in patients, there is another powerful and very direct approach to identify the location of calculation functions in the brain, namely intra-cranial electrical stimulation of cortical surface areas. Electrical stimulation is a routine clinical method that is used in patients who need to undergo brain surgery for surgical evaluation of epilepsy or for the removal of brain tumors.[55] The purpose of this procedure is to gather reliable information about the locations of cognitive functions that must be spared when later removing unhealthy tissue.

To access the brain, parts of the skull need to be removed during a surgery so that specific areas of the brain are exposed. To spare the essential functions that determine the personality of a patient, such as language, the neurosurgeon needs to know precisely where these functions are located on the cortical surface in order not to remove them surgically. For this purpose, the individual locations of essential brain functions are mapped out by applying mild electrical currents directly to the surface of a brain region of interest. Relatively large, blunt electrode pairs with tips spaced about 5 mm apart are used that touch but don't penetrate the cortical surface. During the procedure, the patients are awake and pain-free in order to have their behavior examined to see how it is transiently affected by the procedure. If language areas are being stimulated, the patients are no longer able to speak for the time of stimulation. If counting and calculation areas are electrically stimulated, subjects stop counting and make calculation errors. Therefore, there is a direct behavioral feedback if the proper sites are found, namely a behavioral arrest. During the subsequent surgeries, the neurosurgeon will stay away from such brain areas. Of course, this approach is, first and foremost, a diagnostic tool. The experimental tests that can be applied are therefore severely limited in time. Still, electrical stimulation of larger brain areas provides valuable and causal insights into the locations of brain functions and has been used from the early days of neurosurgery to probe the functional anatomy of the human brain .[56,57]

The available studies that apply direct electrical stimulation all concern calculation tasks that the patient performs during the surgery. The results point toward a separation between subtraction and multiplication, but the cortical areas are, in fact, somewhat mixed. Electrical stimulation of the left IPS primarily disrupted subtraction,[58] which is consistent with the idea that the IPS is implicated in quantity processing, since subtraction involves mental transformations of actual numerical quantities. Sometimes, however, multiplication was also found to be impaired in the IPS.[59] Electrical stimulation of the left angular gyrus as well as the left supramarginal gyrus, on the other hand, primarily led to multiplication disruptions while subtraction was largely spared.[60,61] This result is consistent with the idea that the angular gyrus is more involved in learned mathematical facts, as multiplication is thought to rely on memorized calculations. However, the opposite finding was reported as well, namely that the left angular gyrus is primarily responsible for subtraction.[62,63]

It is interesting to note that all these studies found clear segregations of different calculation types. However, there is no clear picture as to which parietal area is specialized to support a certain mathematical operation. One explanation of these findings would be that specific calculation aspects are parcellated and mixed within the IPS and angular gyrus, which would explain the conflicting results via stimulations. Alternatively, the location of calculation functions might vary between individuals and their specific calculation strategies, so that for some people, one area is responsible for subtraction, while for others, it is responsible for multiplication. One also has to be aware that calculation functions may plastically invade neighboring areas adjacent to the original, but now diseased, brain areas in patients.

10.5 Left versus Right Brain

The idea that calculation functions may be lateralized and preferentially represented in one endbrain hemisphere over the other is a recurring theme in numerical cognition. Functional imaging studies, in particular, often report stronger activation in one hemisphere at the expense of the other.

However, the results between studies are often conflicting. This may partly be due to the fact that functional imaging measures an indirect correlate of brain activity, and thus presents a more qualitative picture of brain locations rather than a quantitative statement about the relative functions of activation sites. Based on the extensive meta-analyses across many imaging studies, Arsalidou and Taylor found systematic differences between the two hemispheres.[64] They report that, on average, addition was left-lateralized, whereas subtraction led to mainly bilateral activations, while multiplication was mainly right-lateralized.

While imaging can only correlate function with brain areas, two other methods explore more causal relations and can inform about whether a particular hemisphere is truly necessary for calculation—transcranial magnetic stimulation (TMS), a non-invasive method, and the previously mentioned direct electrical stimulation of the brain during neurosurgery.

A number of TMS studies investigated the crucial areas for processing the value (or magnitude) of symbolic numbers. Because the spatial precision of TMS is low, almost all of the experiments focused on whether left and/or right posterior parietal areas (PPC) are crucial for quantity processing. While all studies found some effect on the processing of numerical values after TMS-inactivation of the PPC, there is no agreement about the respective contributions of the left and right sides. One explanation for this discrepancy may be because all these studies use different behavioral protocols to test number processing, which may engage the hemispheres differentially. In addition, the studies differ in the electrical parameters used during TMS, and also in the precise brain areas that were stimulated. All of these factors may contribute to the confusion. But what is reassuring is that all the studies support a causal role of the PPC in number processing, irrespective of hemisphere.

When different types of calculations were investigated with TMS, the results concerning left versus right hemispheres were equally inconsistent. The arithmetic operations studied via TMS included addition, subtraction, and multiplication. However, the TMS results are not always in agreement with findings from functional imaging. In confirmation of imaging data, one study found left hemisphere predominance, particularly in the angular gyrus, for exact addition.[65] In contrast, two other TMS studies reported the involvement of the bilateral IPS during addition, subtraction, and multiplication.[66,67] My conclusion from the current TMS literature is that both parietal hemispheres play some role in symbolic number processing and calculation, but this function is dependent on the tasks given.

Direct electrical stimulation studies also support that both cortical hemispheres are involved in calculation. In the past, most of these experiments were focused on the left parietal lobe only, because the language functions that needed to be mapped are usually located in the left hemisphere. For this bias alone, it may seem that only the left parietal lobe is involved in number processing. However, with more recent studies investigating both hemispheres, this conclusion turned out to be premature. One study found that electrical stimulation of both parietal hemispheres in patients impaired performance on simple subtraction problems.[68] However, multiplication was never affected in the right parietal lobe. The authors concluded that the right parietal cortex plays a more significant role in quantity processing (subtraction) than in verbal processing (multiplication).

The importance of the right parietal lobe as well as the left one for calculation is fully supported by Alessandro Della Puppa and colleagues[69] at the University Hospital of Padua. With a total of 13 patients, they showed that stimulation of either parietal lobe specifically impaired calculation.[70] In every single patient, stimulation of either hemisphere, specifically the area around the IPS, could impair multiplication and addition (figure 10.6). Moreover, the differences between hemispheres were nuanced. While the

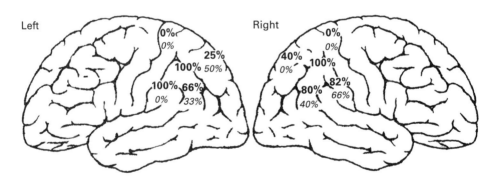

Figure 10.6
Electrical stimulation of either parietal hemispheres specifically impairs calculation. Data from intraoperative mapping in several patients show how often (in percent) multiplication (percentages in bold letters) and addition (percentages in non-bold letters) was impaired after electrical stimulation of different parietal areas of the left and right hemispheres (from Della Puppa et al., 2015).

left angular and supramarginal gyri only represented multiplication but not addition, the right angular and supramarginal gyri represented both. In addition, the superior parietal lobule was inconsistently involved in calculation processing (40% of cases in the left and 75% in the right side). As a control, the somatosensory area on the gyrus postcentralis, which conveys touch sensation, was never involved in calculation. This is striking evidence that calculation is achieved by a necessary involvement of the parietal lobes in both hemispheres.

10.6 Dissociated Brain Networks for Calculation and Language

The question of whether arithmetic is an independent faculty or is rather part of our language abilities has sparked intense debates among scientists for many decades.[71] At first glance, one might expect that linguistic and numerical abilities are, to some extent, tightly associated, if not the same. After all, language and number theory are our full-blown symbol systems that both rely on semantics (meaning) and syntax (rules) for processing symbols. As such, both systems rely on a variety of supporting cognitive functions, such as working and long-term memory. Studies with brain-injured patients provide a very direct way of addressing this issue. The predictions after brain damage are straightforward: If arithmetic and language are hosted in the same brain networks, then damage to these areas should affect both abilities, and patients should show similar deficits in linguistic and calculation tasks. However, if patients exhibit deficits in only one type of task, this would be evidence for segregated faculties that are nested in different brain networks.

Indeed, some studies with brain-damaged patients have noted that arithmetic and linguistic deficits often co-occur in the same individual.[72,73] These studies have suggested that verbal skills play a role in arithmetic ability. However, as shown below, many studies indicate considerable autonomy of the neurocognitive mechanisms underpinning language and calculation and argue for dissociated brain networks. You may remember that this was the very reason why Henschen back in 1925 coined the term *acalculia* for acquired impairments in calculation.[74]

Several studies found that language and arithmetic are actually independent processes. In one case study, a patient with fronto-temporal dementia (Pick's disease) was investigated.[75] Pick's disease is a devastating neurodegenerative disease leading to frontal and temporal lobe dysfunctions that manifest in personality and emotional changes, as well as the deterioration of language (aphasia). CT scans of this patient's brain showed a decrease (atrophy) of neural tissue in the frontal and, particularly, in the temporal lobes. During neuropsychological tests, the patient's verbal skills were hugely compromised. He could only utter a few stereotype phrases (e.g., "I don't know") and jargon utterances (e.g., "millionaire bub"). In addition, his comprehension of both spoken and written language was gravely impaired (called "global aphasic syndrome"). Despite these massive verbal impairments, his calculation abilities were surprisingly well preserved when he was required to write down his responses. He could even add and subtract multidigit numbers. And although his ability to retrieve multiplication facts was not entirely intact, it was at least partially preserved. Obviously, from this case, language skills are not necessary for calculation.

Several patients like the one just discussed showed relatively preserved mathematical skills but severe language impairments.[76,77,78,79,80] This has also been confirmed in a large-scale study with 60 aphasic patients.[81] The authors found better patient performance for numerical tasks compared to linguistic tasks. This again argues for a certain degree of independence of numerical processes from language. Interestingly, the reverse dissociation has also been described. Patients with selective impairments to calculation can show intact or relatively preserved linguistic skills.[82,83,84,85,86] Data like these indicate a dissociation of language and numerical competence and argue for the anatomical and functional independence of the number domain itself.

But let's have a deeper look at the constituents of the number symbol system. As indicated above, both natural language and arithmetic share two fundamental properties characteristic for true symbol systems: semantics and syntax. Both in language and arithmetic, syntax determines the meaning of an entire expression. Therefore, sentences and equations can only be correctly processed if, in addition to the meaning of individual symbols, the structuring rules are also understood.

One such important aspect of syntax that gives specific meaning to an entire expression is order. As mentioned earlier, how we arrange known words in a sentence gives rise to very different meanings: "the child bites the cat" is something very different than "the cat bites the child." Simply by reversing the order of the nouns (child and cat), the meaning of the expression is entirely changed. This is the power of syntax.

In exactly the same way, the computation of arithmetic expressions relies on order. The subtraction "20 – 10" is clearly not the same as "10 – 20." Other syntactical rules we have to obey in arithmetic are "solving multiplication and division before addition and subtraction" or "solving terms in brackets before terms outside of brackets." For example, the result of 5 × (6 + 2) is not the same as (5 × 6) + 2. We therefore learn such arithmetic rules in school. More and more neuropsychological studies have suggested that syntax for math and syntax for language rely on distinct brain substrates.

In 2005, Rosemary Varley, now at University College London, and her team investigated whether syntactic rules for arithmetic and those for natural language are dissociable in patients.[87] The team examined three profoundly aphasic men who suffered from extensive damage throughout the left middle cerebral artery territory. The damage affected the left temporal, parietal, and frontal cortices around the Sylvian (lateral) fissure. All three patients displayed severe language disturbances, particularly in the domain of syntactical rules. For instance, they could no longer understand reversible sentences such as the one described above. Despite such severe impairments in language, the patients surprisingly were able to solve mathematical equations. They mastered reversible subtraction and division problems (e.g., 59 – 13 and 13 – 59). They were even able to solve long, bracketed equations, for example "50 – [(4 + 7) × 4]." Varley and coworkers concluded that the results in these patients

> are incompatible with a claim that mathematical expressions are translated into a language format to gain access to syntactic mechanisms specialized for language.[88]

In other words, syntax for language and syntax for arithmetic are distinct. This may be due to two alternative explanations. One is that both language and mathematics access a general ability for grammar; in this case, arithmetic expressions could directly gain access to a functioning grammar module, whereas damage to the language faculty prevented such access. Alternatively, the syntactic capabilities for language and mathematics may be entirely separated, so that if the syntactic abilities for language are impaired, the ones for mathematics are still functional. Whatever the precise cause of the dissociation, the study provides additional evidence that language and mathematics are independent in the brain.

Two years after the Varley study, the possible dissociation between language and arithmetic syntax was examined by Juliana Baldo and Nina

Dronkers with tasks very similar to those in a study mentioned earlier.[89] A large group of 68 patients who had suffered a single left hemisphere cerebrovascular accident were examined. Depending on the site and broad extension of the lesions, it can of course be expected that both deficits of arithmetic and language comprehension were present, and this is what they found in a subset of the patients. However, two different subsets of patients exhibited either arithmetic impairment with preserved language comprehension, or, conversely, language comprehension impairment with preserved arithmetic ability. When examining the deficits in order to relate them to brain areas, the researchers found that the inferior parietal cortex (more specifically, the supramarginal and angular gyri) was systematically affected in patients that showed arithmetic problems. This confirmed again earlier neuropsychological studies implicating the PPC in calculation. In contrast, patients with isolated language comprehension problems predominantly had the middle and superior temporal gyri affected.[90] Damage to the inferior frontal gyrus, however, affected both arithmetic and language comprehension. While the inferior fontal gyrus, specifically Broca's language production area, is required for both language and arithmetic comprehension, the two systems are segregated upstream in the parietal lobe. Despite a partial overlap of language and arithmetic networks, these two symbol systems are operating independently for the most part.

Together with Nicolai Klessinger, Rosemary Varley went a step further and asked whether higher-order mathematics in the form of elementary algebra could be retained despite severe language impairment.[91] Algebra is a generalization of arithmetic, in which letters representing numbers are combined according to the rules of arithmetic. They report in a case study a patient who showed severely impaired capabilities in the language domain and difficulties with processing both heard and written number words. Despite these impairments, he was able to judge the equivalence of algebraic notation and to transform and simplify mathematical expressions in algebraic notations. He displayed a considerable capacity to retrieve algebraic facts, rules, and principles, as well as an ability to apply them to novel problems. Moreover, the patient demonstrated a similar capacity to solve expressions containing solely either numeric or abstract algebraic symbols (e.g., $8 - (3 - 5) + 3$ versus $b - (a - c) + a$). The results show the retention of elementary algebra despite severe aphasia and provide evidence for the preservation of symbolic capacities in the number faculty independent of language.

More direct evidence for the partial independence of number faculties stems from direct electrical stimulation studies in neurosurgical patients. In one of the most extensive stimulation studies, Franck-Emmanuel Roux and colleagues from the University of Toulouse mapped and compared both language and calculation areas in 16 patients.[92] To test language functions, the patients were asked to name objects or read words while electrical stimulation was applied to different cortical areas. If the patients could not name objects or read during the stimulation, the neurosurgeons knew that the respective cortical site was necessary for language. The tests for calculation, on the other hand, comprised the addition of two-digit numbers that were presented on a paper sheet during electrical stimulation. If the patients could not give an answer or gave the wrong answer, the respective site was marked as relevant for calculation.

In almost half of the cortical sites in the left parietal and frontal lobes, only calculation impairments were registered (figure 10.7). Such calculation-specific sites were located in the angular gyrus and around the IPS of the parietal lobe, as well as in the middle frontal gyrus (F2) of the frontal lobe. This clearly demonstrates that calculation is not integrated in the language system, but separate from it. In addition, it corroborates the finding that the posterior parietal and the prefrontal cortices are essential number-processing areas.

In conclusion, these studies suggest that some aspects of mathematical processing can be sustained despite severe disruptions to language functions. The evidence of this double dissociation between language and calculation skills demonstrates their functional independence. This argues for considerable functional segregation between some linguistic and higher mathematical functions in the adult cognitive system. Of course, these findings in mature brains do not necessarily implicate that the systems operate independently throughout life. One symbol system may help the other to develop early on. Given that we all learn to speak before we learn to count, the language system most likely entails the number system. In adulthood, however, mathematical thoughts are mostly separate from the language system.

On a broader scale, these results in aphasic patients raise an intriguing question: If patients with linguistic impairments can still perform mathematics, doesn't this imply that it is possible to think and reason logically without language? Research over the years has shown that individuals with global aphasia, who have almost no ability to understand or produce

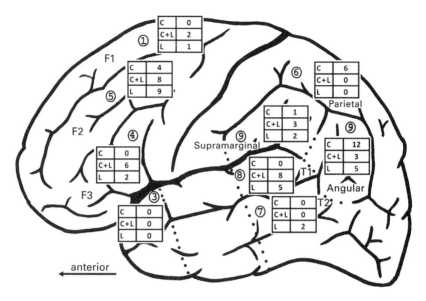

Figure 10.7
Dissociation of calculation and language using electrical stimulation during neurosurgery. The lateral view of a human brain shows the localization of calculation interference sites found in the left hemisphere. Circles with numbers indicate the number of times a cortical region was studied (> 16 brain mappings). C = number of specific calculation interferences found in the region tested; C + L = number of common calculation and language (naming and/or reading) interferences found; L = number of specific naming or reading interferences found (from Roux et al., 2009).

language, are nonetheless able to reason. They are able to add and subtract, solve logic problems, or even think about another person's thoughts.[93] These findings in brain-damaged individuals have also been replicated in healthy adults in neuroimaging studies. When we try to understand a sentence, we strongly engage the brain's language areas, but not when we perform other tasks, such as arithmetic or other rational activities. It therefore seems that many aspects of thought are independent from language because they engage distinct brain regions.

10.7 Professional Mathematicians and Mathematical Prodigies

Most of us are happy in mastering basic arithmetic. To get through the day, addition, subtraction, multiplication, and division are usually sufficient.

However, science and technology requires more elaborate mathematics than these four types of calculation. In school, we experience that not all of us are capable to reach true mathematical heights. It therefore commands our admiration when we hear of particularly gifted people who seem to juggle numbers almost effortlessly. And of course, we want to know how their brains do it.

In fact, the study of gifted brains was the start of what later became scientific neuroscience, and it started with the death of the eminent German mathematician Carl Friedrich Gauss (1777–1855). Gauss, already celebrated during his lifetime as *princeps mathematicorum* ("prince of mathematicians"), made numerous significant contributions to the fields of physics, astronomy, and mathematics. In fact, every student of statistics today knows the famous bell-shaped, or Gaussian, normal distribution. When Gauss died on February 23, 1855, in Göttingen, Rudolph F. J. H. Wagner, a comparative anatomist and physiologist from the University of Göttingen, obtained Gauss's brain for study. Wagner then compared the gross anatomy of Gauss's and other gifted individuals' brains to those of ordinary people.[94] At the time Wagner published his first data, "phrenology" was still influential.[95] Phrenological thinking assumed that character traits, thoughts, and emotions scale with specific localizable parts of the skull and brain. Today, we know of course that this is absolutely false. Thankfully, Wagner, already distancing himself from phrenology, set out to question two of phrenology's propositions (*Lehrsätze*) about the relationship between the brain and mental capabilities. The first proposition states that highly intelligent individuals would have larger brains and hemispheres than ordinary people, while the second claimed that such eminent people showed more convolutions of the cerebral cortex as a way to increase brain surface and thus brain size.

Wagner discovered that Gauss's brain was not particularly large, as one might have hoped for such a titan of mathematics. Gauss's brain weighed 1.492 kg, which was only a bit larger than the average brain. Wagner therefore concluded that

> highly skilled humans possess a well-developed brain, but its total weight does not differ noticeably from the weights of other well-developed and normal people.[96]

Rather disappointingly for those who were looking for simple anatomical mind-brain correlations, Wagner was forced to conclude that neither total brain weight, nor complexity of the cortical convolutions were obvious predicators of giftedness.

Among his contemporaries, however, the idea of a great mind–big brain relationship was too appealing to be abandoned lightheartedly. Amongst the disappointed scholars was the eminent French anatomist Pierre Paul Broca (1824–1880), who discovered the language production area now known as "Broca's area." He tried to save this belief system by doubting that Wagner was looking at such a splendid selection of gifted scientists in the first place. With a biting undertone he remarked:

> A professorial robe is not necessarily a certificate of genius; there may be, even at Göttingen, some chairs occupied by not very remarkable men.[97]

In the wake of Wagner's pioneering work, more and more supposedly eminent individuals made arrangements for the preservation and later study of their own brains. This bizarre competition for the "best brain" led to the foundation of the Society of Mutual Autopsy of Paris in 1876, and later the American Anthropometric Society in 1889. The purpose of these societies was to ensure that the members' bodies, most notably their brains, would be donated to the organization for study upon death.

The American anatomist Edward Anthony Spitzka (1876–1922) conducted some of the most extensive autopsies on the brains of distinguished people. Convinced that "eminent men" exhibit heavier average brain weights than normal controls, he also reexamined C.F. Gauss's brain, which he found *"in many respects the most notable of this series"*:[98]

> The surface configurations of the cerebrum are remarkable for the multiplicity of the fissures and the great complexity of the convolutions. The richness of fissuration is particularly notable in the frontal region while the subparietal regions, especially the marginal and angular gyres, exhibit a relatively enormous expansion.

To complete his enthusiastic picture, both literally and figuratively, he contrasted the brain of Gauss with those of a bushwoman and an ape. His results seemed uncontestable.

> The brain of a first-class genius like Friedrich Gauss is as far removed from that of the savage Bushman as that of the latter is removed from the brain of the nearest related ape.[99]

Of course, Spitzka ignored the simple fact that the relatively small stature of the bushwoman relative to Gauss's body mass was enough to explain the brain weight difference. It is well known that brain volume scales with body size, so that larger humans have bigger brains. Spitzka's conclusion was clearly just wishful and racist thinking. Wagner's original warning that it would be dangerous to make far-reaching generalizations from the gross anatomy of the highly complex cerebrum of Gauss went unheard. The American anatomist Franklin Paine Mall (1862–1917) already got to the heart of this issue when he wrote:

> It certainly would be important if it could be shown that the complexity of the gyri and sulci of the brain varied with the intelligence of the individual, that of genius being the most complex, but the facts do not bear this out, and such statements are only misleading.[100]

As if these scientific problems weren't enough, sloppy handling of the material additionally hampered the quest for genius in the brain. In 2014, a study comparing MRI brain images and Wagner's meticulously depicted original brain drawings from 1860 found that Gauss's brain was mixed up with another scientist's brain.[101] The jar marked "C. F. Gauss" actually contained the brain taken from Conrad Heinrich Fuchs, a medical scholar who, like Gauss, died in 1855, whereas the original brain taken from Gauss was in a jar marked "C. H. Fuchs." The misclassification could be detected because of a very rare anatomical variation in Gauss's brain that is visible in Wagner's original drawings: a divided central sulcus that is found to be missing in C. H. Fuchs's brain. It is suspected that the two scientists' brains were already mixed up many years ago. Consequently, what scientists had examined in recent decades as Gauss's brain was not his brain at all.

Even today, the belief in simple gross-anatomical specializations of talented individuals lives on, as the case of Albert Einstein's brain exemplifies. The theoretical physicist Albert Einstein was born in Ulm, Germany, in 1879. In 1921, he was awarded the Nobel Prize in Physics for his contributions to theoretical physics. Einstein is probably most famous for his development of the theory of relativity, one of the two pillars of modern physics. His mass–energy equivalence formula "$E = mc^2$" is almost common knowledge and has been dubbed "the world's most famous equation."[102] So outstanding were his insights and intellectual achievements that the name "Einstein" is now synonymous with "genius." After Hitler's rise to power, he fled Nazi Germany and, in 1933, became a professor in mathematics at the Institute for Advanced Study in Princeton, where he became friends with Kurt Gödel. Einstein later died in Princeton, New Jersey, in 1955 as an American citizen.

The fate of Albert Einstein's brain has all the characteristics of a mystery novel. After Einstein's death in 1955, Dr. Thomas Harvey, then a pathologist at Princeton Hospital, performed Einstein's autopsy and decided to secure his brain, the hallowed relic. The circumstances of how he got into possession of Einstein's brain are mysterious, and some would say criminal.[103] Nonetheless, in 1955, Harvey photographed Einstein's brain (figure 10.8) before passing it on to dissect parts of it by slicing it into paper-thin histological sections. Over the ensuing years, Harvey struggled with major personal and professional problems, while Einstein's brain was stowed in a beer cooler along with the histological slides and the whole-brain photographs.[104] For decades, Einstein's brain seemed lost and forsaken.[105]

Three decades later, in the 1980s, Marian C. Diamond from the University of California, Berkeley, contacted Harvey to request the brain slices for examination. The tissues she examined were taken from classical association areas, Brodmann area 9 of the medial PFC and area 39 of the angular gyrus in the posterior parietal cortex. Neurons and glia cells (the non-signaling support cells of the brain) were counted and compared to the cell counts of a control group of male human brains.[106] It was found that the ratio of neurons to glia cells was smaller in the left area 39 in Einstein's brain. In other words, his brain contained more glia cells than controls! If anything, one might have expected more neurons, as they are the processing units providing brain power, not glia cells. With a bold rhetorical turn, Diamond and her coauthors argued that more glia cells suggested a higher metabolic need of sugar and oxygen of the neurons and that this "might reflect the enhanced use of this tissue in the expression of his unusual conceptual powers."[107]

In the following years, more neuroanatomists embarked on the quest for genius in Einstein's brain. While some focused on the allegedly greater neuronal density of Einstein's thinner frontal cortex,[108] others scrutinized the atypical absence of the parietal operculum, a part of the lateral fissure that separates the parietal from the temporal lobe.[109] Later, anthropologist Dean Falk from Florida State University and coworkers revisited the original photographs in a paper published in the journal *Brain*.[110] They detected an "extraordinary prefrontal cortex" and "unusual" parietal lobes. Since the PFC is associated with abstract thought, whereas the parietal lobe is implicated in spatial and numerical cognition, such findings were embraced unquestioned. In an attempt to explain Einstein's genius as a

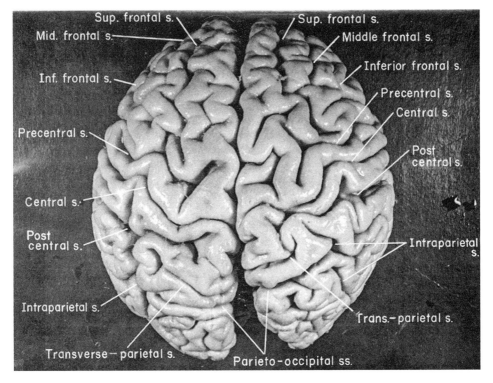

Figure 10.8
The brain of genius Albert Einstein. Before pathologist Dr. Thomas Harvey at Princeton Hospital partitioned Einstein's brain into 240 chunks, he photographed it from various angles. This is the top view. (Courtesy of the Otis Historical Archives, National Museum of Health and Medicine; ID number OHA184.06.001.002.00001.00002.)

consequence of better communication between hemispheres, photographs of sections inside the midline of Einstein's brain were intensely examined. Sure enough, it was found that Einstein's corpus callosum, the big fiber bundle connecting the hemispheres, was thicker than in controls.[111]

Over the years, about half a dozen studies on Einstein's brain have been published, each emphasizing a different anatomical trait as the possible source of his mathematical talent. All have produced great media fanfare, but none have revealed a credible anatomical foundation for Einstein's genius. Terence Hines from Pace University states this clearly when he writes that the results of the studies examining Einstein's brain "do not, in fact, support the claim that the structure of Einstein's brain reflects his intellectual abilities."[112] For him, the quest to find Einstein's genius based on morphological and histological examinations is plain "neuromythology." All these studies are based on the naïve belief that brain structure rather than function indicates intellect. However, explaining mental feats requires an examination of the processes and functions at work in a living brain. In short, it requires an understanding of the brain's physiology.

Today, we can localize mathematical functions in the brains of living people using functional imaging. With this method, we have a chance to understand what distinguishes a mathematician from an average numerate person. But we also need the right subjects to study, because not everyone who seems to be a math wizard is really doing the math. Some individuals with extraordinary calculation skills earn themselves the title of "mathematical prodigies" by being able to perform the most difficult arithmetic problems effortlessly and in no time.

Rüdiger Gamm is one of them. When asked "What is 99 to the power of 20?" Gamm closes his eyes and seconds later he utters the answer flawlessly. He easily calculates two-digit numbers to their twentieth power in his head and rattles off the square and cubic roots of each answer. Or let's take a different set of problems: how many Wednesdays the 13th did the year 1286 have? After brief contemplation, he announces the correct answer confidently.[113] So exceptional are his number skills that he has become a welcome guest on television game shows. However, according to his own admission, he struggled with arithmetic at school. But, one day, he stumbled across a formula that allowed him to calculate what day any date fell on. His interest sparked, he then trained with numbers and calculations for hours each day until he could make a living of his skills.[114]

Mauro Pesenti of the Université Catholique de Louvain in Belgium and colleagues wanted to know how Gamm does it.[115] They studied Gamm's brain in a PET scanner as he calculated away. Next, they compared the activity in his brain with several control people of average calculation skills. The result was surprising. The brain scans show that Gamm uses different regions for calculation than other people do. Instead of juggling and transforming numbers while maintaining them in his working memory buffer, a process we normally would understand as calculation, he derived the route-learned results from long-term memory. To use a computer analogy: While you and I would calculate using the RAM of our brain, Gamm does the trick by storing and retrieving results on a hard disk. For instance, while we would struggle

The Calculating Brain   231

and fail to remember a 12-digit telephone number because our working memory has a limit of only few numbers, Gamm would use the long-term memory that he has specifically trained and elaborated. Rather than being a gifted mathematician, Gamm is a highly trained mnemonist who has learned how to store numerical information in his almost boundless long-term memory. Gamm is exceptional at memorizing, but not at mathematics. And while his brain activity can shed light on memory processing, it doesn't inform us about the mathematical brain. For that, we need professional mathematicians.

Using fMRI, Marie Amalric and Stanislas Dehaene from Paris studied professional mathematicians crunching math problems and indeed found interesting differences in brain activations relative to controls. The researchers compared the fMRI scans of 15 expert mathematicians and 15 controls with comparable academic qualifications.[116] During the scanning session, the subjects listened to mathematical and other statements and had to decide whether the statements were true, false, or meaningless. If you want to test your high-level math skills, here are three of the statements used:

- "A square matrix with coefficients in a principal ideal domain is invertible if and only if its determinant is invertible."
- "Any matrix with coefficients in a principal ideal is equivalent to a companion matrix."
- "Any matrix with cardinality greater than 3 is factorial."

Well, do you know the answers? The first statement is true, the second is false, and the third is meaningless. Here are some examples of non-mathematical statements that you may find easier to answer:

- "The concept of robots and avatars was already present in Greek mythology."
- "The Paris metro was built before the Istanbul one."
- "A poet is a predominantly green tax over the metro."

The answers here are: true, false, meaningless.

After comparing and contrasting fMRI activation patterns of mathematicians and control subjects to a variety of such statements, a special brain network for advanced mathematics was identified, but only in expert mathematicians. Only professional mathematicians activated a reproducible set of bilateral frontal, intraparietal, and ventrolateral temporal regions

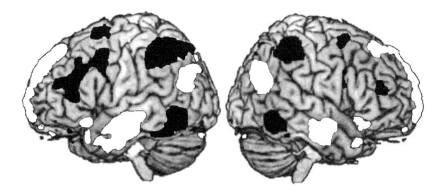

Figure 10.9
Brain networks for advanced mathematics in expert mathematicians. The lateral views of the left and right hemispheres of a human brain shows the distinct areas activated during reflection on mathematical statements (black) versus general knowledge (white; from Amalric and Dehaene, 2016).

in response to mathematical statements (figure 10.9). Moreover, the math network was closely linked to and overlapped with the brain's core number network, consisting of the bilateral PFC, IPS, and inferior temporal cortex. The latter finding replicated early observations of number instinct and number form areas. Crucially, by showing that professional mathematicians do not recruit brain areas associated with language when engaged in mathematical thinking, Amalric and Dehaene again refute the hypothesis that higher-level mathematical abilities are grounded in language systems of the brain. Mathematicians simply employ the independent and highly-trained number network.

11 Space and Number

11.1 Small Numbers on the Left, Large Numbers on the Right

At first glance, number and space are unrelated categories. However, scientists realized early on that numbers and space are somehow linked in our minds. In 1880, Francis Galton noticed that some people imagine "visualized numerals" when they think of numbers.[1] These people experience numbers automatically and consistently as being positioned at a fixed place within a mental spatial structure.[2,3] Such forms of "mental numbers" are usually organized in a line—often arranged in a three-dimensional space—which supports the metaphor of a "mental number line" as a substrate to get access to numbers[4] (figure 11.1).

In 1993, Stanislas Dehaene and colleagues published a seminal paper in which they report a very peculiar and by now classical behavioral finding they termed the SNARC effect.[5] The acronym SNARC, borrowed from Lewis Carroll's nonsense poem "The hunting of the Snarc," stands for "Spatial Numerical Association of Response Codes." When subjects are asked to classify numbers from one to 10 as even or odd, thus making parity judgments, by pressing one of two keys, the responses to smaller numbers (say one or two) are quicker when the responses are made on the left of the body (usually the left hand), whereas the responses to larger numbers are quicker when responses are made on the right side (figure 11.2). This association of numbers and space occurs despite the fact that the task itself has nothing to do with numerical quantity. The SNARC effect is believed to originate from the fact that quantities are spatially organized in the mind by numerical proximity, oriented from left to right, thereby causing a congruity between small numbers and left-side responses, and vice versa. This has led to the

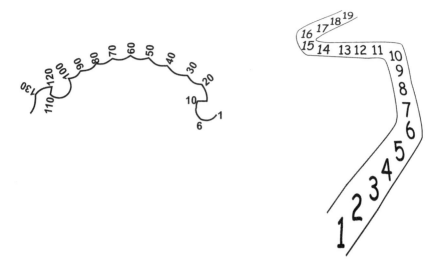

Figure 11.1
Some people experience numbers automatically as being positioned at a fixed place within a mental spatial structure (top image from Galton, 1880; bottom image from Rickmeyer, 2001).

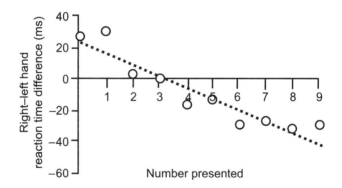

Figure 11.2
Space influences the speed of reaction to numbers. When subjects are asked to classify numbers from onoe to 10 as even or odd by pressing one of two keys, the responses to smaller numbers are quicker when the responses are made on the left side of the body (usually the left hand), whereas the responses to larger numbers are quicker when responses are made on the right side of the body. This finding is termed the "SNARC effect" (from Dehaene et al., 1993).

assumption of a left–to–right oriented mental number line in the case of left-to-right reading cultures.

Even the plain presentation of a numeral automatically draws attention to either the left or right visual field according to the relative value of the number.[6] In a very simple visual detection task, subjects were asked to respond as quickly as possible by pressing a key with a finger whenever they saw a target stimulus flashing either to the left or to the right of a center cross that they were asked to fixate on. When one of four possible Arabic numerals (1, 2, 8, or 9) was briefly shown a few hundred milliseconds before the target appeared, the subjects' reaction times were systematically influenced: when the smaller numbers were presented, the target on the left of the fixation cross was detected faster. In contrast, when the larger ones were shown, the target on the right side of the fixation cross was detected faster. Clearly, our perception of numbers, even though they are completely irrelevant to a task, influences the direction of the spatial allocation of attention. Attention causes us to move up or down on a mental number line.

The mental number line as a spatial representation of number is evidenced in the brain. Patients with brain damage in the parietal lobe and the temporo-parietal junction of the right cerebral hemisphere ignore the left portion of space; they have what is called a "hemi-spatial neglect." When asked to indicate the midpoint of a line, these patients miss the midpoint, because they ignore the left portion of the line, and so perceive the midpoint to the right of the line's center (figure 11.3). Curiously enough, these patients—who otherwise have intact numerical skills—also misplace the midpoint of a numerical interval when asked to bisect it.[7] For example, they would state that six instead of five is the numerical midpoint between one and nine. They are also slower to judge smaller numerals relative to a reference numeral than larger ones.[8] The notion of a mental number line might be more than a simple metaphor; number lines and physical lines might be functionally isomorphic.

Since the discovery of the SNARC effect, scientists have tried to understand how the association between space and number originates. It was clear from early on that the magnitude and direction of a spatial representation of numbers can be influenced by learning, such as the direction of reading and writing. Right-to-left reading cultures associate small numbers with right and large numbers with left,[9] whereas the opposite association

"Where is the midpoint of the line?"

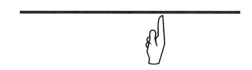

"What number is halfway between 1 and 9?"

"6"

Figure 11.3
The mental number line is spatially represented in the brain. Patients with "hemi-spatial neglect" ignore the left portion of space. When asked to indicate the midpoint of a line, these patients miss the midpoint and place the perceived midpoint to the right of the line's center. They also misplace the midpoint of a numerical interval when asked to bisect it verbally.

is true for left-to-right reading cultures.[10] Because of this, the number-space association appeared to be a purely cultural artifact. However, recent work with newborns indicates a different story and suggests that we are born with an oriented mental number line.

Maria Dolores de Hevia and coworkers from the University of Paris-Descartes in France used a spontaneous looking-time protocol to test whether newborn infants up to 3 days old associated auditory magnitudes with a particular side of a screen.[11] The babies were presented with a line on the screen and either heard six or 18 sounds. Next, the newborns who were primed with six sounds were tested with 18 sounds and two simultaneously shown lines on the left and right side of the screen. If the newborns associate an increase of numerosity with the right space, they are expected to look longer toward the right line on the screen, and this is precisely what they did. Conversely, if they associate a presented decrease of numerosity with the left space, they should look longer toward the line on the left side of the screen. As predicted for this condition, the newborns looked longer toward the left side. It seems that our brain is wired to associate numbers to an oriented space, irrespective of language or culture.

Space and Number 237

This spatial association of numerical quantities in our brain has an evolutionary foundation and can also be seen in animals. Rhesus monkeys show a spatial bias when mapping numerosity onto space as well.[12] The same is true for chicks right after fledging; they spontaneously associate smaller numbers with the left space, and larger numbers with the right space.[13] Just as with numerical quantity itself, it seems that the spatial connotation of numbers is not only ontogenetically wired into our brain, but deeply rooted in phylogeny. This cognitive bias to associate numbers with space is likely to be adaptive; it may provide a scaffold that helps to bolster memory and learning throughout our lives.

11.2 Carried along the Number Line During Calculation

Interestingly, a spatial bias is also present during non-symbolic calculations with sets of items, also called "approximate calculation." In a behavioral study from 2007, Koleen McCrink and Ghislaine Dehaene-Lambertz from Paris demonstrated that we are swept along the mental number line during additions and subtractions with visual numerosities, a phenomenon called *operational momentum*.[14] Similar to studies with preverbal infants, adult subjects viewed videos in which sets of dots disappeared and reappeared behind a blocker. In an addition problem, for instance, first 10 dots moved behind the blocker from one side followed by six more. Once all dots had disappeared to the subject, the blocker disappeared to reveal a proposed outcome set that either showed the correct (i.e., 16 in this example), or an incorrect set with more or fewer dots. The subjects had to judge without counting whether the outcome set showed the correct sum or not. Subtraction problems were shown in the same manner.

During this task, the subjects made many calculation errors and often misjudged the correct outcome as being smaller or larger than it really was. This resulted in a bell-shaped response distribution that was centered around the correct solution, a classic signature of the non-symbolic and approximate number processing based on the ANS we have seen earlier. Surprisingly, however, subjects systematically overestimated the answers to addition problems and underestimated the answers to subtraction problems. In other words, as numbers were mentally added or subtracted, the subjects seemed to be carried further along than necessary on their number line. For example, when given the problem of 24 + 8, they were much more likely to judge 40 as correct rather than 26, despite the fact that both of these incorrect outcomes differed equally from the correct outcome of 32.

The operational momentum is not a specific feature of numerate adults; 9-month-old infants already show it.[15] This indicates a deep developmental continuity across non-symbolic and symbolic calculations and the underlying arithmetic capacities. Whether this effect would also be present in other animals is currently not known.

While a shift of spatial attention is thought to contribute to operational momentum, it cannot explain the entire phenomenon. When subjects have to divide their attention between two competing processes and have less attention available for calculating, operational momentum for addition does not decrease, as expected, but increases.[16] This suggests that a "rule of thumb," or heuristic, contributed to the result: if adding, accept more; if subtracting, accept less. This would correlate with the finding that the use of heuristics is generally increased when attention is decreased.[17]

In 2009, André Knops and Stanislas Dehaene in Paris set out to explore the neural underpinnings of the links between calculation and space.[18] They wondered whether brain activation would indicate that our mind's eye wanders on a horizontal mental number line when we calculate. Because the mental number line is oriented from left to right, our mind may actually move horizontally according to the location of a calculated outcome. For addition operations, we would therefore move rightward, and leftward for subtractions. Could brain activation in functional imaging indicate such a movement along the number line?

To test their hypothesis, they had to do two things. First, they had to find out what activation patterns emerged when participants moved their eyes rightward and leftward in physical space. And second, they had to find out whether the pattern emerging during eye movements was somehow related to the activity pattern elicited by additions or subtractions, respectively. In the first step, the researchers therefore measured the fMRI activation of participants while they moved their eyes. The PPC is well known to be involved in representing eye movements, and unsurprisingly BOLD activity from the posterior SPL was strongly related to eye movements.

Now they needed the fMRI data from calculations for comparison. The participants lying in the scanner saw two successive numbers (operand 1 and operand 2) on a monitor and had to add or subtract according to the instruction. Similar to previous studies, the calculations activated a network of brain areas comprising the bilateral horizontal segment of the IPS, SPL, prefrontal, and premotor areas. Since calculation requires the manipulation and transformation of numerical information in mind, it is not surprising to find the posterior SPL also activated in this task. After all, the superior parietal lobule is known to be critical for the manipulation of information in working memory.[19]

But the important question was still unanswered: is calculation-related brain activation comparable to a mental movement in physical space? Since the different fMRI activation patterns resulting from eye movements and calculation tasks are too complex to be compared by eye, Knops and Dehaene used artificial intelligence to analyze them. They had a statistical classifier learn everything there is to know about the activation patterns from the posterior SPL when the participants made leftward and rightward eye movements. If some activation patterns are shared between eye movements and calculation, the classifier that had learned to discriminate between eye movements should be able to immediately predict the type of operations based on the activation patterns that emerged during calculations. Indeed, when the classifier was fed with the fMRI data from the posterior SPL measured for addition and subtraction instead of eye movements, it significantly predicted whether the correct calculation was performed in 55% of the cases. Simply based on activation for eye movements! One may assume that 5% better than chance level is not much, but remember that two strikingly different conditions were compared. In one case the participants made actual eye movements, while in the other cases, they mentally calculated with fixed eyes. In addition, the posterior SPL was not even the optimal activation area, and both sets of data overlapped only partially there.

But there is more to learn from this study because, during calculation, the numbers were presented as either numerals (symbolic calculation) or as sets of dots (non-symbolic calculation).[20] During calculations, considerable overlap in brain activation between the different formats was detected. Obviously, calculation with symbols is rooted in brain networks of non-symbolic calculation. Moreover, the authors also examined the ability to predict which operation was being performed in one format on the basis of a classifier being trained to sort additions from subtractions in the other format. This cross-format generalization yielded remarkably good results. The classification accuracy for the prediction of non-symbolic calculation from the symbolic format was 61%, and 62% for the prediction of symbolic calculation from the non-symbolic format. This finding indicates that the posterior SPL region is comparably involved in solving mental arithmetic problems in both formats. The calculative capacities of the brain build on shared neural substrates, not only for the representations of numerical values, but also for arithmetic. Obviously, a calculative brain circuitry was in place even before the arrival of arithmetic with culturally determined number symbols.

11.3 Space and Number in the Brain

The study by Knops and Dehaene[21] indicates that the reason why space and numbers are linked in our mind has to do with the workings of the PPC. As a true association cortex, the PPC represents several abstract concepts—space and number, among others—in close anatomical vicinity. The neural circuitry in the posterior parietal lobe, and the IPS in particular, that is crucial for abstract representations of quantity overlaps with the parietal circuitry involved in spatial representations.

Much of the known details about spatial representations in the IPS stem from monkey electrophysiology (see figure 6.2 for anatomical details). To avoid misunderstanding: spatial cognition depends not only on the PPC, but on a larger network of parietal, frontal, and hippocampal (and subcortical) regions. The parietal cortex, however, is specifically dedicated to spatial representations, as can also be seen by spatial neglect after lesions there. In addition, the close proximity of areas representing space and numbers in the PPC favor neural interactions that underlie the behavioral interactions outlined above.

In general, neurons in these areas are selective for the locations of objects in the visual field. The neurons are only excited if a stimulus is displayed in a specific location of the visual field; they have what is called a visual receptive field. This receptive field is anchored relative to different parts of the body. This relationship between physical object location and representation relative to body parts is called a reference frame, and the reference frame changes with cortical processing hierarchy.

Neurons in the lateral intraparietal (LIP) area have the most basic, retina-centered (or eye-centered) receptive fields (see figure 6.2 for neuroanatomy). Just like neurons in the retina or in the primary visual cortex, they encode stimuli relative to the position of the retina; if the eye moves, so does the receptive field of an LIP neuron. However, unlike neurons in the early visual areas, the strength of the neuronal responses in the LIP depends on whether the monkey is paying attention to the stimulus.[22] Without the monkey paying attention to the stimulus, LIP neurons hardly discharge, indicating that the LIP—just like the other IPS regions—is a high-level cognitive brain area. This is also emphasized by the neurons' activations during memory-guided saccades. When the monkey is required to remember a specific location as a target of a subsequent eye movement, neurons fire to a confined target region in the visual field that the monkey is about to make an eye movement to.

Neurons in the medial intraparietal (MIP) area also have retina-centered receptive fields, but fire primarily when the monkey is preparing to reach for a visual target. The MIP therefore is mostly involved in coding the location of an intended reach. For that reason, the MIP is also called the parietal reach region.[23] Hand-eye coordination and reach guidance are functions of the MIP.

I have mentioned the ventral intraparietal (VIP) area several times throughout this book as the key area to represent numerical quantities. This is one big job for VIP neurons, but they encode other features of a stimulus as well. VIP neurons also have visual receptive fields, but some of the neurons have more advanced, head-centered receptive fields.[24] When the head is in a fixed position while the monkey is required to shift its gaze by staring at fixation targets at various locations, the receptive field does not move along with the eyes, but remains at a stable position relative to the fixed head. And unlike area LIP, which responds only to visual stimuli, many VIP neurons process multimodal information and have joint tactile, auditory, and visual receptive fields.[25] While area LIP codes for stimuli remote from the body, area VIP codes for near space within the vicinity of the head or body.

In the anterior intraparietal (AIP) area, finally, neurons usually have a hand-centered frame of reference. Neurons fire when the monkey is looking at or preparing to grasp an object; they are bimodal (visual-tactile) and selective for specific objects.[26] These characteristics highlight the role of area AIP in reaching and grasping three-dimensional objects.[27]

Neuroimaging studies in humans have identified putative homologues of these macaque IPS regions.[28] Of course, such parallel anatomies are always tentative for a few reasons. First, the human association areas experienced rapid expansion during evolution, hampering the identification of corresponding brain regions. And second, symbolic functions that are of paramount importance in humans and have likely been processed in newly evolved sub-regions are simply missing in monkeys. Nevertheless, these putative anatomical homologies can be exploited in human imaging studies.

One of the most extensive fMRI studies to examine the topographical relationship of calculation-related activations to spatial and language-related areas in the human parietal lobe was performed by Olivier Simon and Stanislas Dehaene in France.[29] Consistent with previous data in both humans and monkeys, they found that the PPC was broadly activated by all visuospatial tasks (grasping, pointing, eye movements, and spatial attention). But they also found specific activation patterns. Grasping exclusively activated an anterior IPS region bilaterally that possibly coincided with area AIP in monkeys. Posterior to this was a region deep in the IPS that was activated by calculations alone, putatively corresponding to area VIP. Still posterior to this region fast eye movements (saccades) were represented, suggestive of area LIP. Finally, calculation and language tasks jointly activated a portion of the IPS lying underneath the left angular gyrus. Overall, these results indicate that number tasks activate the bottom of the IPS, a region closely surrounded by a geometrically reproducible array of areas that are involved in manual, visuospatial, and verbal tasks.

The overall similar pattern of parietal organization, with a systematic transformation from sensory to eye and hand-specific neuronal selectivities, presents a striking parallel to monkey physiology.[30] Using neuroimaging, several functions have been identified in human IPS that parallel the above described functions in LIP, VIP, and AIP in the monkey. Crucially, the horizontal segment of the IPS region that has been shown to be consistently activated in numerical tasks in humans roughly coincides with the putative human area VIP.[31] Such a co-localization is consistent with the localization of monkey number neurons of monkey area VIP.

The overlap between numerical and spatial networks in the IPS might account for the behavioral interactions seen between representations of space and numbers. In a number comparison task, for example, interference between number and size can be shown: choosing the numerically larger number takes significantly longer if the numeral is smaller in size compared to the numerically smaller number (e.g., the comparison 2 versus 7).[32] Vincent Walsh from the University College in London has therefore suggested that a common magnitude system for the representation of numerical (discrete) and spatial (continuous) quantities in the parietal cortex might be responsible for behavioral interference phenomena between different types of quantities.[33,34]

To investigate how representations of abstract quantities other than numerosities are encoded by single neurons and how they relate to numerical representations, my PhD student, Oana Tudusciuc, and I trained two rhesus monkeys in a delayed matching-to-sample task to discriminate between both the number (one to four) of items in multiple-dot displays and the length of a line (out of four different lengths) in random trial alternations. When we recorded single-unit activity from area VIP while the animals performed the task, about 20% of anatomically intermingled single neurons encoded numerosity, line length, or both types of quantities. Thus, two partly overlapping populations of neurons within this area seem to give rise to quantity judgments, suggesting "distributed but overlapping" neural coding of quantitative dimensions in the IPS. A statistical classifier analysis showed that the relatively small population of quantity-selective neurons carried most of the categorical information; by exploiting the neurons' firing rates, the classifier was able to accurately and robustly discriminate both line lengths and numerosities in a behaviorally relevant way.

Information about spatial quantities such as length is not only represented by parietal neurons, but also by PFC neurons, which may inherit spatial information from the PPC. PFC neurons are responsive not only for absolute line length,[35] but also for relative distances[36] and even for proportions, which is the relation between two quantities. My PhD student, Daniela Vallentin, and I trained two rhesus monkeys to judge the length ratio between two lines, a reference and a test line.[37] The length ratios between the test and reference lines were 1:4, 2:4, 3:4, and 4:4. Twenty-five percent of the tested PFC neurons were significantly tuned to proportion, irrespective of the absolute lengths of the test and reference bars, and each of the selective neurons preferred one of the four tested proportions. Just as with numerosities or lines, a labeled-line code was found for proportion coding, with neurons exhibiting peaked tuning curves and preferred proportions. The areas where such proportion-selective neurons were found coincided with PFC regions that also house number neurons. These data suggest that the perception of relational quantities is represented by the same frontal-parietal network, and relational magnitudes are coded as absolute quantities in the primate brain.

In conclusion, the finding that numbers and space as abstract magnitude categories interact with one another seems to be related to them being processed in close anatomical vicinity. As a consequence, the neural circuits for spatial and numerical representations overlap in the brain's fronto-parietal quantity network. Moreover, neurons processing magnitude information share coding properties, and sometimes encode different types of magnitudes. From a mechanistic point of view, these commonalities in the brain can explain why space and number are so intimately entwined on a behavioral level.

Part V Development

12 The Developing Number Brain

12.1 Counting in Children

Children learn to count between the ages of two and four. But what does it really mean to say that children can count and understand numbers? Obviously, counting is more than approximately estimating set size. It is a process that allows us to develop a very precise representation of numerical quantity. Parents sometimes say that a toddler who can count up to three or four "knows" those numbers. However, the child might only be reciting a list of arbitrary words without understanding the quantitative meaning of them. It may superficially look like counting, but it is not, at least not yet. How can counting be defined?

The influential Swiss psychologist Jean Piaget (1896–1980) provided one operational definition of counting. He suggested that children understand numbers when they pass the conservation-of-number task, at around 5 or 6 years old. In this task, children typically face two rows of the same number of items; for example, two rows of five candies. They are then asked whether the two rows consist of the same number of items. When the items in one row are more spaced apart than in the other, younger children typically think that this results in a change in number and state the longer row has "more." They obviously fail to understand that both sets have the same number, even if the arrangement of the items differs. For Piaget, the key number concept therefore is "equinumerosity" (sometimes called "exact equality")—the idea that two sets have the same number of items, if and only if their members can be placed in a perfect one-to-one correspondence. This idea goes back to German mathematician Gottlob Frege (1848–1925), who defined integers by capitalizing on the property of exact equality between numbers.[1] But equinumerosity alone may not suffice to characterize counting.

An alternative definition of counting was provided by Rochel Gelman and Charles Ransom Gallistel from Rutgers University in their seminal book from 1978 on toddler number representations. Gelman and Gallistel listed three core principles that characterize counting.[2] First, each item in a set is given a unique label in order, resulting in a one-to-one correspondence between items and tags (the *one-to-one principle*). Second, there must be a stable, ordered list of unique labels, or tags (the *stable order principle*). And third, the label that is applied to the final item determines the absolute quantity of the set, its cardinal value (the *cardinality principle*). This final criterion is of particular importance, for it states that the last word uttered in counting expresses the number of items in the whole set. It is the cardinality principle that gives number words their meanings, by making the cardinal meaning of any number word understood from its ordinal position in the counting list. According to this, children are said to understand numbers when they apply the cardinality principle on "give-a-number tasks." In the give-a-number task, an experimenter simply asks a child to give a certain number of items ("Give me one," "Give me two," and so on). Here, children are asked to establish sets with the numerical value named by the given number word. Successful performance in this task would indicate counting.

But understanding cardinality and linking it to number words takes time. Starting at 2 years old, preschool children need up to two additional years to figure out how counting represents numbers.[3] Between age 2 and 4, an interesting developmental sequence of numerical understanding surfaces in the give-a-number task. Initially, children cannot even give one item when asked for it; they are said to be "no numeral-knowers." Between 24 and 30 months of age, most children can reliably hand over one object as "one-knowers," but otherwise give a random number of objects for any other number word. After about another half-year, children become "two-knowers," then "three-knowers," and rarely also "four-knowers," while showing the same randomness for larger number words.

But then, around three-and-a-half years old, the children exhibit something miraculous: they become knowers of the cardinal principle. They suddenly understand counting and are now able to reproduce sets with cardinalities of any number word in their counting list. At this point, children stop reciting the counting sequence when asked "how many?" and instead use the last word in their count as an answer.[4] A quantitative change in the understanding of counting has happened. Now, children have acquired the counting principles based on knowing the numerical meanings of "one," "two," "three," and "four." They do not need to know counting sequences with larger numbers to understand that this is an infinite numeral list.[5] Once a child masters this list, they can infer the meaning of any newly mastered numeral symbol just from its position in the numeral list.

Susan Carey therefore posits that the numeral list representation of natural numbers is key for an understanding of numbers.[6] She thinks that children who understand cardinality also understand the key numerical concept of succession (often called the *successor principle* or *successor function*). The successor function, the idea that each number is generated by adding one to its predecessor, is at the heart of numeral list representations of positive natural numbers (i.e., integers). A numeral list is therefore an ordered list of numerals such that the first item represents "1" and, for any word on the list that represents the cardinal value n, the next word on the list represents $n + 1$. In mathematics, the successor function or successor operation is a simple recursive function S such that $S(n) = n + 1$ for each natural number n. For example, $S(1) = 2$ and $S(2) = 3$. The successor function is used in the Dedekind–Peano axioms,[7] which define the natural numbers, and are named after the German mathematician Richard Dedekind (1831–1916) and the Italian mathematician Giuseppe Peano (1858–1932).

As with so many cognitive processes, defining "counting" is a surprisingly difficult matter, even though we intuitively know what counting is. Investigating young children that painstakingly learn to enumerate objects is probably our best chance to understand what it takes to become a proficient counter. Based on these studies, equinumerosity and succession have been identified as "two key concepts on the path toward understanding exact numbers."[8]

12.2 Startup Tools for a Symbolic Number System

Research with infants over the past decades shows impressively that the human mind is, in fact, not a blank slate when we are born. On the contrary, human cognition is built on a set of innate intuitions, or systems of "core knowledge."[9] These four systems of core knowledge represent significant aspects of the environment, such as physical objects and their mechanical interactions, space and geometric relationships, living beings and their goal-directed actions—and, importantly, numbers. The systems of core knowledge, which we share with other species, are not acquired throughout life by experience (although they can, of course, be modified by experience), but exist from birth.

While it is well accepted that counting builds on a core knowledge for numbers, it is a matter of intense debate which antecedently available representations serve as the start-up tools for counting.[10] Is it the approximate number system (ANS), the object tracking system (OTS), or both? The problem is that the positive integers as the target of any symbolic number system cannot yet be fully grasped by these two non-symbolic number systems. Neither of the systems can represent numerical magnitude both precisely and infinitely, which is a hallmark of counting.

Several scientists propose that symbolic counting in children arises out of the ANS and that there is some sort of continuity between the non-symbolic and symbolic representations. For instance, Gelman and Gallistel[11] suggest that infants and animals establish numerical representations through a non-verbal counting procedure. According to this view, learning what "five" means requires a mapping between a number symbol and an appropriate cardinal value within the ANS. An important argument for this hypothesis is that only the ANS provides information about what a quantity means, and does so for infinite numerical magnitudes. In addition, this system is available long before language and symbolic understanding develop in children. Such connections between symbolic numbers and approximate number representations are supported by experimental evidence in children.[12,13] However, the ANS cannot explain every phenomenon of symbolic counting. Symbolic counting is exact, whereas the ANS is not. In addition, the ANS does not contain the successor function, which is the method of generating subsequently greater integers.

Other scientists, such as Susan Carey, see a primary role of the OTS as a precursor for counting.[14] Children first use small number words ("one," "two," and "three") as placeholders and then discover the meanings of these words, which they link to numbers given by their OTS. The advantage of the OTS is that it represents small numbers or items relatively precisely. The downside to it is that object tracking provides no representation of cardinal values and is limited to around three or four items. Despite these limitations, the proposed idea is that, after some practice with small number words, which children learn to recite in order, they eventually discover how to order number words in their counting list based on ascending succession. Once they have learned this relationship, they build upon—or bootstrap from—that knowledge to derive a numeral list. At this point, children understand that this rule is valid throughout the entire counting list and proceed to generalize it to the last counting words as well. Thus, children learn the list of counting words first and then give each word a numerical meaning only afterward, as they infer the principle of the successor number from the organization of their list.

Beyond the antecedently available non-symbolic representations, a third scenario features our faculty of language. Linguist Noam Chomsky from MIT and his colleagues view recursion, the capacity to generate an infinite range of expressions from a finite set of elements, as the key ingredient for counting.[15] He postulates that open-ended, precise counting skills with large numbers, including the integer count list, depends on the presence of domain-general (universal) recursion. In that respect, he says, our symbol system for exact numbers parallels the uniquely human faculty of language, which also relies on a recursive computation. Elizabeth S. Spelke, a psychologist at Harvard University, similarly thinks that language mediates natural numbers.[16] She suggests that children discover natural numbers when they learn a natural language. Exact numbers, she says, depend both on our species-specific language faculty as well as our innate core knowledge for numbers. This brief overview of hypotheses on the ontogenetic origins of counting illustrates that the question of how we learn to grasp numbers symbolically will continue to keep us busy in the years to come.

12.3 Out of Approximate Quantity and into Symbolic Number

Despite the intense debate about which available representation serves as the start-up tool for counting, there is a general agreement that, one way or another, approximate number representations must play a key role. The most important argument for this position is that only the ANS provides information about cardinality; there is no other system that could inform what a numerical quantity means. Indeed, ample evidence suggests that symbolic counting is, at least partly, grounded in non-symbolic quantity representations. The remnants of the non-symbolic number system in symbolic number representations are highlighted by behavioral and neural commonalities, as the following paragraphs outline.

The shared behavioral characteristics of the non-symbolic and symbolic number systems are very obvious in numerate adults. When we discriminate numerical symbols, performance shows some key characteristics inherited from approximate quantity judgments: in both cases, we need more time and make more errors when values are close—for instance, when discriminating five from six as opposed to five versus nine. A classical psychophysical experiment by Robert S. Moyer and Thomas K. Landauer in 1967 showed that time to decide which number is larger gets shorter in duration with increasing numerical differences between the two stimulus digits (or numerals).[17] This is of course the numerical distance effect. Moyer and Landauer previously suggested that this process is analogous to judgments on inequality for physical continua, such as lengths or time intervals. Similarly, the numerical size effect is present when we deal with number symbols. For instance, if the discrimination between 11 and 14 is easy, the discrimination of 61 and 64 is harder, even though the numerical distance is the same. The numerical distance and the numerical size effect are captured by Weber's law, and both non-symbolic and symbolic number judgments follow it, albeit the symbolic system in much more subtle ways.[18,19,20]

If adults show a link between non-symbolic and symbolic numbers, then the commonalities should be even more obvious during number symbol acquisition in childhood. As we have seen earlier, infants just moments after birth are sensitive to the numerical attributes of the world around them. Infants rely on the ANS to make numerosity judgments, and their precision improves continuously throughout childhood. At some point in their development and education, symbolic number representations emerge and take over, while the ANS gets overshadowed but remains fully functional. If these two systems are linked or partly depend on each other, one would expect that precision in one system would predict accuracy in the other. Indeed, several studies demonstrate a positive correlation between acuity in discriminating non-symbolic set sizes and symbolic math ability. Justin Halberda and Lisa Feigenson from Johns Hopkins University demonstrated this by exploiting the finding that large individual differences exist in the non-verbal approximation abilities of 14-year-old children.[21] They showed that these individual differences in the present correlated with the children's past scores on standardized achievement tests, extending all the way back to kindergarten. In short, children who can discriminate small differences of set sizes, on average, are also better at symbolic math tasks later in life. Importantly, this finding has nothing to do with individual differences in other cognitive capabilities, such as intelligence and verbal skills, but is specific only to numbers and mathematics. Overall, these results show that individual differences in achievement in mathematics are related to individual differences in the acuity of the evolutionarily ancient and innate ANS. This will be brought up later when discussing disabilities in learning math, termed dyscalculia.

Earlier, I reported how Karen Wynn could demonstrate that infants are not only able to discriminate numerosities, but can also perform rudimentary arithmetic calculations when watching addition and subtraction events.[22] This non-symbolic calculation skill interacts with symbolic calculation in young children, as Camilla Gilmore and Elizabeth Spelke from Harvard University showed.[23] Five-year-old children, who had mastered verbal counting but were ignorant of formal arithmetic, spontaneously used non-symbolic system processes to manipulate quantities presented symbolically. In other words, they could build on their non-symbolic number system to perform symbolic calculations that they had never learned.

Both of these number systems also communicate with each other during learning. Training studies in children show that exercises that aim at enhancing addition and subtraction with numerosities have a positive effect for symbolic arithmetic, but not for reading as a control task.[24] Of course, these studies ensure that general training effects unrelated to numbers do not cause this effect. The learning effect also works in the other direction: learning to count with number symbols also enhances precision during numerosity judgments.[25] This holds true not only for children in the West, but also for indigenous people with a limited counting system. Members of the Munduruku in Brazil who had some formal schooling in math were able to discriminate numerosities better than members with less formal schooling.[26]

Even in numerate adults, calculation with dot numerosities transfers to symbolic mathematics. In one study, adults were first trained to add or subtract, without counting, large quantities of visually presented dots in two arrays.[27] On addition trials, participants observed two arrays of dots each disappear behind a centrally located block. On subtraction trials, participants observed a single array move behind a block and then a subset of items float out from behind the block and leave the screen. The participant's task was to estimate the total number of dots behind the block in these addition or subtraction problems. Participants with only this sort of non-symbolic calculation training could later solve more two- and three-digit addition and subtraction problems than people in a control group.

While many studies have reported a correlation between numerosity acuity and math performance, some individual studies could not find this effect.[28,29,30] However, when combined data from several studies are taken into account, a relationship is present. A total of three meta-analyses found support for a modest but significantly positive relation between approximate numerosity estimation and math ability.[31,32,33] In general, it is therefore safe to say that the ANS and the symbolic number system interact and rely on each other.

Given that the ANS originates early both in development and evolution, these data inform the "symbol grounding problem"—the question of how symbols acquire their meaning. The data support an initial and at least partial grounding of the symbolic system in the non-symbolic ANS. In other words, number symbols acquire their meaning by linking with pre-existing, innate representations of numbers. Even after learning, these two systems remain intimately linked and mutually interact with each other.

12.4 Brain Activity in the Developing Brains of Children

While neuronal correlates of adult number cognition are relatively well described, only a few functional imaging studies have looked at the developing brains of children. This is because it is often more challenging to work with children than with adults. The scanning environment with its noises and large tubes may prove frightening to children. They have to keep very still during the scans, and, depending on their age, the test protocols have to be carefully adapted so that the children understand what they need to be doing while lying in the scanner. Normally, children answer questions by speaking. However, these verbal responses may partly overshadow numerical brain regions, not to mention create artifacts in the data sets because of mouth movements. As a consequence, children—like adults—are instructed to press keys to the right or left of them depending on a situation, which they may find difficult to comply with. Even adults sometimes struggle with left versus right if they are under time pressure.

To avoid active responses during measurements, imaging studies with young children employ passive viewing protocols that exploit fMRI adaptation effects mentioned in previous chapters. The first fMRI study with 4-year-old children used precisely such an adaptation protocol.[34] The children lay inside an MRI scanner and passively viewed a stream of visual arrays. To avoid habituation to non-numerical cues, the cumulative surface area, density, size, and spatial arrangement of the items varied among stimuli. For habituation, 16 dot arrays were repeated as adaptation stimuli, causing a decrease of activity in specific brain areas. Occasionally, an array was shown that deviated in the number of items (32 items instead of 16). This change caused a release from adaptation, reflected by a relative increase in BOLD activity. To ensure that an observed BOLD signal increase was due to numerosity changes and not to any other change of the visual stimulus, some deviant stimuli showed only changes in shape (squares instead of dots). An increase of the BOLD signal for 32 items relative to the last presentation of 16 items—in the absence of a signal increase for changes in shape—indicated sensitivity to number. The numerical deviants showed activation in children's parietal and frontal lobes (figure 12.1). In the parietal lobe, the right IPS, right superior parietal lobule, and left inferior parietal lobule were activated. In the frontal lobe, stronger activation was observed in the left precentral gyrus, the left superior frontal gyrus, and the right middle frontal gyrus. Children's number-related activity in the parietal and frontal lobes was strikingly similar to activity in adult participants tested in the same study under identical conditions. This demonstrates that very similar cortical areas are recruited for non-symbolic number processing early in development, and before formal schooling has begun.

After this study in 4-year-old children, the race began to test younger and younger children. Greater sensitivity for number deviants in the parieto-frontal network has later been confirmed with other imaging techniques in six-month-old,[35] and even in 3-month-old, children.[36] This suggests that parieto-frontal adaptation is specific to visual numerosities, and reflective of a very primordial number-processing capacity that precedes explicit learning and language.

A Children:

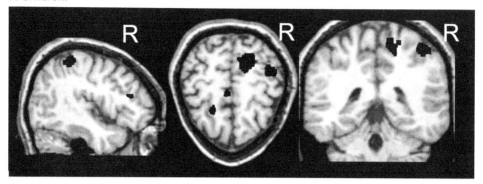

B Adults:

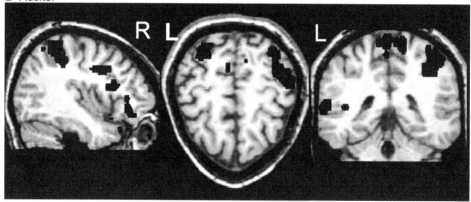

Figure 12.1
Number-related activity in 4-year-old children is strikingly similar to activity in adults. A) FMRI activation in the parietal and frontal cortices in 4-year-old children. Black regions were more active during the presentation of number compared to shape deviants (left to right: sagittal, horizontal, and frontal planes). B) Similar brain regions are activated in adults (from Cantlon et al., 2006).

While these studies clearly showed that the parieto-frontal number network is functioning early on in our lives, the design of these earlier studies could not inform about how numerical information is encoded in children's brains. In adults, evidence from fMRI[37] and single-cell recordings[38] confirmed results in non-human primates[39] and showed tuning to preferred cardinal values. Would this number tuning already be present in children? Alyssa Kersey and Jessica Cantlon from the University of Rochester have demonstrated that this is the case.[40] The researchers used the aforementioned fMRI adaptation protocol in three- to four-year-old children, but in contrast to previous studies, they not only presented one deviant numerosity occasionally mixed with a habituated one, but a range of both smaller (eight and 12 items) and larger deviant numerosities (24 and 32 items). This allowed them to explore systematic BOLD signal release from adaptation as a function of numerical distance from habituated numerosity of 16. And as always, controls for non-numerical factors were included. Indeed, the young children exhibited fMRI tuning to cardinal numerosities in the IPS: the more distant the deviant numerosities were relative to the habituated stimulus, the stronger the BOLD recovery signal became in the IPS. In other words, the BOLD recovery response to numbers resulted in an upside-down, bell-shaped tuning function. These tuning functions were even best described on a logarithmic number scale, just as in monkey number neurons and fMRI data from humans.

Kersey and Cantlon also presented convincing evidence that the neural tuning in the IPS was related to children's numerosity discrimination performance. They found that the sensitivity of children's neural tuning to number in the right IPS was comparable to their behavioral discrimination sensitivity observed outside of the scanner. Children with sharp neural tuning curves in the right IPS were better at differentiating numbers. This extensive study strongly argues in favor of a developmental continuity in the neural representation of cardinal numbers over development. Clearly, the right IPS in particular underlies early development of numerical perception.

When children grow older, they learn to represent numbers in a symbolic format. How might non-symbolic and symbolic representations in the brain change with development? In a meta-analysis, Liane Kaufmann from the General Hospital in Hall, Austria, and coworkers analyzed 19 fMRI studies with children and compared findings with those from adults.[41] They found that the two number formats are more segregated in the brains of children compared to adult brains. In the parietal lobe, activations in response to non-symbolic numbers (items of dot patterns) were confined to the right (bordering the IPS), whereas symbolic number magnitudes produced bilateral posterior parietal activations. While children activate the core of the IPS for symbolic numbers just like adults, for non-symbolic numbers, they surprisingly engaged somato-motor related areas, such as the anterior part of IPS and parts of the somatosensory cortex on the postcentral gyrus. It is thought that activations in the postcentral gyrus and the neighboring anterior IPS may reflect a link between finger counting and number processing.[42,43] As children first learn to enumerate non-symbolic numbers by using fingers, they also recruit brain regions that represent the sensation and movements of fingers.

Both number formats not only activate different regions of the parietal lobe in children, but also different areas of the frontal lobe. Only non-symbolic presentations engaged the inferior and middle frontal gyrus. The exact functional role of these areas is still debated today. It could be related to the semantic retrieval of numerical magnitude, or, alternatively, it could be a matter of working memory.

The respective contributions of number brain areas vary quite fundamentally with development and age. For processing Arabic numerals, a developmental shift from frontal to parietal activations is observed from childhood to adulthood. The first study to report this shift was performed by Daniel Ansari and coworkers from the University of Western Ontario in Canada.[44] They had 12 ten-year-old children and 12 adults lay in the scanner and judge the relative magnitudes of two single-digit numerals by selecting the numerically larger one. Next, they identified brain areas that were more activated during judgments of number pairs with varied numerical distances. The group of children primarily engaged frontal regions, such as the left inferior frontal gyrus of the lateral PFC. In the adult group, however, this numerical distance modulated bilateral parietal regions. The conclusion, then, is that the frontal number activation in children gets replaced by parietal activation in adulthood. This frontal-to-parietal shift in number processing with age, as well as changes within the parietal lobe, were later confirmed by other studies.[45,46,47]

The above effects were not only seen for simple number comparisons, but also for calculations. This was first reported by Susan M. Rivera and Vinod Menon, who asked 8- to 19-year-old participants in the scanner to judge whether displayed results for arithmetic equations (two-digit additions or subtractions) were correct.[48] The older subjects in the group showed greater activation in the left parietal cortex, while younger subjects showed greater activation in the PFC and the hippocampus. This prefrontal involvement in children's number processing may partly reflect effortful number processing that places high demands on working memory, attention, and cognitive control.[49] As children's arithmetic competencies are yet not automatized but rather require effortful processing, calculation may be supported by prefrontal regions known to mediate overarching cognitive functions such as attention, working memory, and monitoring. With age and proficiency, activation of the PFC decreases and is shifted more and more to the PPC, which in adults is seen as the major processing site for symbolic numbers.

The greater activation in the hippocampus in young children could be explained by greater demands placed on memory. Indeed, the hippocampal system is involved in learning to count and arithmetic skill acquisition, specifically during childhood.[50,51] Hippocampal-frontal circuit reorganization plays an important role in children's progression from effortful counting to efficient memory-based problem solving.[52,53] It seems that increased expertise with age leads to increased activations in parietal number areas as a sign of a more automated processing of mental arithmetic.[54]

These results point to the PFC as the cardinal structure in acquiring symbolic number concepts during ontogeny. In agreement with this hypothesis, and as described earlier, we only find numerosity-sign association neurons in the PFC, not in the IPS, of monkeys.[55] Perhaps the neural correlates of non-verbal number processing at a phylogenetically more ancient stage in non-human primates parallel those early in human ontogeny.

12.5 Abstractness of Number Representations in the Brain

We have seen that brain activation patterns for symbolic number processing change quite dramatically during development in childhood. But the question remains of how numbers are represented in a mature and adult human brain. Numerical representations are said to be abstract if they are signaled by the same neuronal substrate irrespective of the way the numbers are depicted relate to format (non-symbolic versus symbolic), notation (number words versus numerals), and modality (seen versus heard).

Let us first consider the format of number representations. The first direct demonstration that numerals and dot numerosities are both represented in parietal and frontal cortex according to a common quantity metric was provided by Manuela Piazza and Stanislas Dehaene in 2007.[56] In this study, they used fMRI adaptation to study whether the cardinality of non-symbolic quantities (i.e., sets of dots) and digits are processed in the same neural circuits in the IPS. They found that number adaptation in the

IPS transferred from non-symbolic quantities to numerals, and vice versa. Later, Daniel Ansari's group also found that the right inferior parietal lobule was significantly active in both symbolic and non-symbolic numerical comparison.[57] This suggests that both formats are processed in the same neural circuits in IPS, a finding that was confirmed and extended in later studies. Across dozens of imaging studies, assessment of both formats of numbers resulted in overlapping activations in the posterior parietal lobe (SPL, IPS, and IPL), but also in the superior, medial, and inferior frontal gyri, the precentral gyrus, the cingulate gyrus, the insula, and the left fusiform gyrus (see figure 9.2).[58] Overall, the same brain areas respond to numerosities and number symbols in adults.

Not only the anatomical locations but also the functional activation patterns in these brain areas show striking similarities. Eger and Kleinschmidt[59] trained a statistical classifier to predict the numerical value a participant had seen based on fMRI activation patterns in the PPC. The classifier trained on the activation patterns caused by numerals nicely predicted the corresponding dot numerosities. However, this did not work the other way around from numerosities to numerals. One possible interpretation of this asymmetric generalization might be a less precise tuning for non-symbolic format compared with symbolic numerals, which may lead to impoverished generalization across formats.

If both formats of numbers share a common neural substrate, both are expected to become influenced when these areas are temporarily inactivated. This is suggested by experiments using transcranial magnetic stimulation (TMS). When TMS is applied to the IPS, both representation formats are affected.[60] Curiously, both number representations were impaired after TMS to the left IPS, but enhanced by it on the right side. In contrast, TMS stimulation of the left and right angular gyrus did not cause impairments.

Patient studies are another valuable source of information about the relationship between non-symbolic and symbolic magnitudes. If both formats share a common cortical substrate, discrimination of both types of these quantities is expected to suffer after permanent damage to specific regions. Exactly this has been reported in case studies. A patient who had a focal lesion in the left parietal lobe exhibited problems in processing numbers, both with arrays of dots and numerals.[61] Another patient who suffered from acalculia with number symbols following an infarct restricted to the left IPS also had difficulties in processing visual numerosities one to nine.[62] Finally, in a patient with cortical atrophy in parietal (including the IPS) and frontal regions, simple judgments of numerical values in both representation formats were impaired.[63] All studies suggest the IPS as a key area of the brain's system for different quantity formats. The results from many studies and from different methods imply that number representations share, at least to some extent, a format-independent component.

Do we also find abstract representations with respect to the symbolic number notation? The most common symbolic number notations are numerals and number words. Would the brain generalize across these symbolic notations? To find out, Lionel Naccache and Stanislas Dehaene used a subliminal priming protocol.[64] Subliminal priming consists of a prime (a number) that is presented so briefly that it cannot be seen consciously, but can nevertheless facilitate the recognition of a subsequently shown target (another number). In other words, even though we cannot report on the value of the prime, its value is unconsciously processed in the brain and helps in deciding more quickly about the value of the target. In this study, participants had to decide as quickly as possible whether a target number in different notations (1, 4, 6, or 9 as numerals or spelled-out number words) was greater than or less than five. The notation of target and prime numbers was changed independently of each other, so that a prime numeral could be followed be either a numeral or a number word. The main finding was a notation-independent neural priming effect: a decrease of BOLD activity in the bilateral IPS when the prime and target were numerically identical (e.g., 4–4, or 6–six), as opposed to non-identical prime-target pairs (e.g., 1–4, or six–9). This indicates that the numerical meaning of a number symbol is encoded irrespective of its notation as a word or numeral. As an extension of this cross-notational priming study, the neural priming effect was analyzed as a function of numerical prime-target distance (close: one–2 or 8–nine; far: 1–eight or two–9) in another study.[65] The authors observed a distance-dependent modulation of IPS activity for numeral-to-number word prime-target pairs, but not for the reverse symbol order. Finally, Roi Cohen Kadosh and coworkers presented participants with numerals or number words and found a notation-independent adaptation in the left IPS, whereas adaptation in right IPS related with notation.[66] Despite some variation across these studies, and hemispheric differences in particular, these studies expand the idea of a notation-independent magnitude representation in the IPS.

The final remaining question in the context of abstract number representations relates to the sensory modality of the number stimulus, whether number is seen, heard, or felt on the skin. The abstract concept of number implies that it should be represented independent of the modality in which it is displayed. Although this is an obviously crucial question, surprisingly few studies have investigated which brain regions become activated by numbers across sensory modalities. One of the first studies was performed by Evelyn Eger and Andreas Kleinschmidt.[67] They presented spoken numbers and written visual numerals to passively perceiving participants. Compared to spoken or written letters and color words, the IPS was more active for numbers in both auditory and visual contexts. This was the first evidence for a supramodal number representation. A later study corroborated this finding for non-symbolic numbers and reported a right-lateralized network of frontal and parietal areas that was active during the estimation of sequentially presented items in both the visual and auditory modality.[68] However, these findings could not always be replicated.[69]

Saudamini Damarla and Marcel Just took a much closer look and showed that a statistical classifier trained on the fMRI activation patterns resulting from the presentation of a set of sequential visual dots could recognize the number of events based on patterns emerging from auditory trials, and vice versa.[70] The representations that were common across modalities were mainly right-lateralized in frontal and parietal regions. This suggests that BOLD activation patterns produced by numbers of visual and auditory elements are very similar. In addition, the authors found that the neural patterns in parietal cortex that represent quantities were common across participants. These findings demonstrate impressively that a common neuronal foundation for the representation of quantities across sensory modalities resides in the brain, and in particular in the IPS.

So, are number representations abstract? Based on human studies, it seem that at least some parts of the core number network respond in an abstract way. But we need to be careful. Functional imaging does not measure neuron activity and has a limited spatial and temporal resolution; claims about abstract number representations derived from BOLD signals might also be explained by functionally segregated neuron ensembles that overlap at the macroscopic level in certain brain regions. The mere overlap of activations does not necessarily imply the common recruitment of identical neurons for all conditions. Indeed, the idea that numerical representations are abstract has previously been rejected as premature based on behavioral differences found for the processing of different number formats, and based on BOLD response differences in number tasks.[71] Of course, whether the issue of abstract representation can ever be resolved based on behavioral outcome effects and the methodologically limited BOLD signals is a question in itself. We need to know what single neurons have to say in this matter.

If "abstract representation" is operationalized as "neuronal populations that code numerical quantities and are insensitive to the form of input in which the numerical information was presented,"[72] then the notion of the abstractness in numbers is indeed supported by single-cell recordings. In the monkey brain, the PFC, not the IPS, hosts the most abstract number representations. Number neuron responses in PFC were generalized across spatial features in visual item arrays, spatio-temporal visual presentation formats, visuo-auditory presentation formats, and in numbers of dots and associated number signs.[73] Of course, if neurons operate on lower, more sensory levels of the cortical hierarchy, such generalized responses cannot (and should not) be expected for all selective neurons. Abstract number information could also be extracted from the activity of segregated neuronal populations. This seems to be the case in the human MTL. Here, we find neurons that respond abstractly within the non-symbolic and symbolic domain separately, but not across formats.[74] MTL neurons therefore contain format-dependent populations of number neurons. Whether this also holds true in the core parietal-frontal number network of humans remains to be explored.

13 Developmental Dyscalculia

13.1 Developmental Dyscalculia and How It Affects Life

Seven is greater than five, right? What sounds trivial for most of us turns out to be a struggle for people diagnosed with developmental dyscalculia, a brain difficulty in comprehending numbers and arithmetic. The term "dyscalculia" stands for weak or poor ("dys-") calculation abilities. More specifically, developmental dyscalculia is a severe disability in learning and, as a result, performing arithmetic. This does not mean that dyscalculic people are generally less intelligent than average people, or simply too lazy to learn math. Rather, no matter how hard they try and how much effort they put into it, numbers remain barely comprehensible to them.

Brian Butterworth, a professor emeritus from University College in London and champion in the field of dyscalculia research, reports some impressive example statements by dyscalculic individuals[1]:

Emma King, cosmologist: I can't add up, subtract or multiply in my head ... I calculate 4 + 3 by counting.

Vivienne Parry, broadcaster, science journalist: I was in the top set for all other subjects. ... No matter how hard I tried or how much homework I did, I just didn't get it.

Articulate 8-year-old, good at all other school subjects, expert on dinosaurs, but "the only subject I don't like is math":

  –**Teacher:** What do you need to add to 8 to make 30?
  –**Child:** Two.

BD, a 23-year-old reading English at an Ivy League university:

–**Experimenter:** Can you please tell me the result of nine times four?

–**BD:** Yes, well, looks difficult. Now, I am very uncertain between fifty-two and forty-five ... I really cannot decide: it could be the first but could be the second as well.

–**Experimenter:** Make a guess then.

–**BD:** Okay ... uhm ... I'll say forty-seven.

–**Experimenter:** Good, I'll write down forty-seven. But you can still change your answer, if you want. For example, how about changing it with thirty-six?

–**BD:** Bah, no ... it does not seem a better guess than forty-seven, does it? I'll keep forty-seven.

A journalist once caught me by surprise when he insisted that dyscalculia might be used as a convenient technical term in the field of numerical cognition, but was not a real and accepted disorder. However, it is one according to the official diagnosis catalogue of the American Psychiatric Association. The *Diagnostic and Statistical Manual of Mental Disorders*, or *DSM* for short, edited by the American Psychiatric Association, serves as a universal authority for psychiatric diagnoses in America, Australia, and many European countries. The manual specifies the diagnostic criteria for each recognized mental health disorder and provides a systematic and reliable approach to diagnosis. In its fifth edition, from 2013, the DSM-5 uses the umbrella term "specific learning disorder" for impairments in reading, written expression, and mathematics. The DSM-5 diagnostic code 315.1 (F81.2) describes the specific learning disorder for impairments in mathematics. The diagnostic criteria include difficulties mastering number sense, number facts, or calculation (e.g., individual has poor understanding of numbers, their magnitude, and relationships; counts on fingers to add single-digit numbers instead of recalling the math fact, as peers do; gets lost in the midst of arithmetic computation and may switch procedures). In addition, difficulties with mathematical reasoning are evaluated (e.g., individual has severe difficulty applying mathematical concepts, facts, or procedures to solve quantitative problems). To diagnose dyscalculia, standardized tests of arithmetic are used. Substantial underachievement in such tests relative to the level expected considering age, education, and overall intelligence is taken as an objective criterion for developmental dyscalculia.[2]

Dyscalculic people are by no means a negligible minority. The estimated prevalence of developmental dyscalculia is between 5% to 7%.[3] In other words, at least one in every 20 people has dyscalculia. This is about the same prevalence as developmental dyslexia, a much more recognized disability in reading.[4] A major report by the UK government therefore warns, "Developmental dyscalculia is currently the poor relation of dyslexia, with a much lower public profile. But the consequences of dyscalculia are at least as severe as those for dyslexia."[5]

One may dismiss this disorder as some children being better at math than others. However, this learning disability has severe consequences for individuals and for society at large. The ability to understand and process numbers is essential to functional living, and low numeracy affects many daily activities, from calculating accounts to following medication instructions. Believe it or not, a large UK cohort study found that low numeracy was more of a handicap for an individual's life than low literacy was. This means that dyscalculic people earn less, spend less, and are more likely to be sick, be in trouble with the law, and need more help in school.[6] Therefore, improving education could not only help the individuals that suffer from it, but also dramatically improve economic performance.[7]

13.2 Domain-General and Domain-Specific Impairment in Dyscalculia

To help dyscalculic people, it will be essential to understand where this disability originates. Despite significant advancement, the neural causes of dyscalculia are not fully understood. Most scientists working in the number field think that counting, calculation, and arithmetic need both domain-general and domain-specific capabilities.

Above all, we need domain-general capabilities to learn anything at all. For instance, we need working memory and long-term memory to encode, store, and retrieve what we have learned—not just numbers, but all sorts of things, such as the meaning of words, our birth data, or the circumstances of our first kiss. Consistent with the idea of domain-general impairments is the finding that people with dyscalculia have a higher chance of also suffering from other developmental disorders, such as reading disorders.[8] One study in children in grades 2 to 4 found that the rates of deficits in arithmetic, reading, or spelling were four to five times higher in those children already experiencing marked problems in one of those abilities compared to the full population.[9] This co-occurrence is particularly detrimental because good language abilities appear to be needed for the typical development of counting, calculation, and arithmetical principles.[10] It is also known that 11% of children with attention-deficit hyperactivity disorder also suffer from dyscalculia.[11] Such disabilities in several cognitive domains are best explained by deficits of domain-general capabilities that simultaneously affect specific skills.

In addition, however, we also possess domain-specific capabilities that apply only to numbers. In fact, dyscalculia can be highly selective, meaning that it affects learners who have otherwise normal cognitive capabilities, such as normal intelligence or working memory.[12] This indicates that only the part of the brain necessary to learn counting and arithmetic is inefficient in this case. The arising difficulties are often so specific that they only apply to numerical tasks, but leave tasks involving other types of quantities spared.

This finding is demonstrated in an experiment by Marinella Cappelletti and coworkers[13] in which dyscalculic participants and control subjects were tested on tasks requiring either the discrimination of continuous spatial and temporal quantities or judgments of numerical stimuli. The continuous quantity tasks involved the differentiation of the duration of two stimuli or the lengths of two lines. In the numerical tasks, the participants had to select the larger of two numbers, compare a number with a set of dots, and evaluate a simple calculation (figure 13.1). Adults with severe dyscalculia were significantly worse than controls only when judging numbers, but not when evaluating length or time. Findings like these argue that it is important and justified to distinguish numerical abilities from quantitative abilities in general.

Brian Butterworth calls the domain-specific number capabilities the *"numerosity tool"* or the *"number module."* This tool or module is supposed to support the normal development of arithmetic competence. Consequentially, normal math development is seriously handicapped when this tool is inefficient. While originally coined for calculation disabilities, dyscalculia is now known to affect even more fundamental number skills, such as those as simple as assessing small sets of objects[14] or comparing the numerosities of two-dot arrays.[15] This indicates that deficits in higher-level mathematical

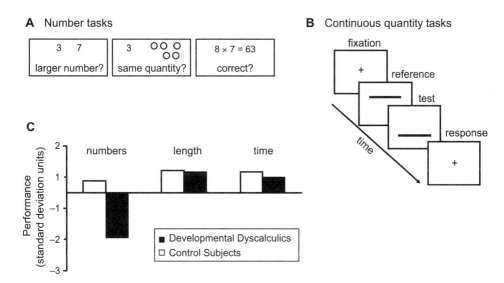

Figure 13.1
Dyscalculia can selectively cause difficulties with numbers while sparing other magnitudes. A) Three examples of number tasks used by Cappelletti et al., 2014. B) In the continuous quantity tasks, participants indicated whether the upper or the lower line was longer in either duration or length, in different blocks. C) Dyscalculic participants show impaired performance on numerical tasks, but not on spatial or temporal tasks, compared to control participants (from Cappelletti et al., 2014).

skills may stem from impaired representations and processing of basic numerical magnitudes.

Evidence supporting this idea stems from the finding that problems with discriminating numerosities in preschool children can predict trouble with learning arithmetic later in school. Manuela Piazza and coworkers compared the acuity to differentiate the number of items in dot displays in dyscalculic and control participants.[16] For both groups, three age ranges were tested: kindergarteners, school-age children, and adults. This allowed the authors to assess how number acuity and dyscalculia develop with age. They found that the acuity to discriminate the number of dots was severely impaired in dyscalculic children in comparison to controls matched for age and general intelligence. In dyscalculic children, the ability to discriminate dot numerosities was already severely impaired. Ten-year-old dyscalculic children scored at the level of 5-year-old typically developing children.

Moreover, the severity of the impairment predicted the defective performance on tasks involving the manipulation of symbolic numbers.

Similar correlations between numerosity discrimination and symbolic number processing were found by Robert Reeve and Brian Butterworth in a six-year longitudinal study with 159 children who were tested for dot enumeration and number comparison from age 5 on.[17] Instead of discrimination accuracy, the authors measured how quickly the children could solve the tasks. The slower they solved the task, the more difficulties they had with numbers. Remarkably, children who were slow at age 5 remained slow throughout the six-year testing period. This deficit was independent of domain-general capacities, such as non-verbal intelligence or symbol understanding. To conclude, low efficiency with numerosities in kindergarten is a reliable predictor of children who will struggle with learning arithmetic in school.

Moreover, children's precision in discriminating the values of numerosities and number symbols is correlated.[18,19] Children who can discriminate small differences of set sizes are also better, on average, at dissociating small differences in number symbols. Moreover, these differences early in childhood are also predictive of future performance in mathematical tasks later on. Children who are sensitive to fine-magnitude differences early in life tend to score better on math tests when they are older, and this effect is independent of intelligence, language skills, or other cognitive factors. Of course, the importance of the ANS that supports numerosity discrimination seems to be largely overshadowed by other basic numerical and cognitive abilities once children begin formal math education. Therefore, the processing of symbolic numbers, such as choosing numerals with larger numerical values, becomes more predictive of arithmetic performance.[20] But, as mentioned previously, by now it is well accepted that the analog number system informs symbolic math skills.[21,22,23]

13.3 Tracing Dyscalculia Back to Brain Anatomy

In an attempt to find a neural correlate of dyscalculia, anatomical changes in the brains of people with severe numerical learning disabilities have been discovered. Looking at anatomical cross-sections of the brain, one can see that brain tissue can generally be distinguished into dark and light parts called, respectively, "gray matter" and "white matter." Gray matter consists mainly of the densely packed bodies of nerve cells, or somata; these are the regions of the brain where information is processed and transformed. High gray-matter volume therefore implies intensive processing power. White matter, on the other hand, consists of thick fiber bundles and comprises the axons of thousands of neurons. Because axons are the communication lines of nerve cells, white matter represents the connections between brain areas. High white-matter volume therefore indicates the presence of intense connections. Gray and white matter nowadays can clearly be seen not only in histological slices manufactured postmortem, after death, but also on structural MRI images of living brains. The volumes of gray and white matter in the brains of people with developmental disabilities in number processing can therefore be compared with the volumes of brain matter in normally developing people.

Voxel-based morphometry is a method that detects anatomical differences in living brains. It is a neuroimaging analysis technique that examines the relative amount of brain mass present in specific brain areas between two groups of participants. Voxel-based morphometry adjusts every individual brain to a "standard" brain, thus removing the large differences in brain anatomy among people. Next, the brain images are segmented into gray matter, white matter, and cerebrospinal fluid cavities. Finally, the volume of gray or white matter in distinct brain areas is compared between the two groups of participants. This allows a glimpse into potential anatomical changes in dyscalculic brains.

In typically developing children of about 7 to 10 years, math competence specifically correlates with gray-matter volume in the left IPS.[24,25] The more gray matter, and therefore neurons, a child had in the left IPS, the better they scored in a standard arithmetic test battery. Also, the opposite is true. Children with low numerical proficiency had less gray-matter volume in the parietal areas (particularly in the left IPS and bilateral angular gyri) than those with high numerical proficiency.[26]

Based on this finding in normally developing children, one might expect reduced gray matter in the IPS of dyscalculics. Indeed, several studies confirm that impairments of arithmetical abilities correlate with abnormalities in the organization of parieto-frontal networks. In a pioneering study, Elizabeth Isaacs and coworkers from University College London compared the volume of grey matter in adolescents who were born at equally severe grades of prematurity.[27] For this study, this meant the participants had been born preterm at 30 weeks' gestation or less. Half of these subjects in this study (with otherwise normal IQ) suffered from dyscalculia. When comparing the brains of those with dyscalculia to those of normally developing preterm adolescents, the left IPS showed reduced gray matter. In other words, preterm children without a deficit in arithmetic ability had more grey matter in the left IPS than those who had a deficit.

Later it was found that not only the IPS, but also other brain areas related to numerical competence, particularly those in the frontal lobe, can show gray-matter differences. Stephanie Rotzer and coworkers[28] investigated structural differences in brain scans of 9-year-old children with developmental dyscalculia and compared them to scans of children of the same age who showed normal math development. Compared to controls, children with developmental dyscalculia showed significantly reduced gray-matter volume in the right IPS, the anterior cingulum, the left inferior frontal gyrus, and the bilateral middle frontal gyri. This study by Rotzer and coworkers also found differences in white matter. Dyscalculic children had significantly less white-matter volume in the left frontal lobe and in the right parahippocampal gyrus; because the parahippocampal cortex is an important part of the declarative memory system, the latter finding might explain problems with arithmetic fact retrieval in dyscalculia. Overall, these studies in dyscalculic subjects indicate that fewer neurons in the fronto-parietal number network, and less connections between the network regions, make people prone to impaired arithmetic processing skills.

While growing older, typically developing children show increases in white-matter volume, a structural correlate of increased connectivity in the frontal and parietal lobes, especially in the angular gyrus and the supramarginal gyrus. This suggests that the frontal lobes may be connecting increasingly to parietal areas in the fronto-parietal network known to be involved in arithmetic. In contrast, this age trend of increasing white matter was absent in dyscalculic children.[29] This suggests that one of the problems in dyscalculia is a relative failure to link up the relevant brain regions adequately.

An MRI technique called diffusion tensor imaging enables researchers to non-invasively measure the dimensions and integrity of white matter fiber tracts.[30] In a study that examined fiber tracts in children with dyscalculia and in typically achieving children, evidence was found for impaired fiber tracts connecting different brain areas of the fronto-parietal network known to be responsible for number processing and calculation.[31] Deficient fiber connections therefore are also implicated as a neurological correlate of dyscalculia.

All of these findings strongly suggest that dyscalculia in children is the result of specific disabilities in basic numerical processing.[32] Since core number-processing networks cannot work in isolation, more widespread changes are also to be expected. Some studies suggest that dyscalculia stems from more extensive functional aberrations in a distributed network of brain areas. This distributed network encompasses not only posterior parietal but also prefrontal as well as ventral temporal-occipital cortices that are known to serve multiple cognitive functions necessary for successful numerical problem solving.[33,34]

13.4 Functional Differences in Brain Activation of Dyscalculic Children

Not only anatomical, but also functional, differences in brain activation can be witnessed in dyscalculic children. The fronto-parietal fMRI responses of children with and without dyscalculia are quite different.[35] The functional differences are complex: dyscalculic children showed decreased activation in some number-related brain areas as well as stronger activation in others.

The few developmental fMRI studies published thus far suggest two main differences. Firstly, compared to controls, those children with dyscalculia tend to have less robust number-related activations in the IPS.[36] This could indicate a dysfunctional intraparietal number system and suggest that the neural basis of core numerosity is disrupted only during pre-adolescent cortical maturation. Secondly, children with dyscalculia recruit more distributed brain regions, particularly in the frontal lobe. The latter finding is interpreted in such a way that low achievers might compensate for the deficient number areas by engaging detour strategies requiring executive brain areas in the frontal lobe. And with different tasks—whether simple numerosity comparisons or more complex arithmetic tasks—the strategies and engaged brain areas seem to change in dyscalculic subjects.

In addition, growing evidence suggests that deficits in working memory—the capacity to maintain and manipulate information actively—can contribute to dyscalculia.[37] Both complex arithmetic problem solving and basic quantity representation are dependent on working memory.[38,39] No child would be able to calculate without working memory. It requires rule-based manipulation and an update of the contents of stored information.[40] Only with increased proficiency and a switch to fact retrieval strategies later in life does the need for working memory diminish. As a consequence, functional neuroimaging research has revealed significant overlap in multiple parietal and PFC regions involved in working memory and numerical problem solving. Overlapping patterns of activation have most prominently been detected in the supramarginal gyrus and IPS in the PPC, the premotor cortex, and in the ventral and dorsal aspects of the lateral PFC.

Studies of typical development provide the first evidence for the changing role of working memory with age. As mentioned earlier, Susan Rivera and colleagues found that relative to adolescents and young adults, children engage the PPC less, and the PFC more, when solving arithmetic problems.[41] This may partly reflect the increased role of visuospatial working memory processes, and a concurrent decrease in demands on cognitive control with age. In children between 7 and 9 years old, visuospatial working memory is the strongest predictor of mathematical ability. The importance of visuospatial working memory and associated fronto-parietal processing during arithmetic is further highlighted by neuroimaging studies in dyscalculic children. Compared to typically developing children, children with low math abilities had lower visuospatial abilities and lower activity levels in the right anterior IPS, inferior frontal gyrus, and insular cortex during a visuospatial working memory task.[42]

Finally, there is also evidence that immature prefrontal control systems may contribute to weaker math skills in children. More specifically, immature prefrontal control systems may contribute to weaknesses in the ability to inhibit irrelevant information, such as arithmetic facts or operations during numerical problem solving. In agreement with this idea, a study with dyscalculic subjects identified weaknesses in visuospatial working memory and inhibition control as key cognitive factors contributing to this learning disorder.[43]

13.5 It's (Partly) in the Genes

It's without question that individual differences in relation to math proficiency exist among people. Not everybody has the cognitive means to become a Gauss or an Einstein. And of course, there are many reasons for these differences. Good teaching of and a satisfying experience with mathematics surely is an important factor. But could there also be a genetic reason why some are bad at math while others excel? Behavioral geneticists consistently find moderate to high heritability for all sorts of cognitive and behavioral traits.[44] Why should numerical competence be any different? After all, the non-symbolic foundations of numerical skills seem to be innate. How else than based on genetic predispositions would newborn infants without an opportunity to learn about numbers be able to discriminate abstract numerosities?[45]

Given the complexity and hierarchical structuring of the 20,000 or so protein-coding genes in the human genome, it would be naïve to assume that a few of those genes do nothing else than ensure our number skills. Whenever behavioral geneticists explore the heritability of different sorts of cognitive skills, they find that all cognitive traits are "polygenic," that is influenced by many genes, with each gene contributing small effects. Consequently, any learning disorder will also be affected by many genes. In addition, the same gene can influence multiple cognitive traits (an effect called "pleiotropy"). And to complicate things even further, the effect of one gene may depend on the presence of one or more other genes (a process called "epistasis").

Despite such complications, it is now known that a substantial proportion of about 30% of genetic variance is specific to mathematics.[46] But how can one find such genes? After all, we all differ in our genome, and searching for the "math genes" is like searching for a needle in the haystack. Fortunately, nature itself provides a helping hand in the form of twins.

Twins come in two types: monozygotic twins ("identical twins"), who develop from one fertilized egg that splits and later forms two individuals with identical genes; and dizygotic twins ("fraternal twins"), who develop from two separate eggs forming individuals that share 50% of their genes. However, just like identical twins, and in contrast to usual siblings, fraternal twins are born at the same time and experience a very similar environment while growing up. Twin studies therefore allow researchers to disentangle genetic from environmental sources in math performance. In such twin studies, the performance correlation between pairs of identical twins is compared to the performance correlation between pairs of fraternal twins. Higher performance similarity for identical as compared to fraternal twins reflects genetic sources of variance (i.e., heritability), whereas equal performance between the twin pairs points to shared environmental sources (e.g., socio-economic status, home environment, or school factors) as a cause for performance variance.

In 1997, Bruce Peddington's group from the Colorado Learning Disabilities Research Center presented the first twin study to estimate the heritability of mathematic disabilities.[47] The researchers investigated 40 pairs of identical twins, and 23 same-sex fraternal twins, in which at least one member of each twin pair had math disabilities. They found that if one twin individual had a math deficit, then the other twin shared this deficit in 58% of the cases for identical twins, but only in 39% for fraternal twins. Due to the relatively small number of participants, this difference in math disability concordance between identical and non-identical twins did not reach statistical significance, but the study provided a first hint that math capabilities and disabilities might be heritable.

The most extensive twin study by Yulia Kovas and coworkers from the University of London consisted of a sample of thousands of pairs of identical and fraternal twins at 7, 9, and 10 years of age.[48] They tested heritability in three domains: mathematical skills, science skills, and English language skills, as well as their relationship to general intelligence. The study found astonishingly high genetic influence of about 60% for individual differences in early academic performance. This points to a substantial genetic influence and modest shared environmental influence. The authors were also surprised to find such consistently high heritability of academic performance in all three domains (math, science, English) at these years, despite the major changes in content across these years.

A further unexpected result was the high heritability between the learning domains. About two-thirds of the learning abilities between mathematics, science, and English were mediated genetically. Such genes that affect several other cognitive domains are called "generalist genes."[49] Obviously, generalist genes largely determine mathematical, science, and literary learning abilities and disabilities. As a consequence, this suggests only a very mild environmental influence. However, the study also postulates specialist genes that contribute to predisposing children to perform better in one domain than in another. About 30% of the genetic variation in this large twin sample was specific to only mathematics performance.[50] In other words, there are some genes that can make children better at understanding mathematics than science, or vice versa. Because genetic influence on learning abilities is so substantial, such specialist genes contribute importantly to differences in learning abilities, even though most genes are generalists. The authors sum up the genetic influence in this way[51]: Most genes found to be associated with a particular learning ability or disability (such as reading) will also be associated with others (such as mathematics). In addition, most of these generalist genes for learning abilities will also be associated with other cognitive abilities (such as memory).

As we have seen, part of math learning difficulty has to do with a deficient core "number sense," that is, our number instinct. In a more recent study, Yulia Kovas's team focused on the heritability of number sense by testing how accurately teenagers can non-symbolically discriminate the number of dots in dot arrays.[52] This time, they tested 836 pairs of identical and 1,422 pairs of fraternal 16-year-old twins. Genetic analysis of the twin data revealed that the number sense is modestly heritable (32%), with individual differences being largely explained by differing environmental influences (68%). The number sense's modest heritability may come as a surprise, but the authors point out that evolutionarily useful traits are not always necessarily heritable. Fear, for example, is considered an evolutionary useful trait, but individual differences in fear processing are mostly explained by environmental influences.[53] The authors suggest that in the case of the number sense, one set of genes may provide a blueprint for the development of this ability across many species, whereas a different set of genes may contribute to variation in the trait between individuals in any population. Such variation-causing genes may, for example, affect perceptual processes, speed of processing, and other cognitive functions relevant to perform numerosity estimations.

When talking about genetic differences in learning, the putative gender differences are usually of great interest to the public and create some media hype. It is no secret that males and females differ in certain aspects of body anatomy and function. Of course, an individual's sex has a genetic basis, namely differences in sex chromosomes (gonosomes): in all mammals, females have two X chromosomes (XX), whereas males have an X and a Y chromosome (XY). Many rumors exist that one gender may have superior number skills compared to the other. Usually, boys are assumed to have better math skills. However, scientific studies find no or only minimal sex differences in number skills. In one of the most extensive meta-analyses of gender differences in the USA, no significant gender difference for numbers and arithmetic could be observed.[54] Moreover, the same genetic and environmental factors affect both males and females in similar ways in both typical and low achievers.[55,56] Despite all the gossip and prejudices about gender differences, there really isn't much to it. Males and females are similarly genetically endowed to perform mathematics.

However, this does not mean that sex chromosomes would be unimportant for number skills. If the number of sex chromosomes does deviate from the normal pair due to genetic accidents during early development, pathological conditions can arise that cause arithmetic deficits. One such condition is Turner's syndrome, a condition in which a female is partly or completely missing an X chromosome (X monosomy). People with Turner's syndrome suffer from a variety of anatomical and functional aberrations but have normal intelligence levels. Turner's syndrome is linked with dyscalculia, but not with any other specific learning disorders.[57] Another genetic abnormality resulting in math deficits is fragile X syndrome. It is caused by a genetic mutation on the X chromosome and mostly affects males, leading to significant intellectual impairment. In both conditions, atypical fMRI activation was found in the IPS and wider parieto-frontal networks.[58,59,60] Turner syndrome people, for example, showed an abnormal intraparietal hypo-activation (decreased BOLD activation) during calculation. In addition, the right IPS showed abnormal length, depth, and geometry that point to a disorganization of this region in Turner syndrome. Because this genetic form of developmental dyscalculia can be related to both functional and structural anomalies of the right IPS, a crucial role of this region in the development of arithmetic abilities is further strengthened.

But let us return to the question of how much of arithmetic learning ability is caused by genes compared to the environment. While genes mediate a stable endowment for learning abilities, the extensive twin study by Yulia Kovas and coworkers from 2007 also pointed out interesting differences in learning abilities that cannot be explained by genes, even between identical twins.[61] If genes cannot explain these differences, then the environment must be causing them. According to the study, it is the varied environment of twins that largely contributes to change. In other words, when identical twins show differences in math learning performance, it is because one twin of the pair experiences a somewhat different environment than the other. And not only do varied environmental influences on learning abilities make two children in the same family different from one another, but they also make children different from themselves at different ages.

Differing environments are a major mystery for learning abilities and disabilities, because usually twins live in the same family, attend the same school, and are often even in the same classroom. If the family environment is identical, then educational influences might have their greatest impact on remediating discrepant performances among learning abilities. This insight provides a unique opportunity to help children at risk of developing learning disabilities. As Kovas and coworkers point out:

> The most important benefit of identifying genes that put children at risk for developing learning disabilities is that the genes can be used as an early warning system to predict problems before they occur..[62]

Whatever the exact causes, the finding that the environment plays a role in math skills means that early interventions, particularly those informed by neuroscience findings, can ameliorate performance in low achievers. Over the past few years, a number of adaptive software programs, or computer games, have been designed that target the inherited ANS in the IPS, which seems to support early arithmetic. All of these games share a design that improves precision in discriminating between sets of dots. While such games significantly improve number comparison, this effect unfortunately does not seem to generalize to counting or arithmetic.[63]

But there is hope for dyscalculic children, because the brain can be tutored. In a study by Teresa Iuculano and Vinod Menon from Stanford University, the researchers wanted to know if and how specific math training could remedy mathematical learning disabilities.[64] After they had acquired functional brain scans from groups of dyscalculic and normally developing children, all the children took part in an eight-week math tutoring program. This program was geared at strengthening arithmetic knowledge in such topics as cardinality and relations between addition and subtraction. Indeed, the authors found a normalization of math performance in the dyscalculic group. Moreover, they also found that tutoring elicited extensive functional brain changes in dyscalculic children. Their brain activity became equal to those of typical achievers. The prominent differences in brain activation between dyscalculic and normally developing children in prefrontal, parietal, and temporal cortices that were evident before tutoring vanished entirely after tutoring. In other words, math tutoring resulted in significant reductions of widespread fMRI overactivation in brain areas important for numerical problem solving. Intervention has beneficial effects, and one of the important research agendas for the future is to explore exactly which strategy is most helpful for dyscalculic children.

Part VI   The Brain Departing from Empirical Reality

14 The Magical Number Zero

> What is crooked cannot be straightened; what is lacking cannot be counted.
> (Ecclesiastes 1:15)

14.1 A Special Number

So far, we have dealt exclusively with the positive natural numbers. But we have neglected one number that is usually associated with them, namely zero. Zero is indeed a very special number. In terms of mathematics, zero is somewhat of a double-edged sword, simultaneously making some mathematical aspects possible in the first place, but others a bit harder.

The pleasant side is very obvious. Arithmetic works much better if you think of zero as a number. With only positive integers, we can do calculations like three minus two, which yields an obvious, positive result. But what about three minus three? It is an arithmetic operation we can actually do, and it demands an answer that only zero can provide. We need zero in order to represent "nothing." But that's not all zero does for our calculation abilities. It is also the doorway to negative numbers. On a mental number line for natural numbers, zero is the smallest, and all positive integers follow on a nicely ordered scale in one direction. But what about the numbers on the other side of the number line? Without zero as an anchor point, there are no negative numbers. This is a big leap into a mental area that can be called transcendental, because it leaves real things behind. Our lack of representation of negative numbers represents this; we can't tangibly have "negative three" of something. Still, negative numbers are imperative for all sorts of mathematics. Because of this, we can develop a full-fledged number system, and all of this is because of the number zero.

However, the irritating side of zero becomes visible when using it in calculations. Zero causes unprecedented arithmetic paradoxes. Whenever zero is added to or subtracted from a number, this number remains unchanged. Multiplying any number with zero results in zero. This is remarkable, but not yet paradoxical. But try dividing by zero—it is not possible. Whenever you punch numbers into your calculator and try dividing it by zero, the result is "ERROR." And just as divisions by zero are not possible, zero to the power of zero (0^0) is also not a possible operation. The bottom line is: We can't perform some standard arithmetic operations with zero.

Another fundamental issue is that some operations with zero are not reversible. With positive integers, multiplications are fully reversible. For instance, $3 \times 2 = 6$. The multiplication doubles the number, and if it gets divided by the same number, the multiplication is undone. That's a reversible operation. With zero, this doesn't work. Multiplying by zero is not a reversible operation, as the result is always zero and division becomes impossible. Somehow, in these calculations, logic breaks down and paradoxes proliferate. And the reason is the zero.

Zero is a very demanding numerical concept, indeed, both from a mental and a neurobiological point of view. Why is it so special? After all, any number is an abstraction. The number 2 is the common property of all sets containing a pair, the number 3 of all sets that contain a triple, and so on. For any set, the brain needs to abstract the empirical properties of the elements comprising it, and that is cognitively demanding. But challenging as it is, positive integers at least correspond to real items that can be enumerated. As a result, we first learn to count small numbers of items and later use this counting procedure to comprehend infinite positive numbers.[1,2]

With zero, however, it is entirely different. Zero is an empty set, "nothing," and has no elements that can be counted. In fact, it is defined by the absence of any countable item. Understanding that zero is nonetheless a collection (even if empty) and a numerical concept requires ultimate abstract thinking. "Nothing" needs to become "something." The absence of elements needs to become a mental category, a mathematical object. This demands thinking about objects that can no longer be experienced with the senses in the real world.

As a reflection of this mental challenge, zero is a true latecomer in human history, development, evolution, and neural processing. As shown in subsequent sections, it took a long stretch of human history for zero to be recognized and appreciated.[3,4] This cultural hesitation is mirrored in a protracted ontogenetic understanding of numerosity zero in children. In the animal kingdom, only species with advanced cognitive skills exhibit a rudimentary grasp of zero numerosity. Finally, for a brain that has evolved to process sensory stimuli ("something"), conceiving of empty sets ("nothing") as a meaningful category requires abstraction beyond what is perceived and experienced. Neuronal recordings in monkeys provide a first hint of how this is accomplished in the primate brain.

14.2 Zero in Human History

The number zero is a surprisingly recent development in human history. There was a time, not too long ago, when mankind only possessed positive natural numbers. The word, the symbol, and the very concept of zero had not yet been discovered. Even today, the origin of zero is still mysterious, partially because it was intially used in different ways and with different meanings.

Zero was first used as a sign to indicate an empty place in number notation. Empty places are of great significance in a special type of number notation system that we also use today: the positional notation systems, also called the place value system. The positional notation system is a universal notation system for infinite numbers. In a positional notation system, one and the same numeral adopts different numerical values according to its position in a record—for example, tens versus hundreds depending on a number's position.

An effective place value system requires a sign to denote the absence of a positive digit, and for this purpose, zero was first used.[5] Without a sign that indicates that a position remains empty, any place value number record is highly ambiguous. In China around 200 B.C., for instance, rods were placed on the columns of a counting board to represent digits in a base-10 decimal system and perform calculations. Each column represented a decimal order: the first and rightmost column was for units, the second for tens, the third for hundreds, and so on. Zeroes were represented by an empty space.[6] The problem with this notation is that an entry such as ||| | may represent any one of several numbers: 31, 301, or even 310, as well as others. In order to avoid this ambiguity, it is essential to have some method of representing the void—that is, a sign for an empty column. One must have a zero to have a way to say "don't put any number in this particular position." The discovery of zero as a positional placeholder is therefore entwined with the invention of positional notation systems.

The number notation systems in the West originally did not rely on place values, and therefore had no need for zero. In ancient Egypt, Greece, and Rome, specific number values were denoted by dedicated signs. In the Roman numeral system, for instance, units up to three were denoted by corresponding amounts of I, five was represented by V, 10 by X, 50 by L, 100 by C, 500 by D, and 1 million by M. Such signs were simply added to represent higher or intermediate numbers, without the need for zero. More importantly, calculations couldn't be performed properly with these cumbersome systems; mathematics couldn't progress without zero.

Zero as a sign for an empty column in positional notations appears to have first been used around 400 BCE in ancient Mesopotamia by the Babylonians,[7] who used two slanted wedges as a placeholder in a base-60 (sexagesimal) system (figure 14.1A).[8] Slightly later, the Greeks used a circle (∘) as a placeholder, probably based on Babylonian influences.[9] Independent of the Babylonians, the ancient Mayan civilization introduced a shell-like sign as zero in a base-20 (vigesimal) positional numeral system around the beginning of the Christian era[10] (figure 14.1B). "Nothing," the absence of a numeral in a positional notation system, was now first realized as a meaningful category, a void placeholder, and denoted by a sign. So zero became an irreducible component of this notation system. Importantly, however, these signs for zero invented by the Babylonians and Mayans had no own numerical value. Therefore, they cannot be interpreted as representations of the *concept* or the *number* zero.[11]

But once mankind had a sign to denote emptiness in a notation system, it was only a small step to realize that this sign represents an empty set. This insight of an empty set is important, because only with it does zero acquire a quantitative meaning. Zero becomes a set that is empty and is adjacent to a set that contains only one element. Now this sign denotes zero as part of a mental number line. When exactly this discovery was made in history is difficult to determine. However, this realization marks a cognitive turning point, because it requires the insight that even if a set is empty, it still is a quantitative set.

Once zero was associated with a quantitative meaning, with time it turned into a real mathematical concept: the number zero. Zero first

A Babylonian numerals

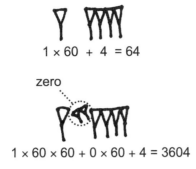

B Mayan numerals

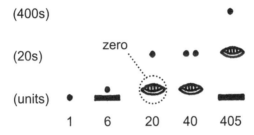

C Gwalior inscription, India

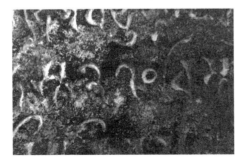

Figure 14.1
Signs of zero. (A) Babylonian base-60 (sexagesimal) number system, read from left to right. Because they used a sexagesimal positional notation number system, a new place value begins at 60. The number 64 (*top*) would be written with one wedge to the left (1 sixty), and four to the right (4 ones). The Babylonians used an oblique and superscript double wedge as sign for the empty place. In the second number (*bottom*), the second place is empty, so this sexagesimal number in decimal writing is 3,604. (B) Mayan base-20 (vigesimal) positional numeral system, read from bottom to top. After the number 19, larger numbers were written in a vertical place value format using powers of 20. (C) Ninth-century inscription in Gwalior, India: The number 270 is in the middle of the image. Photograph courtesy of Alex Bellos.

became associated with an elementary concept of number in India around the seventh century. The first written record of the use of zero as a number in its own right comes from the Indian mathematician Brahmagupta.[12] In his book *Brahmasphutasiddhanta* (628 CE), written in Sanskrit verse, Brahmagupta was the first to provide rules to compute with *zero*, a clear sign of it as a number. For instance, he writes, "When zero is added to a number or subtracted from a number, the number remains unchanged; and a number multiplied by zero becomes zero."[13] He even thought about division. But there, he got it wrong and and erroneously claimed that "zero divided by zero is zero."[14]

Zero as a true number used for calculations also appeared in the so-called Bakhshali manuscript, one of the oldest mathematical texts from India. Recent radiocarbon dating showed that the manuscript consists of three birch-bark folios of different ages dated to roughly 300 CE, 700 CE, and 900 CE. The treatise on arithmetic have been written at the time of the latest of the manuscript's leaves, around 900 CE.[15]

The numeral for zero as we know it today ("0") first appeared in an Indian inscription on the wall of a temple in Gwalior (Central India), dated 876 CE.[16] A small circle in engraved numbers denotes the measures of a flower garden made as a gift to the temple (figure 14.1C). It is soothing to know that the numeral for zero first served such a peaceful purpose. Now, zero was no longer a humble empty set, but a number that became part of a complex number theory and a combinatorial symbol system, enabling arithmetic operations. In utter admiration about this discovery, Mike Hockney writes

> The Indians—by embracing zero—created a mathematical revolution. Zero is India's greatest achievement and constitutes one of the greatest leaps forward in human history.[17]

Zero was called *sunya* ("empty") in Sanskrit. When the Arabs, the intermediaries between India and the West, became acquainted with the zero in the ninth century, they translated the Indian name *sunya* literally into the Arabic *as-sifr* ("the empty"). In the West, zero arrived circa 1200 CE when the Italian mathematician Fibonacci (a.k.a. Leonardo of Pisa) brought it back from his travels in North Africa, along with the rest of the Indo-Arabic numerals 1 to 9 and a base-10 (decimal) positional numeral system. When the West took on the numeral system, it also adopted the symbol and the name for zero from Arabic, and in the thirteenth century transformed the name into the Latin forms *cifra* and *cephirum*. The two Latin words then gradually worked their way into idioms. In Italian, *cephirum* was changed to *zero,* which was to be maintained in contemporary French and English. *Cifra* in fourteenth-century France became *chiffre,* and later in Germany became *Ziffer.*

After arriving in the West, zero created quite a bit of confusion and insecurity in the minds of thinkers in the Middle Ages. Initially, theologians, philosophers, and mathematicians were reluctant to deal with zero. "Nothing" was chaos, the state out of which God created the world. The ancient Greek χάος (*chaos*) originally meant "yawning void" and described a condition of utter disorder that contrasts against κόσμος (*kósmos*), "world order." No wonder that the "nothing" from which the world escaped was not really something they wanted to deal with. The symbol for the mysterious zero was therefore also used as a secret sign. The English verb *"to decipher"* that similarly also exists in other languages with the meaning to unlock a secret, remains as a remembrance of this usage. It is difficult to imagine such irrational fear nowadays, but one has to remember that during the Middle Ages, education was overseen by the church all over Europe. The church's doctrine therefore permeated all disciplines, including mathematics. As late as the fifteenth century, zero was regarded as *umbre et encombre*, "dark and encumbered," and its German name, *Null*, is derived from the idea that it is *nulla figura*—not a "real" number.[18]

However, merchants, in particular, soon learned to appreciate the power of the place value notation system, including zero, to do business. Interestingly enough, in 1299, zero was temporarily banned in Florence, along with all Arabic numerals, because the new numerals were said to encourage fraud. And one can see why: simply adding zeros to an entry in an account book makes it look like there is 10 or even a hundred times more money. As a remnant of this worry over falsified written records, we still write out numbers with number words, not numerals, when writing on bank checks. And because zero was also the gateway to negative numbers, the concept of debt and money lending obtained unprecedented validity.

Ultimately, however, fascination for zero prevailed. Although it itself is nothing, it miraculously changes numerical values in the place-value notation system. As Alex Bellos, author of the book *Alex's Adventures in Numberland,*[19] wonderfully describes:

The Renaissance was really sparked by the arrival of the Arabic number system, containing zero. And when that happened, the black and white world of arithmetic suddenly became glorious and technicolour.[20]

The magic of zero not only enthralled mathematicians but also greatly fascinated artists. The great poet William Shakespeare (1564–1616), for instance, was heavily influenced by the idea of nothingness, and zero played a prominent role in his work. In his play *King Lear* alone, he used the term "nothing" approximately 40 times in different contexts. The tragedy unfolds from "nothing" when King Lear decides to divide his kingdom among his three daughters, and his youngest daughter, Cordelia, who loves him most, cannot find the words to articulate her love ("my love's more ponderous than my tongue").[21]

Lear: What can you say to draw
A third more opulent than your sisters? Speak.
Cordelia: Nothing, my lord.
Lear: Nothing?
Cordelia: Nothing.
Lear: How? Nothing will come of nothing. Speak again.
(*King Lear*, Act 1, Scene 1)

In Shakespeare's time, the idea of nothingness as a number embedded in the new number notation system was a rather new idea that had gradually worked its way from Italy to England.[22] In the second half of the sixteenth century, William Shakespeare belonged to the first generation of children in England to have learned about zero[23] from Robert Recorde's *The Ground of Artes* (1543).[24] From this standard arithmetic textbook of the time, Shakespeare also knew the arithmetic meaning of a number's "place" in a place-value system.[25] Shakespeare uses this place-value idea of zero in his plays, for instance when Polixenes in *The Winter's Tale* explains:

And therefore, like a cipher [zero]
(Yet standing in rich place), I multiply
With one "We thank you" many thousands more
That go before it
(*The Winter's Tale*, Act 1, Scene 2)

Shakespeare seems to be the first poet to use Arabic numeral "0" metaphorically. He does so not only with zero as a place holder, but also as a sign to denote "a mere nothing" in a quantitative meaning. The zero becomes

The Magical Number Zero 291

a signifier of utter abandonment and annihilation when the Fool in *King Lear* speaks:

> Now thou art an O without a figure. I am better than thou art now. I am a fool. Thou art nothing.
>
> (*King Lear*, Act 1, Scene 4)

Here Lear is compared to an isolated zero that simply indicates empty quantity, not a zero that could increase the value of a number ("without a figure").

Just as numbers inspired poets like Shakespeare, mathematicians also see beauty in zero and the role it plays in arithmetic equations. One of these astonishing equations is Euler's identity,

$$e^{i\pi} + 1 = 0,$$

named after Swiss mathematician and physicist Leonhard Euler (1707–1783). Euler's identity is often judged as the most beautiful equation in mathematics. Keith Devlin, a British mathematician at Stanford University writes

> Like a Shakespearean sonnet that captures the very essence of love, or a painting that brings out the beauty of the human form that is far more than just skin deep, Euler's equation reaches down into the very depths of existence.[26]

The American theoretical physicist Richard Feynman (1918–1988) called it a "jewel" and "one of the most remarkable, almost astounding, formulas in all of mathematics."[27] The reason for this utter admiration is that this equation contains many basic concepts of mathematics—once and only once—in a single expression, such as Euler's number e, the imaginary unit i; the constant π, one and zero. Mike Hockney called Euler's identity "The God Equation."[28] In the introduction to his book, he praises Euler's equation:

> If there were such a thing as a Creator, he would be a mathematician, and the "God Equation" would be the equation he invoked to make the world. ... One Equation to rule them all, One Equation to find them, One Equation to bring them all and in the darkness bind them.

When we hear about zero and the positional notation system, we usually think of the decimal, base-10 system that is widespread in the West. However, this is just a cultural convention; other base systems may be equally important. The hexagesimal, base-60 system of the ancient Babylonians was used until relatively modern times, particularly in astronomy. Indeed, we still use it to measure angles by degrees, minutes, and seconds and to measure time by hours, minutes, and seconds.[29]

However, our technological world wound be utterly inconceivable without the modern binary number system, a base-2 system, as devised in 1679 by German mathematician Gottfried Wilhelm Leibniz,[30] who was often considered the last universal scholar. In a binary system, all numerical values are expressed using only two different symbols: 0 and 1. A binary digit, or bit, can have only one of these two values. The advantage of the binary system is that it can be easily implemented in electronics. Digital electronic circuitry adopts only two voltage states termed "high" or "low," or simply 1 and 0. Therefore, the binary system is at the heart of all modern computers and digital devices. The world's first programmable computer, the Z1, was a binary electrically driven mechanical computer designed by German engineer Konrad Zuse (1910–1995) in 1936 and built by him in 1938. Because of this and the later technological explosion of the years, we are, quite literally, surrounded by ones and zeros.

14.3 Development of Zero-like Concepts in Children

The cultural hesitation to appreciate zero as a number is mirrored in a protracted understanding of zero relative to positive integers in developing children. Positive integers are already grasped early in life. As we have seen, newborns and infants have the capacity to represent the number of items in a set and to solve basic addition and subtraction operations in puppet shows that demonstrate the calculations. Surprisingly, however, eight-month-old infants do not discriminate operations of $1 - 1 = 1$ versus $0 + 1 = 1$.[31] This has been interpreted as the inability of infants to conceive of a null quantity, even though they are able to understand the absence of entities.[32] Interestingly, children during their first years of development then pass the same stages that we have seen in culture: first zero as "nothing" versus "something," next as an empty set, and finally as a number.[33]

Beginning at around 3 years old, children start to realize that "nothing" can be a meaningful category that is different from all other categories that comprise things. In a counting task in which children are asked to count backward as long as possible from a given number of cubes, they comprehend that the condition arrived at by taking away the last cube is "none," "nothing," or "zero." Thus, zero becomes a sign for absence. However, zero is not yet integrated in their quantitative knowledge of other small integers. For example, when asked, "which is smaller, zero or one?" children often insist that "one" is smaller.[34] This finding has also been reported in a study in which children were asked to distribute cookies, note the quantity of cookies on a Post-It note, and read their notation later.[35] Interestingly, the most common way of denoting zero was to leave the Post-It note blank. When asked about leaving the paper blank, children frequently said "it means no cookies," or "because there are no cookies." The blank paper serves as an iconic representation of absence, much like the empty column denoting an empty position in a positional notation system.

The next leap forward is for children to understand that zero is a quantity and can be placed on a numerical continuum along with other positive integers. When faced with tasks employing number words and numerals, children realize that zero is the smallest number in the series of (non-negative) integers by about age 6.[36] However, since verbal counting in itself is already demanding for children, the difficulty may be more related to dealing with symbolic number tasks in general than an actual lack of understanding of zero. Maybe even younger children could comprehend empty sets as a void quantity if they don't have to use symbols. This is exactly what Dustin Merritt and Elizabeth Brannon from the University of Pennsylvania wondered when they devised a more direct, non-verbal numerosity ordering test.[37] They came up with the idea to test 4-year-old children with numerosities represented by the number of dots in displays. In this numerical ordering task, children were required to select the quantitatively smaller of two numerosities by touching it on a touch screen with a finger (figures 14.2A and 14.2B). As was expected based on earlier studies, Merritt and Brannon found that children were more accurate at ordering countable numerosities compared to ordering pairs that contained an empty set. In addition, however, they also found evidence for a rudimentary grasp of empty sets in these children. To figure out whether children treat empty sets as quantitative stimuli rather than a category beyond numbers, the numerical distance effect was exploited. This means that if children represent empty sets with a quantitative meaning in relation to countable numerosities, they should have the most difficulty to discriminate empty sets from the smallest numerosity 1. In other words, they should mix up the empty set with sets of one more often than with other sets. This is precisely what happened. Children who could reliably order countable numerosities

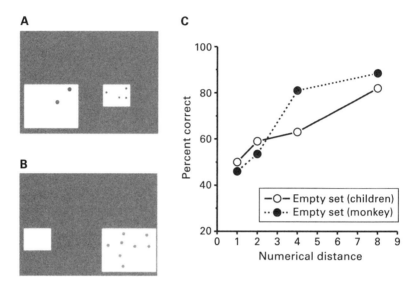

Figure 14.2
Small children and monkeys represent empty sets. (A) Children and monkeys learned in standard trials to order numerosities (dot collections on white background) in ascending order by selecting them on a touch sensitive screen. In this example image, they first had to select numerosity 2 (*left*) and then numerosity 4 (*right*). (B) In intermingled empty-set probe trials, one of the sets contained no item. Again, both sets had to be touched in ascending order. (C) Performance accuracy of 4-year-old children and two rhesus monkeys for empty sets as a function of numerical distances of 1, 2, 4, and 8. For instance, accuracy at distance 1 represents performance for empty sets and numerosity 1, accuracy at distance 2 represents performance for empty sets and numerosity 2, and so on (after Merritt et al., 2013).

indeed showed a numerical distance effect with empty sets (figure 14.2C). This suggests that around 4 years of age, children begin to include a nonquantitative representation of "nothing" as an empty set into their mental number line.

Later in life, children from ages 6 to 9 acquire a concept of zero as a number. They are increasingly able to correctly affirm and deny simple algebraic rules, particularly if they involve zero (such as, "If you add 0 to a number, it will be that number"). From age 7 onward, children typically understand three general rules, namely: $0 < n$, $n + 0 = n$, and $n - 0 = n$. Their ability to justify and reason about such rules, again with zero, also develops dramatically. Henry Wellman and Kevin Miller from the University of Michigan concluded that young elementary school children possess some understanding of simple algebraic rules, and that zero holds a special status in fostering their reasoning about such rules.[38]

In conclusion, learning the meaning of zero is hard work for children. It may therefore not be entirely surprising to realize that zero remains a special number even for adults. Psychophysical experiments indicate that the representation of zero is based on principles other than those used for positive integers. For instance, while adult humans' reading time for numbers from one to 99 grows logarithmically with the number magnitude, zero takes consistently more time to read than expected based on the logarithmic function.[39] In parity judgment tasks, zero is not judged as a typical even number and is suggested not to be investigated as part of the mental number line.[40,41] Even educational experts may not do better. When elementary school teachers were tested on their knowledge about zero, they exhibited confusion as to whether or not zero is a number, and showed stable patterns of calculation error using zero.[42] This clearly shows that zero stands out among other integers, even in adulthood.

14.4 Zero-like Concepts in Animals

Do animals have zero-like concepts? They do, but it is important to take a closer look. Just as it is important to differentiate the developmental progression children show when they develop an understanding of zero, it is necessary to classify the types of zero-precursors in animals. Before children understand that zero refers to the null quantity of an empty set, they need to grasp that "nothing" is a behavioral concept that opposes "something."

Indeed, animals can be trained to report not only the presence, but also the absence of stimuli ("nothing"). Rhesus monkeys, for instance, can be taught to press one of two buttons to indicate the presence or absence of a light touch,[43] or to report the presence or absence of a faint visual stimulus.[44] Clearly then, animals are able to represent "nothing" not just as the absence of a stimulus, but also as a behaviorally relevant category.

The next question, then, is whether animals can also ascribe a quantitative meaning to "nothing" as an empty set. In many earlier studies, the situation is far from clear. Animals have been trained to associate set sizes with visual or vocal labels, including a sign for "no item," in an attempt to mimic "symbolic" number processing. The previously mentioned African grey parrot Alex, trained by Irene Pepperberg, used human speech sounds to report the absence or presence of same/different relationships between objects.[45] For instance, when asked "What's different?" when faced with two identical objects, he correctly responded with "none."[46] However, he failed to respond similarly when asked how many items were hidden under two empty cups in a follow-up study.[47] Maybe the parrot used "none" to signify the absence of object attributes rather than the absence of the objects themselves. Alternatively, his response may have simply indicated a failed search. In the absence of reports of a numerical distance effect, it would be premature to attribute a quantitative meaning to the parrot's response.

The data situation is not better in non-human primates, and similar interpretational limitations arise for monkey and ape studies that were trained to associate sets with specific visual signs. A female chimpanzee trained by Sarah Boysen at Ohio State University in Columbus learned to match a 0-sign with an empty tray.[48] Moreover, when the chimp was shown a pair of numeral signs, she was able to select the sign that indicated the arithmetic sum of the two signs (e.g., shown 0 + 2, she selected 2). Importantly, however, the chimp would also have succeeded if she had simply ignored the 0-sign that denoted nothing. Whether she learned to interpret the zero-sign in a quantitative way therefore remains an open question. In another study, squirrel monkeys were trained to choose between all possible pairs of numerals 0, 1, 3, 5, 7, and 9 to receive that many peanuts in return.[49] The monkeys always chose the larger sum between two sets of numeral signs (e.g., 1 + 3 versus 0 + 5). However, the easiest interpretation of why the monkeys succeeded in this task is that they chose the sign that was associated with the greater amount of reward and therefore went for hedonic value. Obviously, discriminating reward magnitude is not the same as choosing between numerical quantities. But even if the monkeys performed numerical operations in this task, the 0-sign most likely meant "nothing" to the squirrel monkeys rather than the null quantity.

The famous female chimpanzee "Ai," studied by Tetsuro Matsuzawa at Kyoto University in Japan, is yet another aspirant for an understanding of a null quantity. As shown in chapter 8, Ai learned to successfully discriminate numerosities and signs associated with numerical values.[50] She had also been trained to match certain numbers of dots to signs using a computer-controlled setup.[51] And finally, she also mastered matching blank squares containing no dots to the sign for zero. Had Ai learned a concept of zero quantity that she could immediately transfer to a related numerical task? Unfortunately, no, because when she was later tested on her ability to order the signs in ascending order, she failed to transfer the zero sign from the matching task (cardinal domain) to the ordinal task without further training.[52]

The aforementioned studies allow only limited conclusions about zero-like concepts in animals. First, it cannot be ruled out that animals associate the 0-sign with "nothing" instead of with "null quantity." Second, even though animals learn to associate the absence of items with arbitrary shapes, transfer to novel contexts (e.g., from cardinal to ordinal tasks) is severely limited.

However, one behavioral effect that turned out to be most informative in children has not been addressed so far in animals: the numerical distance effect. Just as with children, evidence of a distance effect would demonstrate that animals represent empty sets quantitatively with other numerosities along a "mental number line." More precisely, empty sets should be more often mixed up with numerosity one than with any larger numerosity. Elizabeth Brannon's laboratory that had already shown a distance effect for empty sets in 4-year-old children, demonstrated the same effect in rhesus monkeys that were proficient at matching and ordering dot numerosities.[53] The monkeys were trained to touch pairs of variable dot numerosities shown on a touch screen in ascending order, for instance, first two dots, then five dots (figure 14.2A). Obviously, the monkeys understood the ordinal sequence of numerosities, but would they spontaneously incorporate the empty set as the smallest numerosity in this sequence? To find out, transfer trials with empty sets as an occasional member of the numerosity pairs were presented to the monkeys (figure 14.2B). Because they were arbitrarily rewarded for responses to numerosity pairs containing empty sets, they did not have a chance to learn the correct response. This turned out to be an unnecessary concern, because the monkeys were immediately able to order empty-set stimuli of variable appearance as the smallest numerosity, thus indicating a conceptual understanding of null quantity (figure 14.2C). And not only did the monkeys show a numerical distance effect with empty sets in ordering tasks that examine an ordinal grasp of numerical quantity, but also in numerosity matching tasks that tap into a cardinal understanding. My PhD student, Araceli Ramirez-Cardenas, and I replicated the same effect for empty sets when we trained rhesus monkeys to perform a visual delayed match-to-numerosity task for numerosities zero to four in preparation for neurophysiological investigations mentioned later.[54] Both monkeys mistakenly matched empty sets to numerosity one more frequently than to larger numerosities. Distance effects for empty sets were also observed in rhesus monkeys trained to perform numerical additions or subtractions on visual dot displays in the laboratory of Hajime Mushiake at Tohoku University, Sendai, Japan.[55] I already reported about this study in chapter 10. Macaque monkeys were trained to assess the numerosity of a target display that showed zero to four items and, by manipulating a hand device, add or subtract items in a second display to match the target numerosity. As an indication of a numerical distance effect with empty sets, the monkeys erroneously chose quantities two and zero more often than larger numerosities when the target quantity one was presented.

While we may expect that non-human primates, our animal cousins on the tree of life, grasp the concept of an empty set, we might be surprised to learn that honeybees also belong to the elite club of animals that comprehend the empty set as the conceptual precursor of zero. Earlier in the book, I reported an experiment by Scarlett Howard and Adrian G. Dyer from the Monash University in Australia in which honeybees were shown to rank numerical quantities according to the rules of "greater than" and "less than."[56] However, what I concealed so far is that these insects also could extrapolate the less-than rule to place empty sets next to one at the lower end of the mental number line. You may remember from chapter 3 that one group of bees was trained on the less-than rule and rewarded for landing at the display presenting fewer items. The bees learned to master that task with displays consisting of one to four items. Surprisingly, the bees obeying the less-than rule spontaneously landed upon an occasionally inserted and unrewarded displays showing no item (i.e., an empty set). In doing so, bees understood that the empty set was numerically smaller than sets of one, two, or more items. Further experiments confirmed that this behavior was related to quantity estimation and not a product of learning history. What is even more remarkable is that the bees' performance accuracy improved as the magnitude difference between two respective numerosities increased. Bees had a hard time judging whether an empty set was smaller than one, but were progressively better when larger numbers had to be compared to empty sets. With this, the bees demonstrated the numerical distance effect with empty sets, a hallmark of number discrimination. These series of experiments therefore demonstrate that bees grasp the empty set as a quantitative concept.

Collectively, these studies provide clear evidence that animals, when faced with a range of numerosities, can judge the empty set as a quantity that is related to other numerical values. I am quite sure that we will hear more of these empty-set representations also in other animal species in the future. Is there hope that someday animals may even transcend empty set representation in order to arrive at representations that satisfy number theory? This would require animals to comprehend zero as a true number symbol embedded in a combinatorial system. In my opinion, there is no evidence that any animal will ever grasp true numbers or linguistic symbols the way we humans do. By this logic, the number zero therefore will remain a uniquely human concept. But this should not come as a disappointment. Rather, it is fascinating to see that the earlier and more basic concepts of empty sets we first need to understand as children are already within the reach of animals. The foundations or precursors of a zero concept can therefore be investigated in monkeys, and in the next part of this chapter we shall get a first glimpse of how neurons in the brain deal with this demanding concept of "nothing" and the empty set.

14.5 Neuronal Representations of "Nothing" and Empty Sets

So far, I have exclusively discussed behavioral representations of zero-like concepts. But how does the brain deal with such concepts? Clearly, representing "nothing," empty sets, or the number zero must be a challenge for the brain. If anything, the brain and its sensory neurons have evolved to always represent something—an obstacle, a predator, a mate—all of them can be perceived because of the sensory energy they radiate that is detected and processed in the brain. And in the very same way, a collection of items is realized by the energy of the stimuli that constitute this set. In the absence of stimulus energy, neurons are inactive and at a resting state.

However, cognitively advanced animals can learn that the absence of stimulation is also a meaningful condition and relevant for behavior. And like other categories that are learned based on shared properties of elements, the lack of sensory stimulation can also become a meaningful category and as such become encoded by neurons. Indeed, my PhD student, Katharina

Merten, and I found evidence that the absence of a stimulus is actively represented by neurons in the frontal association cortex.[57] At the behavioral level, we made sure that "nothing" was an important category by training rhesus monkeys to report both the presence and the absence of a stimulus (figure 14.3A). This was done in a simple detection task in which half of the trials were visible stimuli that were flashed on a monitor, whereas in the other half of the trials, no stimulus at all was presented. The intensity of the stimuli was varied so that the faint ones were barely visible. This was a difficult task for the monkeys, and they were often uncertain whether a stimulus was shown or not. But this was precisely what we wanted, because if a sensory stimulus is faint enough that neurons have difficulty sensing it, the monkeys are forced to "make up their minds" and subjectively judge whether they had or had not seen a barely visible stimulus. As a result, the monkeys respond from trial to trial in opposite ways to a faint stimulus that had the same physical energy. The internal status of the monkeys determined whether they judged a stimulus as being present or absent, allowing us to investigate how neurons encode both the "present" and "absence" categories irrespective of the stimulus energy, which was identical in these situations. Before the monkeys were prompted by a rule cue on how to respond to their decision, they first had to wait for three seconds. During this waiting period, they had to memorize their decision, but could not yet know how they would report their decision. This allowed us to dissociate the decision about stimulus absence or presence from motor-preparation processes.

When we recorded from the PFC while the monkeys performed this task, we found two groups of neurons that signaled their decisions about the stimulus even before they could report it. One group of neurons increased their discharge rates whenever the monkeys later reported to have seen the stimulus (figure 14.3B). Such a response might be expected for neurons that respond to the energy from a present stimulus, and similar responses in the frontal lobe have also been reported for monkeys responding to touch.[58]

Surprisingly, however, a second group of neurons increased their discharges whenever the monkeys decided that they had *not* seen a stimulus, but remained silent whenever the monkeys voted for a "presence"[59] (figure 14.3C). This finding was entirely unexpected, but the neuronal responses were highly correlated with the monkeys' subjective reports. Analyses of

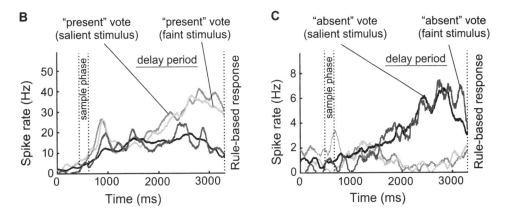

Figure 14.3
"Something" and "nothing" are represented in the brain. (A) In behavioral tasks, monkeys report that they have seen a salient visual stimulus (*left*) or that they have not seen such a stimulus if none was shown (*middle*). If a very faint stimulus is presented that can hardly be seen (*right*), the monkeys are uncertain and sometimes vote for "present," but other times for "absent." (B,C) Neurons in the prefrontal cortex of performing monkeys signal the monkey's decision about whether or not a stimulus was present. The neuron in (B) fires more action potentials after the monkey made a "stimulus present" decision and is waiting during a delay period before it is prompted to report its decision. The firing rate is equally high for salient and faint stimulus conditions, which emphasizes that this neuron encodes the subjective judgment of the monkey, not the stimulus intensity. The neuron in (C) also represents the monkey's subjective "present versus absent" vote, but it increases its firing rate whenever the monkey judges that no stimulus, or "nothing," was on the screen (B, C from Merten and Nieder, 2012).

trials in which the monkeys failed to detect a faint stimulus showed that the activity of these neurons predicted the monkeys' judgments even before a response could be planned. Notably, the active coding of the "stimulus absence" decision was not a visual response, but emerged during the delay phase when the monkey needed to decide whether it had seen the stimulus. The best explanation for this finding is that the categorical stimulus-absence signal in the brain arises during a cognitive processing stage following the sensory period. The new discovery therefore was that behaviorally relevant "stimulus absent" decisions are not encoded by resting-state neuronal responses. Rather, the brain seems to internally translate the lack of a stimulus into a categorical and active "stimulus absent" representation signified by increased firing. In other words, "nothing" as a meaningful condition does excite the brain.

Of course, this sort of activity does not have to bear any quantitative meaning. As a signature for a quantitative representation, neurons are rather expected to represent not only the "absence of stimulus," but also signal empty sets as a quantity at the lower end of a numerical continuum. Therefore, my PhD student Araceli Ramirez-Cardenas and I recorded from monkeys we knew had treated empty sets as a quantity based on the numerical distance effect in the numerical matching task I mentioned earlier.[60] As demonstrated in previous chapters, number neurons in the PFC and area VIP of primates (figure 7.2) are tuned to preferred numerosities by maximum activity and show such a distance effect by a progressive decline of activity relative to the preferred numerosity. Maybe, we reasoned, some of the monkeys' neurons would also be tuned to zero based on their experiences with empty sets as void numerosity. Indeed, we found populations of single neurons tuned to empty sets, just as other neurons were tuned to countable numerosities, and therefore representing them as conveying a quantitative null value.

However, the way such neurons encoded empty sets differed in interesting ways between the VIP in the parietal lobe and the PFC in the frontal lobe. Neurons in the VIP, at the input to the cortical number network, discharged vigorously to empty sets but not at all to other countable numerosities; in other words, they encoded empty sets as categorically different from all other numerosities (figure 14.4A). For VIP neurons, the empty set obviously was not part of a numerical continuum, but simply "nothing," as opposed to the something of all countable numerosities.

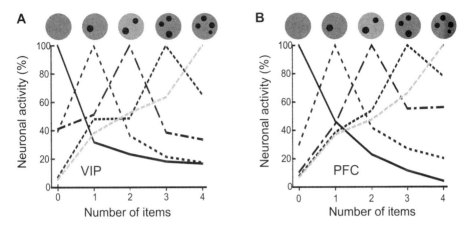

Figure 14.4
Number neurons in the frontal and parietal association cortices represent zero quantity. Besides neurons that preferred countable numerosities from one to four, some neurons in the ventral intraparietal area (**A**) and prefrontal cortex (**B**) were also tuned to zero—that is, empty sets, and had quantity zero as preferred numerosity. While empty set neurons in the ventral intraparietal area responded categorically (A), empty set neurons in the prefrontal cortex showed a graded response characteristic of a neuronal distance effect (B) (from Ramirez-Cardenas et al., 2016).

Higher up the cortical hierarchy, however, PFC neurons were also tuned to empty sets, but they represented zero numerosity more similarly to numerosity one than to larger numerosities; in other words, PFC neurons exhibited a numerical distance effect that, as we have seen earlier, is indicative of empty sets being incorporated together with countable numerosities along the number line (figure 14.4B). Moreover, prefrontal neurons represented empty sets abstractly and irrespective of stimulus variations. Compared to the VIP, the activity of numerosity neurons in the PFC also better predicted the successful or erroneous behavioral outcome of empty-set trials.

These findings indicate a cortical hierarchy of processing from the VIP to PFC. While the VIP is still discriminating "nothing" from "something," empty sets are steadily detached from visual properties and gradually positioned in a numerical continuum on their way to the highest cortical integration center, the PFC. Interestingly, this brain-internal sequential process nicely mirrors the timeline of the cultural and ontogenetic advancement described above. The brain transforms the absence of countable items ("nothing") represented at brain areas lower in the hierarchy like the VIP, into an abstract quantitative category ("zero") in areas higher in the hierarchy, such as the PFC.

Both stimulus-absence and empty-set representations require a transformation from a sensory non-event to an internally generated categorical activity. Understanding the physiological mechanisms behind this process will be challenging. In contrast to countable numerosities, which are represented spontaneously, coding the absence of stimuli and null quantity requires explicit training. As a behaviorally relevant category, zero-like representations need to develop over time as the result of trial-and-error reinforcement learning. If behavioral feedback is provided, reward-prediction error signals arising from the dopamine system could modulate reward-dependent plasticity.[61,62] Reinforcement learning could refine functional connectivity between parietal and frontal neurons, or recurrent connections within associative cortical areas to support neuronal selectivity and sustained working memory coding.[63,64]

Tatiana Engel and Xiao-Jing Wang from New York University devised a network model that showed how category selectivity could arise from reinforcement learning.[65] According to this model, weak but systematic correlations between trial-to-trial fluctuations of firing rates and the accompanying reward after appropriate behavioral choices led neurons to gradually become category-selective. An interesting property of this model is that it does not require the initial tuning of the neurons for successful learning; even non-selective neurons developed categorical tuning, as long as they carried neuronal fluctuations that correlated with behavioral choices. Therefore, when a subject learns to respond appropriately to absent stimuli or empty sets in order to receive a reward, such a mechanism might suffice to produce empty-set-tuned neurons from originally untuned ones.

This neuronal scenario outlined above indicates an effortful cognitive and neuronal process to arrive at a concept of the number zero. The studies mentioned in this chapter from human history, developmental psychology, animal cognition, and neurophysiology provide evidence that the emergence of zero passes through four stages (figure 14.5). In the first and most primitive stage, the absence of a stimulus ("nothing"), corresponds to a (mental/neural) resting state lacking a specific signature. In the second stage, stimulus absence is grasped as a meaningful behavioral category, but its representation is still devoid of quantitative relevance. In the third stage,

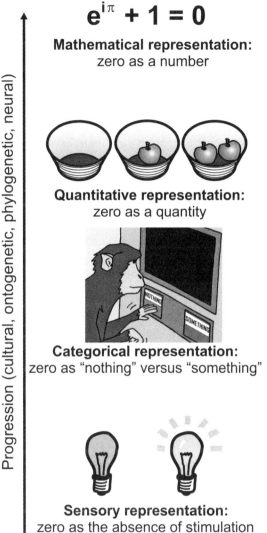

Figure 14.5
Four stages of zero-like concepts appearing in human culture, ontogeny, phylogeny, and neurophysiology. At the most primitive stage (*bottom*), sense organs register the presence of stimuli, such as light. In the absence of stimulation, sense organs are in an inactive resting state. At the next higher level, "nothing" is conceived as a behaviorally relevant category, as exemplified by a monkey judging the absence of a stimulus. With the advent of quantitative representations, a set containing no elements is realized as an empty set or a null set. Finally, zero becomes the number 0 used in number theory and mathematics. For instance, 0 is the additive identity in Euler's identity. Each higher representation encompasses the previous, lower one: The conception of zero as a number requires a quantitative understanding of empty sets, which in turn necessitates a grasp of "nothing" as an abstract category (from Nieder, 2016).

"nothing" acquires a quantitative meaning and is represented as an empty set at the lower end of a mental number line. Finally, the representation is extended to become the number zero, thus becoming part of a combinatorial number symbol system used for calculation and mathematics.

Since cognitive capabilities originate from the workings of neurons, the historical and ontogenetic struggle of mankind to arrive at a concept of zero may at least partly, and in addition to socio-cultural factors, be a reflection of the neurobiological challenge for the brain to leave the sensory world behind and start pondering about concepts that can no longer be experienced. As Mike Hockney wrote:

> Mathematics has been held back whenever mathematicians have tried to match mathematics to observables, "sensory" reality, and has made huge strides whenever mathematics has completely ignored empirical reality.[66]

The mind and its neurons departing from experiential reality: This is what the concept of zero is all about. It shows how our brains, originally evolved to represent sensory objects and events, detach from empirical properties to achieve ultimate abstract thinking. Because of this, the story of zero tells us a great deal about the mind and brain leaving empirical grounds and rising to new intellectual heights.

Epilogue

For the inquisitive human mind, gaining new insights into the brain mechanisms and evolutionary paths of numerical capabilities is intellectually intriguing. Valuable insights, both in animals and humans, about how the brain acquires numbers have been collected over recent decades to arrive at a more complete picture on how our brain "does it." These scientific insights are also the bedrock for more applied endeavors in the future.

One area in which neuroscientific knowledge can be put to good use is the problem of low numeracy. The ability to understand and process numbers is essential to functional living. Low numeracy, either as a consequence of developmental dyscalculia or of acquired acalculia, affects many daily activities, from checking one's finances to determining the dose of a medication. As pointed out, low numeracy seems to be more of a handicap for an individual's life than low literacy is. Low numeracy therefore is not only a private impediment, but also a matter of the economic performance of a society. Improving diagnostic tools and devising educational and medical interventions based on neuroscientific knowledge would be a blessing for individuals and society alike. Of course, several attempts have been made in recent years to come up with better educational interventions to specifically help dyscalculic children. So far, these attempts have yielded rather modest improvements. Currently it is thought that practice with approximate and non-symbolic arithmetic may be an effective way to enhance the math skills of young children, but that hopeful theory awaits confirmation. In addition, more educational intervention programs focus on tablet applications. Here as well, the future will show whether computer-assisted interventions may prove to be superior to traditional approaches.

Another area that will rely on these basic research insights is medicine. Already today, research on number skills helps neurosurgeons to localize and spare brain regions essential for calculation and arithmetic. However, currently there are no means to compensate for calculation abilities after brain damage. Once neurons in an affected area die, they unfortunately will not re-grow. And if the restricted plastic changes in the brain cannot restore counting functions, they are lost forever. Neuroscientists in collaboration with engineers and computer scientists therefore strive to compensate such losses via machines, or brain-machine interfaces (BMIs). In the motor and sensory domain, BMIs aim to (partly) recover a loss of motor or sensory functions by neuroprosthetic devices. While such technologies are still in their infancy and are primarily used in restricted therapeutic contexts, this may change in the future once their performance has improved and BMIs become more widely applicable and accessible. Already today, attempts are being made to use BMIs for cognitive functions as well. Perhaps it will even be possible in the future to restore lost counting functions using BMIs and other therapies.

Whatever blissful application we can envision in the domain of number competence, one thing is certain: None of these applications will be goal-directed and fruitful if they do not originate from solid scientific inquiries. Research in the field of numerical competence is an ongoing process that will require the continued efforts from different biological disciplines. I, for one, am curious to learn about the future discoveries in the fascinating field of counting and calculation.

Notes

Chapter 1

1. Davis, P., Hersh, R., & Marchisotto, E. A. (2012). *The mathematical experience* (study ed.). Boston, MA: Birkhäuser.

2. Livio, M. (2010). *Is God a mathematician?* New York, NY: Simon & Schuster.

3. Davis, Hersh, & Marchisotto, *The mathematical experience.*

4. Hardy, G. H. (1940). *A mathematician's apology.* Cambridge, UK: Cambridge University Press.

5. Kasner, E., Newman, J. R. (1940). *Mathematics and the imagination.* New York, NY: Simon and Schuster.

6. Davis, Hersh, & Marchisotto, *The mathematical experience.*

7. Davis, Hersh, & Marchisotto, *The mathematical experience.*

8. Euclid. (2002). *Euclid's elements: All thirteen books complete in one volume* (T. Heath, trans.). Santa Fe, NM: Green Lion Press.

9. Berkeley, G. (1956). Letter to Molyneux [1709]. In A. Luce & T. Jessop (Eds.), *The works of George Berkeley, Bishop of Cloyne*, Vol. 8. London, UK: Nelson.

10. Cantor, G. (1955). *Contributions to the founding of the theory of transfinite numbers* [1895], transl. P. Jourdain. New York, NY: Dover Publications.

11. Giaquinto, M. (2015). Philosophy of number. In R. Cohen Kadosh & A. Dowker (Eds.), *The Oxford handbook of numerical cognition.* Oxford, UK: Oxford University Press.

12. Locke, J. (1690). *An essay concerning humane understanding.* London, UK: Thomas Bassett.

13. Barrow, J. D. (2011). *The book of universes.* London, UK: Random House.

14. Gödel, K. (1947). "What is Cantor's continuum problem?" *Am. Math. Mon.*, *54*, 515–525. Revised and expanded version in: P. Benacerraf & H. Putnam (Eds.). (1964). *Philosophy of mathematics* (pp. 470–485). Englewood Cliffs, NJ: Prentice-Hall.

15. Dantzig, T. (1930). *Number—The language of science*. New York, NY: Free Press.

16. Dehaene, S. (2011). *The number sense* (2nd ed.). Oxford, UK: Oxford University Press.

17. Spelke, E. S., & Kinzler, K. D. (2007). Core knowledge. *Dev. Sci.*, *10*, 89–96.

18. Quine, W. V. O. (1969). Epistemology naturalized. In W. V. O. Quine, *Ontological relativity and other essays* (pp. 69–90). New York, NY: Columbia University Press.

19. Campbell, D. T. (1974). Evolutionary epistemology. In P. A. Schilpp (Ed.), *The philosophy of Karl R. Popper* (pp. 412–463). LaSalle, IL: Open Court.

20. Thomson, P. (1995). Evolutionary epistemology and scientific realism. *J. Soc. Evol. Syst.*, *18*, 165–191.

21. Vollmer, G. (2003). *Wieso können wir die Welt erkennen?* Neue Beiträge zur Wissenschaftstheorie. Stuttgart, Germany: S. Hirzel.

22. Simpson, G. G. (1963). Biology and the nature of science. *Science*, *139*, 81–88.

23. Ruse, M. (1989). The view from somewhere: A critical defence of evolutionary epistemology. In K. Hahlweg & C. A. Hooker (Eds.), *Issues in evolutionary epistemology* (pp. 185–228). Albany, NY: State University of New York Press.

Chapter 2

1. Fuson, K. C., & Hall, J. W. (1983). The acquisition of early number word meanings: A conceptual analysis and review. In H. P. Ginsburg (Ed.), *The development of mathematical thinking* (pp. 49–107). New York, NY: Academic Press.

2. Wiese, H. (2003). *Numbers, language, and the human mind*. Cambridge, UK: Cambridge University Press.

3. Trick, L. M., & Pylyshyn, Z. W. (1994). Why are small and large numbers enumerated differently—a limited-capacity preattentive stage in vision. *Psy. Rev.*, *101*, 80–102.

4. Treisman, A. (1992). Perceiving and reperceiving objects. *Am. Psychol.*, *47*, 862–875.

5. Pylyshyn, Z. W. (2001). Visual indexes, preconceptual objects, and situated vision. *Cognition*, *80*, 127–158.

6. Kaufman, E. L., Lord M. W., Reese T. W., & Volkmann, J. (1949). The discrimination of visual number. *Am. J. Psychol.*, *62*, 498–525.

7. Mandler, G., & Shebo, B. J. (1982). Subitizing: An analysis of its component processes. *J. Exp. Psychol. Gen.*, *111*, 1–22.

8. Trick, L. M., & Pylyshyn, Z. W. (1993). What enumeration studies can show us about spatial attention: Evidence for limited capacity preattentive processing. *J. Exp. Psychol. Hum. Percept. Perform.*, *19*, 331–351.

9. Simon, T. J. (1997). Reconceptualizing the origins of number knowledge: A "nonnumerical" account. *Cogn. Dev.*, *12*, 349–372.

10. Palmeri, T. J. (1997). Exemplar similarity and the development of automaticity. *J. Exp. Psychol. Learn. Mem. Cogn.*, *23*, 324–354.

11. Lassaline, M. E., & Logan, G. D. (1993). Memory-based automaticity in the discrimination of visual numerosity. *J. Exp. Psychol. Learn. Mem. Cogn.*, *19*, 561–581.

12. Wolters, G., Vankempen, H., & Wijlhuizen, G. J. (1987). Quantification of small numbers of dots—Subitizing or pattern recognition. *Am. J. Psychol.*, *100*, 225–237.

13. Logan, G. D., & Zbrodoff, N. J. (2003). Subitizing and similarity: Toward a pattern-matching theory of enumeration. *Psychon. Bull. Rev.*, *10*, 676–682.

14. Weber, E. H. (1850). Der Tastsinn und das Gemeingefühl. In R. Wagner (Ed.), *Handwörterbuch der Physiologie*, Vol. 3, Part 2 (pp. 481–588). Braunschweig, Germany: Vieweg Verlag.

15. Moyer, R. S., & Landauer, T. K. (1967). Time required for judgements of numerical inequality. *Nature*, *215*, 1519–1520.

16. Dehaene, S. (1992). Varieties of numerical abilities. *Cognition*, *44*, 1–42.

17. Jordan, K. E., & Brannon, E. M. (2006). Weber's Law influences numerical representations in rhesus macaques (*Macaca mulatta*). *Anim. Cogn.*, *9*, 159–172.

18. Merten, K., & Nieder, A. (2009). Compressed scaling of abstract numerosity representations in adult humans and monkeys. *J. Cogn. Neurosci.*, *21*, 333–346.

19. Ditz, H. M., & Nieder, A. (2016). Numerosity representations in crows obey the Weber–Fechner law. *Proc. Biol. Sci.*, *283*, 20160083.

20. Burr, D. C., Anobile, G., & Arrighi, R. (2017). Psychophysical evidence for the number sense. *Philos. Trans. R. Soc. Lond., B, Biol. Sci.*, *373*(1740), 20170045.

21. Dantzig, T. (1930). *Number—The language of science*. New York, NY: Free Press.

22. Dehaene, S. (2011). *The number sense* (2nd ed.). Oxford, UK: Oxford University Press.

23. Gallistel, C. R., & Gelman, R. (2000). Non-verbal numerical cognition: From reals to integers. *Trends Cogn. Sci.*, *4*, 59–65.

Chapter 3

1. Urry, L. A., Cain, M. L., Wasserman, S. A., Minorsky, P. V., & Reece, J. B. (2017). *Campbell biology* (11th ed.). London, UK: Pearson.

2. Wray, G. A. (2015). Molecular clocks and the early evolution of metazoan nervous systems. *Philos. Trans. R. Soc. Lond., B, Biol. Sci., 370*, 1684.

3. Fox, D. (2016). What sparked the Cambrian explosion? *Nature, 530*, 268–270.

4. Sperling, E. A., Frieder, C. A., Raman, A. V., Girguis, P. R., Levin, L. A., & Knoll, A. H. (2013). Oxygen, ecology, and the Cambrian radiation of animals. *Proc. Natl. Acad. Sci. U.S.A., 110*, 13446–13451.

5. Darwin, C. (1859). *On the origin of species by means of natural selection, or the preservation of favoured races in the struggle for life*. London, UK: John Murray.

6. Pfungst, O. (1911). *Clever Hans (The horse of Mr. von Osten): A contribution to experimental animal and human psychology*. New York, NY: Henry Holt and Company.

7. Uller, C., & Lewis, J. (2009). Horses (*Equus caballus*) select the greater of two quantities in small numerical contrasts. *Anim. Cogn., 12*, 733–738.

8. Hassenstein, B. (1974). Otto Koehler—his life and his work. *Z. Tierpsychol., 35*, 449–464.

9. Koehler, O. (1941). Vom Erlernen unbenannter Anzahlen bei Vögeln. *Naturwissenschaften, 29*, 201–218.

10. Koehler, O. (1951). The ability of birds to "count." *Bull. Anim. Behav., 9*, 41–45.

11. Koehler, The ability of birds to "count."

12. Davis, H., & Perusse, R. (1988). Numerical competence in animals: Definitional issues, current evidence, and a new research agenda. *Behav. Brain. Sci., 11*, 561–615.

13. Agrillo, C., & Bisazza, A. (2014). Spontaneous versus trained numerical abilities. A comparison between the two main tools to study numerical competence in nonhuman animals. *J. Neurosci. Methods, 234*, 82–91.

14. Leibovich, T., Katzin, N., Harel, M., & Henik, A. (2016). From "sense of number" to "sense of magnitude"—The role of continuous magnitudes in numerical cognition. *Behav. Brain. Sci., 40*, 1–62.

15. Agrillo & Bisazza, Spontaneous versus trained numerical abilities.

16. Agrillo, C., Miletto Petrazzini, M. E., & Bisazza, A. (2017). Numerical abilities in fish: A methodological review. *Behav. Processes, 141*(2), 161–171.

17. Hager, M. C., & Helfman, G. S. (1991). Safety in numbers: Shoal size choice by minnows under predatory threat. *Behav. Ecol. Sociobiol., 29*, 271–276.

18. Agrillo, C., Dadda, M., Serena, G., & Bisazza, A. (2009). Use of number by fish. *PLoS ONE, 4*(3), e4786.

19. Agrillo, C., Piffer, L., & Bisazza, A. (2010). Large number discrimination by mosquitofish. *PLoS ONE, 5*(12), e15232.

20. Agrillo, C., Miletto Petrazzini, M. E., Piffer, L., Dadda, M., & Bisazza, A. (2012). A new training procedure for studying discrimination learning in fish. *Behav. Brain Res., 230*, 343–348.

21. Bisazza, A., Tagliapietra, C., Bertolucci, C., Foa, A., & Agrillo, C. (2014). Non-visual numerical discrimination in a blind cavefish (*Phreatichthys andruzzii*). *J. Exp. Biol., 217*, 1902–1909.

22. Uller, C., Jaeger, R., Guidry, G., & Martin, C. (2003). Salamanders (*Plethodon cinereus*) go for more: Rudiments of number in an amphibian. *Anim. Cogn., 6*, 105–112.

23. Krusche, P., Uller, C., & Dicke, U. (2010). Quantity discrimination in salamanders. *J. Exp. Biol., 213*, 1822–1828.

24. Stancher, G., Rugani, R., Regolin, L., & Vallortigara, G. (2015). Numerical discrimination by frogs (*Bombina orientalis*). *Anim. Cogn., 18*, 219–229.

25. Miletto Petrazzini, M. E., Fraccaroli, I., Gariboldi, F., Agrillo, C., Bisazza, A., Bertolucci, C., & Foà, A. (2017). Quantitative abilities in a reptile (*Podarcis sicula*). *Biol. Lett., 13*(4), 20160899.

26. Miletto Petrazzini, M. E., Bertolucci, C., & Foà, A. (2018). Quantity discrimination in trained lizards (*Podarcis sicula*). *Front. Psychol., 9*, 274.

27. Dyke, G., & Kaiser, G. (2011). *Living dinosaurs: The evolutionary history of modern birds*. Oxford, UK: Wiley-Blackwell.

28. Lyon, B. E. (2003). Egg recognition and counting reduce costs of avian conspecific brood parasitism. *Nature, 422*, 495–499.

29. White, D. J., Ho, L., & Freed-Brown, G. (2009). Counting chicks before they hatch: Female cowbirds can time readiness of a host nest for parasitism. *Psychol. Sci., 20*, 1140–1145.

30. Hunt, S., Low, J., & Burns, K. C. (2008). Adaptive numerical competency in a food-hoarding songbird. *Proc. R. Soc. B Biol. Sci., 275*, 2373–2379.

31. Garland, A., Low, J., & Burns, K. C. (2012). Large quantity discrimination by North Island robins (*Petroica longipes*). *Anim. Cogn., 15*, 1129–1140.

32. Bogale, B. A., Aoyama, M., & Sugita, S. (2014). Spontaneous discrimination of food quantities in the jungle crow, *Corvus macrorhynchos*. *Anim. Behav., 94*, 73–78.

33. Ujfalussy, D., Miklósi, A., Bugnyar, T., & Kotrschal, K. (2013). Role of mental representations in quantity judgments by jackdaws (Corvus monedula). *J. Comp. Psychol.*, *128*, 11–20.

34. Templeton, C. N., Greene, E., & Davis, K. (2005). Allometry of alarm calls: Black-capped chickadees encode information about predator size. *Science*, *308*, 1934–1937.

35. Pepperberg, I. M. (2010). Evidence for conceptual quantitative abilities in the African grey parrot: Labeling of cardinal sets. *Ethology*, *75*, 37–61.

36. Xia, L., Emmerton, J., Siemann, M., & Delius, J. D. (2001). Pigeons (*Columba livia*) learn to link numerosities with symbols. *J. Comp. Psychol.*, *115*, 83.

37. Smirnova, A. A., Lazareva, O. F., & Zorina, Z. A. (2000). Use of number by crows: Investigation by matching and oddity learning. *J. Exp. Anal. Behav.*, *73*, 163–176.

38. Bogale, B. A., Kamata, N., Mioko, K., & Sugita, S. (2011). Quantity discrimination in jungle crows, *Corvus macrorhynchos*. *Anim. Behav.*, *82*, 635–641.

39. Ditz, H. M., & Nieder, A. (2016). Numerosity representations in crows obey the Weber–Fechner law. *Proc. Biol. Sci.*, *283*(1827), 20160083.

40. Scarf, D., Hayne, H., & Colombo, M. (2011). Pigeons on par with primates in numerical competence. *Science*, *334*, 1664.

41. Rugani, R., Regolin, L., & Vallortigara, G. (2008). Discrimination of small numerosities in young chicks. *J. Exp. Psychol. Anim. Behav. Process.*, *34*, 388–399.

42. Rugani, R., Fontanari, L., Simoni, E., Regolin, L., & Vallortigara G. (2009). Arithmetic in newborn chicks. *Proc. Biol. Sci.*, *276*, 2451–2460.

43. Vallortigara, G. (2012). Core knowledge of object, number, and geometry: A comparative and neural approach. *Cogn. Neuropsychol.*, *29*, 213–236.

44. Hublin, J. J., Ben-Ncer, A., Bailey, S. E., Freidline, S. E., Neubauer, S., Skinner, M. M., Bergmann, I., Le Cabec, A., Benazzi, S., Harvati, K., & Gunz, P. (2017). New fossils from Jebel Irhoud, Morocco, and the pan-African origin of *Homo sapiens*. *Nature*, *546*, 289–292.

45. Schlebusch, C. M., Malmström, H., Günther, T., Sjödin, P., Coutinho, A., Edlund, H., Munters, A. R., Vicente, M., Steyn, M., Soodyall, H., Lombard, M., & Jakobsson, M. (2017). Southern African ancient genomes estimate modern human divergence to 350,000 to 260,000 years ago. *Science*, *358*, 652–655.

46. West, R. E., & Young, R. J. (2002). Do domestic dogs show any evidence of being able to count? *Anim. Cogn.*, *5*, 183–186.

47. Pisa, P. E., & Agrillo, C. (2009). Quantity discrimination in felines: A preliminary investigation of the domestic cat (*Felis silvestris catus*). *J. Ethology*, *27*, 289–293.

48. Vonk, J., & Beran, M. J. (2012). Bears "count" too: Quantity estimation and comparison in black bears (*Ursus americanus*). *Anim. Behav.*, *84*, 231–238.

49. McComb, K., Packer, C., & Pusey, A. (1994). Roaring and numerical assessment in contests between groups of female lions, *Panthera leo*. *Anim. Behav.*, *47*, 379–387.

50. Benson-Amram, S., Heinen, K., Dryer, S. L., & Holekamp, K. E. (2011). Numerical assessment and individual call discrimination by wild spotted hyaenas, *Crocuta crocuta*. *Anim. Behav.*, *82*, 743–752.

51. Abramson, J. Z., Hernandez-Lloreda, V., Call, J., & Colmenares, F. (2011). Relative quantity judgments in South American sea lions (*Otaria flavescens*). *Anim. Cogn.*, *14*, 695–706.

52. Davis, H. (1984). Discrimination of the number 3 by a raccoon (*Procyon lotor*). *Anim. Learn Behav.*, *12*, 409–413.

53. Kilian, A., Yaman, S., von Fersen, L., & Güntürkün, O. (2003). A bottlenose dolphin discriminates visual stimuli differing in numerosity. *Learn. Behav.*, *31*, 133–142.

54. Jaakkola, K., Fellner, W., Erb, L., Rodriguez, M., & Guarino, E. (2005). Understanding of the concept of numerically "less" by bottlenose dolphins (*Tursiops truncatus*). *J. Comp. Psychol.*, *119*, 296–303.

55. Abramson, J. Z., Hernández-Lloreda, V., Call, J., & Colmenares, F. (2013). Relative quantity judgments in the beluga whale (*Delphinapterus leucas*) and the bottlenose dolphin (*Tursiops truncatus*). *Behav. Processes*, *96*, 11–19.

56. Uller, C., & Lewis, J. (2009). Horses (*Equus caballus*) select the greater of two quantities in small numerical contrasts. *Anim. Cogn.*, *12*, 733–738.

57. Perdue, B. M., Talbot, C. F., Stone, A. M., & Beran, M. J. (2012). Putting the elephant back in the herd: Elephant relative quantity judgments match those of other species. *Anim. Cogn.*, *15*, 955–861.

58. Fernandes, D. M., & Church, R. M. (1982). Discrimination of the number of sequential events. *Anim. Learn. Behav.*, *10*, 171–176.

59. Davis, H., & Albert, M. (1986). Numerical discrimination by rats using sequential auditory stimuli. *Anim. Learn. Behav.*, *14*, 57–59.

60. Mechner, F. (1958). Probability relations within response sequences under ratio reinforcement. *J. Exp. Anal. Behav.*, *1*, 109–121.

61. Lewis, K. P., Jaffe, S., & Brannon, E. M. (2005). Analog number representations in mongoose lemurs (*Eulemur mongoz*): Evidence from a search task. *Anim. Cogn.*, *8*, 247–252.

62. Thomas, R. K., & Chase, L. (1980). Relative numerousness judgments by squirrel monkeys. *Bull. Psychon. Soc.*, *16*, 79–82.

63. Judge, P. G., Evans, T. A., & Vyas, D. K. (2005). Ordinal representation of numeric quantities by brown capuchin monkeys (*Cebus apella*). *J. Exp. Psychol. Anim. Behav. Process*, *31*, 79–94.

64. Beran, M. J., Evans, T. A., Leighty, K. A., Harris, E. H., & Rice, D. (2008). Summation and quantity judgments of sequentially presented sets by capuchin monkeys (*Cebus apella*). *Am. J. Primatol.*, *70*, 191–194.

65. Hicks, L. H. (1956). An analysis of number-concept formation in the rhesus monkey. *J. Comp. Physiol. Psychol.*, 49, 212–218.

66. Brannon, E. M., & Terrace, H. S. (2000). Representation of the numerosities 1–9 by rhesus macaques (*Macaca mulatta*). *J. Exp. Psychol. Anim. Behav. Process*, *26*, 31–49.

67. Beran, M. J. (2007). Rhesus monkeys (*Macaca mulatta*) enumerate sequentially presented sets of items using analog numerical representations. *J. Exp. Psychol. Anim. Behav. Process*, *33*, 42–54.

68. Smith, B. R., Piel, A. K., & Candland, D. K. (2003). Numerity of a socially housed hamadryas baboon (*Papio hamadryas*) and a socially housed squirrel monkey (*Saimiri sciureus*). *J. Comp. Psychol.*, *117*, 217–225.

69. Anderson, U. S., Stoinski, T. S., Bloomsmith, M. A., & Maple, T. L. (2007). Relative numerousness judgment and summation in young, middle-aged, and older adult orangutans (*Pongo pygmaeus abelii* and *Pongo pygmaeus pygmaeus*). *J. Comp. Psychol.*, *121*, 1–11.

70. Anderson, U. S., Stoinski, T. S., Bloomsmith, M. A., Marr, M. J., Smith, A. D., & Maple, T. L. (2005). Relative numerousness judgment and summation in young and old Western lowland gorillas. *J. Comp. Psychol.*, *119*, 285–295.

71. Beran, M. J. (2001). Summation and numerousness judgments of sequentially presented sets of items by chimpanzees (*Pan troglodytes*). *J. Comp. Psychol.*, *115*, 181–191.

72. Beran, M. J., & Beran, M. M. (2004). Chimpanzees remember the results of one-by-one addition of food items to sets over extended time periods. *Psychol. Sci.*, *15*, 94–99.

73. Penn, D. C., Holyoak, K. J., & Povinelli, D. J. (2008). Darwin's mistake: explaining the discontinuity between human and non-human minds. *Behav. Brain Sci.*, *31*, 109–130;

74. Nelson, X. J., & Jackson, R. R. (2012). The role of numerical competence in a specialized predatory strategy of an araneophagic spider. *Anim. Cogn.*, *15*, 699–710.

75. Rodríguez, R. L., Briceño, R. D., Briceño-Aguilar, E., & Höbel, G. (2015). *Nephila clavipes* spiders (Araneae: Nephilidae) keep track of captured prey counts: Testing for a sense of numerosity in an orb-weaver. *Anim. Cogn.*, *18*, 307–314.

76. Karban, R., Black, C. A., & Weinbaum, S. A. (2000). How 17-year cicadas keep track of time. *Ecol. Lett.*, *3*, 253–256.

77. Carazo, P., Font, E., Forteza-Behrendt, E., & Desfilis, E. (2009). Quantity discrimination in *Tenebrio molitor*: Evidence of numerosity discrimination in an invertebrate? *Anim. Cogn.*, *12*, 463–470.

78. Reznikova, Z., & Ryabko, B. (1996). Transmission of information regarding the quantitative characteristics of an object in ants. *Neurosci. Behav. Physiol.*, *26*, 397–405.

79. Wittlinger, M., Wehner, R., & Wolf, H. (2006). The ant odometer: Stepping on stilts and stumps. *Science*, *312*, 1965–1967.

80. Chittka, L., & Geiger, K. (1995). Can honeybees count landmarks? *Anim. Behav.*, *49*, 159–164.

81. Dacke, M., & Srinivasan, M. V. (2008). Evidence for counting in insects. *Anim. Cogn.*, *11*, 683–689.

82. Gross, H. J., Pahl, M., Si, A., Zhu, H., Tautz, J., & Zhang, S. (2009). Number-based visual generalisation in the honeybee. *PLoS ONE*, *4*, e4263.

83. Howard, S. R., Avarguès-Weber, A., Garcia, J. E., Greentree, A. D., & Dyer, A. G. (2018). Numerical ordering of zero in honey bees. *Science*, *360*, 1124–1126.

84. Zhang, S. W., Bock, F., Si, A., Tautz, J., & Srinivasan, M. V. (2005). Visual working memory in decision making by honey bees. *Proc. Natl. Acad. Sci. U.S.A.*, *102*, 5250–5255.

85. Giurfa, M., Zhang, S., Jenett, A., Menzel, R., & Srinivasan, M. V. (2001). The concepts of "sameness" and "difference" in an insect. *Nature*, *410*, 930–933.

86. Loukola, O. J., Perry, C. J., Coscos, L., & Chittka, L. (2017). Bumblebees show cognitive flexibility by improving on an observed complex behavior. *Science*, *355*, 833–836.

87. Mechner, F. (1958). Probability relations within response sequences under ratio reinforcement. *J. Exp. Anal. Behav.*, *1*, 109–121.

88. Brannon, E. M., Terrace, H. S. (1998). Ordering of the numerosities 1 to 9 by monkeys. *Science*, *282*, 746–749.

89. Brannon, E. M., & Terrace, H. S. (2000). Representation of the numerosities 1–9 by rhesus macaques (*Macaca mulatta*). *J. Exp. Psychol. Anim. Behav. Process.*, *26*, 31–49.

90. Fechner, G. T. (1860). *Elemente der Psychophysik*, Vol. 2. Leipzig, Germany: Breitkopf & Härtel.

91. Nieder, A., & Miller, E. K. (2003). Coding of cognitive magnitude: Compressed scaling of numerical information in the primate prefrontal cortex. *Neuron, 37*, 149–157.

92. Nieder, A., & Miller, E. K. (2004). Analog numerical representations in rhesus monkeys: Evidence for parallel processing. *J. Cogn. Neurosci., 16*, 889–901.

93. Merten, K., & Nieder, A. (2009). Compressed scaling of abstract numerosity representations in adult humans and monkeys. *J. Cogn. Neurosci., 21*, 333–346.

94. Merten & Nieder. Compressed scaling of abstract numerosity representations.

95. Merten & Nieder. Compressed scaling of abstract numerosity representations.

96. Nieder, A., Diester, I., & Tudusciuc, O. (2006). Temporal and spatial enumeration processes in the primate parietal cortex. *Science, 313*, 1431–1435.

97. Jordan, K. E., Maclean, E. L., & Brannon, E. M. (2008). Monkeys match and tally quantities across senses. *Cognition, 108*, 617–625.

98. Nieder, A. (2012). Supramodal numerosity selectivity of neurons in primate prefrontal and posterior parietal cortices. *Proc. Natl. Acad. Sci. U.S.A., 109*, 11860–11865.

99. Scarf, D., Hayne, H., & Colombo, M. (2011). Pigeons on par with primates in numerical competence. *Science, 334*, 1664.

100. Smirnova, Lazareva, & Zorina. Use of number by crows.

101. Ujfalussy, D., Miklósi, A., Bugnyar, T., & Kotrschal, K. (2013). Role of mental representations in quantity judgments by jackdaws (*Corvus monedula*). *J. Comp. Psychol., 128*, 11–20.

102. Bogale, B. A., Aoyama, M., & Sugita, S. (2014). Spontaneous discrimination of food quantities in the jungle crow, *Corvus macrorhynchos*. *Anim. Behav., 94*, 73–78.

103. Tornick, J. K., Callahan, E. S., & Gibson, B. M. (2015). An investigation of quantity discrimination in Clark's nutcrackers (*Nucifraga columbiana*). *J. Comp. Psychol., 129*, 17–25.

104. Ditz, H. M., & Nieder, A. (2015). Neurons selective to the number of visual items in the corvid songbird endbrain. *Proc. Natl. Acad. Sci. U.S.A., 112*, 7827–7832.

105. Ditz & Nieder. Numerosity representations in crows obey the Weber–Fechner law.

106. Cantlon, J. F., & Brannon, E. M. (2006). Shared system for ordering small and large numbers in monkeys and humans. *Psychol. Sci., 17*, 401–406.

107. Beran, M. J. (2007). Rhesus monkeys (*Macaca mulatta*) enumerate large and small sequentially presented sets of items using analog numerical representations. *J. Exp. Psychol. Anim. Behav. Proc., 33*, 42–54.

108. Barnard, A. M., Hughes, K. D., Gerhardt, R. R., DiVincenti, L., Jr., Bovee, J. M., & Cantlon, J. F. (2013). Inherently analog quantity representations in olive baboons (*Papio anubis*). *Front. Psychol.*, 4(253).

109. Rugani, R., Regolin, L., & Vallortigara, G. (2013). One, two, three, four, or is there something more? Numerical discrimination in day-old domestic chicks. *Anim. Cogn.*, 16, 557–564.

110. Ditz & Nieder. Numerosity representations in crows obey the Weber–Fechner law.

111. Hauser, M. D., Carey, S., & Hauser, L. B. (2000). Spontaneous number representation in semifree-ranging rhesus monkeys. *Proc. R. Soc. Lond. B Biol. Sci.*, 267, 829–833.

112. Bonanni, R., Natoli, E., Cafazzo, S., & Valsecchi, P. (2011). Free-ranging dogs assess the quantity of opponents in intergroup conflicts. *Anim. Cogn.*, 14, 103–115.

113. Hunt, S., Low, J., & Burns, C. K. (2008). Adaptive numerical competency in a foodhoarding songbird. *Proc. R. Soc. Lond. B Biol. Sci.*, 10, 1098–1103.

114. Agrillo, C., Dadda, M., Serena, G., & Bisazza, A. (2008). Do fish count? Spontaneous discrimination of quantity in female mosquitofish. *Anim. Cogn.*, 11, 495–503.

115. Van Oeffelen, M. P., & Vos, P. G. (1982). A probabilistic model for the discrimination of visual number. *Percept. Psychophys.*, 32, 163–170.

116. Vetter, P., Butterworth, B., & Bahrami, B. (2008). Modulating attentional load affects numerosity estimation: Evidence against a pre-attentive subitizing mechanism. *PLoS ONE*, 3, e3269.

117. Gallistel, C. R., & Gelman, R. (1992). Preverbal and verbal counting and computation. *Cognition*, 44, 43–74.

118. Mandler, G., & Shebo, B. J. (1982). Subitizing: An analysis of its component processes. *J. Exp. Psychol.*, 111, 1–22.

Chapter 4

1. Waters, C. M., & Bassler, B. L. (2005). Quorum sensing: Cell-to-cell communication in bacteria. *Annu. Rev. Cell Dev. Biol.*, 21, 319–346.

2. Cronin, A. L. (2014). Ratio-dependent quantity discrimination in quorum sensing ants. *Anim. Cogn.*, 17, 1261–1268.

3. Chittka, L., & Geiger, K. (1995). Can honeybees count landmarks? *Anim. Behav.*, 49, 159–164.

4. Dacke, M., & Srinivasan, M. V (2008). Evidence for counting in insects. *Anim. Cogn.*, *11*, 683–689.

5. Krebs, J. R., Ryan, J. C., & Charnov, E. L. (1974). Hunting by expectation or optimal foraging? *Anim. Behav.*, *22*, 953–964.

6. Uller, C., Jaeger, R., Guidry, G., & Martin, C. (2003). Salamanders (*Plethodon cinereus*) go for more: Rudiments of number in an amphibian. *Anim. Cogn.*, *6*, 105–112.

7. Krusche, P., Uller, C., & Dicke, U. (2010). Quantity discrimination in salamanders. *J. Exp. Biol.*, *21*, 1822–1828.

8. Stancher, G., Rugani, R., Regolin, L., & Vallortigara, G. (2015). Numerical discrimination by frogs (*Bombina orientalis*). *Anim. Cogn.*, *18*, 219–229.

9. Panteleeva, S., Reznikova, Z., & Vygonyailova, O. (2013). Quantity judgments in the context of risk/reward decision making in striped field mice: First "count," then hunt. *Front. Psychol.*, *4*, 53.

10. Nelson, X. J., & Jackson, R. R. (2012). The role of numerical competence in a specialized predatory strategy of an araneophagic spider. *Anim. Cogn.*, *15*, 699–710.

11. MacNulty, D. R., Tallian, A., Stahler, D. R., & Smith, D. W. (2014). Influence of group size on the success of wolves hunting bison. *PLoS ONE*, *9*, e112884.

12. Hager, M. C., & Helfman, G. S. (1991). Safety in numbers: Shoal size choice by minnows under predatory threat. *Behav. Ecol. Sociobiol.*, *29*, 271–276.

13. Buckingham, J. N., Wong, B. B. M., & Rosenthal, G. G. (2007). Shoaling decisions in female swordtails: How do fish gauge group size? *Behaviour*, *144*, 1333–1346.

14. Mehlis, M., Thünken, T., Bakker, T. C. M., & Frommen, J. G. (2015). Quantification acuity in spontaneous shoaling decisions of three-spined sticklebacks. *Anim. Cogn.*, *18*, 1125–1131.

15. Foster, W. A., & Treherne, J. E. (1981). Evidence for the dilution effect in the selfish herd from dish predation on a marine insect. *Nature*, *293*, 466–467.

16. Landeau, L., & Terborgh, J. (1986). Oddity and the confusino effect in predation. Animal *Behaviour*, *34*, 1372–1380.

17. Pulliam, H. R. (1973). On the advantages of flocking. *J. Theor. Biol.*, *38*, 419–422.

18. Templeton, C. N., Greene, E., & Davis, K. (2005). Allometry of alarm calls: Black-capped chickadees encode information about predator size. *Science*, *308*, 1934–1937.

19. McComb, K., Packer, C. & Pusey, A. (1994). Roaring and numerical assessment in contests between groups of female lions, *Panthera leo*. *Anim. Behav.*, *47*, 379–387.

20. Wilson, M. L., Hauser, M. D., & Wrangham, R. W. (2001). Does participation in intergroup conflict depend on numerical assessment, range location, or rank for wild chimpanzees? *Anim. Behav.*, *61*, 1203–1216.

21. Wilson, M. L., Britton, N. F., & Franks, N. R. (2002). Chimpanzees and the mathematics of battle. *Proc. Biol. Sci.*, *269*, 1107–1112.

22. Benson-Amram, S., Heinen, K., & Dryer, S. L., et al. (2011). Numerical assessment and individual call discrimination by wild spotted hyaenas, *Crocuta crocuta*. *Anim. Behav.*, *82*, 743–752.

23. Carazo, P., Font, E., Forteza-Behrendt, E., & Desfilis, E. (2009). Quantity discrimination in *Tenebrio molitor*: Evidence of numerosity discrimination in an invertebrate? *Anim. Cogn.*, *12*, 463–470.

24. Carazo, P., Fernández-Perea, R., & Font, E. (2012). Quantity estimation based on numerical cues in the mealworm beetle (*Tenebrio molitor*). *Front. Psychol.*, *3*, 502.

25. Shifferman, E. M. (2012). It's all in your head: The role of quantity estimation in sperm competition. *Proc. Biol. Sci.*, *279*, 833–840.

26. Bonilla, M. M., Zeh, D. W., White, A. M., & Zeh, J. A. (2011). Discriminating males and unpredictable females: Males bias sperm allocation in favor of virgin females. *Ethology*, *117*, 740–748.

27. Lyon, B. E. (2003). Egg recognition and counting reduce costs of avian conspecific brood parasitism. *Nature*, *422*, 495–499.

28. White, D. J., Ho, L., & Freed-Brown, G. (2009). Counting chicks before they hatch: Female cowbirds can time readiness of a host nest for parasitism. *Psychol. Sci.*, *20*, 1140–1145.

29. Hoover, J. P., & Robinson, S. K. (2007). Retaliatory mafia behavior by a parasitic cowbird favors host acceptance of parasitic eggs. *Proc. Natl. Acad. Sci. U.S.A.*, *104*, 4479–4483.

Chapter 5

1. Starkey, P., & Cooper, R. G., Jr. (1980). Perception of numbers by human infants. *Science*, *210*, 1033–1035.

2. Izard, V., Sann, C., Spelke, E. S., & Streri, A. (2009). Newborn infants perceive abstract numbers. *Proc. Natl. Acad. Sci. U.S.A.*, *106*, 10382–10385.

3. Xu, F., & Spelke, E. S. (2000). Large number discrimination in 6-month-old infants. *Cognition*, *74*, B1–B11.

4. Xu, F. (2003). Numerosity discrimination in infants: Evidence for two systems of representations. *Cognition, 89*, B15–B25.

5. Lipton, J. S., & Spelke, E. S. (2003). Origins of number sense. Large-number discrimination in human infants. *Psychol. Sci., 14*, 396–401.

6. Wood, J. N., & Spelke, E. S. (2005). Infants' enumeration of actions: Numerical discrimination and its signature limits. *Dev. Sci., 8*, 173–181.

7. Xu & Spelke. Large number discrimination in 6-month-old infants.

8. Xu, F., & Arriaga, R. I. (2007). Number discrimination in 10-month-old infants. *Br. J. Dev. Psychol., 25*, 103–108.

9. Halberda, J., Mazzocco, M. M. M., & Feigenson, L. (2008). Individual differences in nonverbal number acuity correlate with maths achievement. *Nature, 455*, 665–668.

10. Carey, S. (2001). Cognitive foundations of arithmetic: Evolution and ontogenesis. *Mind Lang., 16*, 37–55.

11. Feigenson, L., Dehaene, S., & Spelke, E. (2004). Core systems of number. *Trends Cogn. Sci., 8*, 307–314.

12. Starkey & Cooper. Perception of numbers by human infants.

13. Feigenson, L., Carey, S., & Hauser, M. (2002). The representations underlying infants' choice of more: Object-files versus analog magnitudes. *Psychol. Sci., 13*, 150–156.

14. Feigenson, L., & Carey, S. (2003). Tracking individuals via object-files: Evidence from infants' manual search. *Dev. Sci., 6*, 568–584.

15. Pylyshyn, Z. W. (2003). *Seeing and visualizing: It's not what you think*. Cambridge, MA: MIT Press.

16. Sathian, K., Simon, T. J., Peterson, S., Patel, G. A., Hoffman, J. M., & Grafton, S. T. (1999). Neural evidence linking visual object enumeration and attention. *J. Cogn. Neurosci., 11*, 36–51.

17. Piazza, M., Mechelli, A., Butterworth, B., & Price, C. J. Are subitizing and counting implemented as separate or functionally overlapping processes? *Neuroimage, 15*, 435–446.

18. Szkudlarek, E., & Brannon, E. M. (2017). Does the approximate number system serve as a foundation for symbolic mathematics? *Lang. Learn. Dev., 13*, 171–190.

19. Gordon, P. (2004). Numerical cognition without words: Evidence from Amazonia. *Science, 306*, 496–499.

20. Frank, M. C., Everett, D. L., Fedorenko, E., & Gibson, E. (2008). Number as a cognitive technology: Evidence from Pirahã language and cognition. *Cognition*, *108*, 819–824.

21. Pica, P., Lemer, C., Izard, V., & Dehaene, S. (2004). Exact and approximate arithmetic in an Amazonian indigene group. *Science*, *306*, 499–503.

22. Merten, K., & Nieder, A. (2009). Compressed scaling of abstract numerosity representations in adult humans and monkeys. *J. Cogn. Neurosci.*, *21*, 333–346.

23. Whalen, J., Gallistel C. R., & Gelman, R. (1999). Nonverbal counting in humans: The psychophysics of number representations. *Psychol. Sci.*, *10*, 130–137.

24. Cordes, S., Gelman, R., Gallistel, C. R., & Whalen, J. Variability signatures distinguish verbal from nonverbal counting for both large and small numbers. *Psychon. Bull. Rev.*, *8*, 698–707 (2001).

25. Barth, H., Kanwisher, N., & Spelke, E. (2003). The construction of large number representations in adults. *Cognition*, *86*, 201–221.

26. Dehaene, S., Izard, V., Spelke, E., & Pica, P. (2008). Log or linear? Distinct intuitions of the number scale in Western and Amazonian indigene cultures. *Science*, *320*, 1217–1220.

27. Siegler, R. S., & Booth, J. L. (2004). Development of numerical estimation in young children. *Child Dev.*, *75*, 428–444.

Chapter 6

1. Azevedo, F. A., Carvalho, L. R., Grinberg, L. T., Farfel, J. M., Ferretti, R. E., Leite, R. E., Jacob Filho, W., Lent, R., & Herculano-Houzel, S. (2009). Equal numbers of neuronal and nonneuronal cells make the human brain an isometrically scaled-up primate brain. *J. Comp. Neurol.*, *513*, 532–54110.

2. Broca, M. P. (1861). Remarques sur le siége de la faculté du langage articulé, suivies d'une observation d'aphemie (Perte de la Parole). *Bull. Mem. Soc. Anat. Paris*, *36*, 330–357.

3. Finger, S. (1994). *Origins of neuroscience: A history of explorations into brain function*. Oxford, UK: Oxford University Press.

4. Fritsch, G., & Hitzig, E. (1870). Über die elektrische Erregbarkeit des Grosshirns. *Arch. Anat. Physiol. Wissen.*, *37*, 300–332.

5. Ferrier, D. (1876). *The functions of the brain*. London, UK: Smith, Elder and Company.

6. Brodmann, K. (1909). *Vergleichende Lokalisationslehre der Großhirnrinde in ihren Prinzipien dargestellt auf Grund des Zellenbaues*. Leipzig, Germany: Johann Ambrosius Barth Verlag.

7. Hitzig, E. (1874). *Untersuchungen über das Gehirn. Abhandlungen physiologischen und pathologischen Inhalts.* Berlin, Germany: Hirschwald.

8. Ferrier. *The functions of the brain.*

9. Bianchi, L. (1895). The functions of the frontal lobes. *Brain, 18,* 497–522.

10. Flechsig, P. (1896). *Gehirn und Seele.* Leipzig, Germany: Verlag von Veit & Comp.

11. Flechsig, P. (1927). *Meine myelogenetische Hirnlehre.* Berlin, Germany: Springer.

12. Guillery, R. W. (2005). Is postnatal neocortical maturation hierarchical? *Trends Neurosci., 28,* 512–517.

13. Flechsig. *Gehirn und Seele.*

14. Flechsig. *Meine myelogenetische Hirnlehre.*

15. Van Essen, D. C., & Dierker, D. L. (2007). Surface-based and probabilistic atlases of primate cerebral cortex. *Neuron, 56,* 209–225.

16. Donahue, C. J., Glasser, M. F., Preuss, T. M., Rilling, J. K., & Van Essen, D. C. (2018). Quantitative assessment of prefrontal cortex in humans relative to non-human primates. *Proc. Natl. Acad. Sci. U.S.A., 115,* E5183–E5192.

17. Clarke, D. L., Boutros, N. N., & Mendez, M. F. (2010). *The brain and behavior: An introduction to behavioral neuroanatomy.* Cambridge, UK: Cambridge University Press.

18. Wise, S. P., Boussaoud, D., Johnson, P. B., & Caminiti, R. (1997). Premotor and parietal cortex: Corticocortical connectivity and combinatorial computations. *Annu. Rev. Neurosci., 20,* 25–42.

19. Blakemore, S.-J. (2008). The social brain in adolescence. *Nat. Rev. Neurosci., 9,* 267–276.

20. Fuster, J. M. (2008). *The prefrontal cortex* (4th ed.). New York, NY: Academic Press.

21. Baddeley, A. (1992). Working memory. *Science, 255,* 556–559.

22. Miller, E. K., & Cohen, J. D. (2001). An integrative theory of prefrontal cortex function. *Annu. Rev. Neurosci., 24,* 167–202.

23. Rorden, C., & Karnath, H. O. (2004). Using human brain lesions to infer function: A relic from a past era in the fMRI age? *Nat. Rev. Neurosci., 5,* 813–819.

24. Lewandowsky, M., & Stadelmann, E. (1908). Über einen bemerkenswerten Fall von Hirnblutung und über Rechenstörungen bei Herderkrankung des Gehirns. *Z. Psychol. Neurol., 11,* 249–265.

25. Peritz, G. (1918). Zur Pathopsychologie des Rechnens. *Dtsch. Z. Nervenheilkd, 61,* 234–340.

26. Henschen, S. E. (1919). Über Sprach-, Musik- und Rechenmechanismen und ihre Lokalisation im Gorßhirn. *Z. Gesamte Neurol. Psychiatr.*, *52*, 273–198.

27. Henschen, S. E. (1925). Clinical and anatomical contributions on brain pathology. *Arch. Neurol. Psychiatry*, *13*, 226–249.

28. Goldstein, K. (1948). *Language and language disturbances*. New York, NY: Grune & Stratton, p. 133.

29. Lemer, C., Dehaene, S., Spelke, E., & Cohen, L. (2003). Approximate quantities and exact number words: Dissociable systems. *Neuropsychologia*, *41*, 1942–1958.

30. Ashkenazi, S., Henik, A., Ifergane, G., & Shelef, I. (2008). Basic numerical processing in left intraparietal sulcus (IPS) acalculia. *Cortex*, *44*, 439–448.

31. Delazer, M., Karner, E., Zamarian, L., Donnemiller, E., & Benke, T. (2006). Number processing in posterior cortical atrophy—A neuropsycholgical case study. *Neuropsychologia*, *44*, 36–51.

32. Koss, S., Clark, R., Vesely, L., Weinstein, J., Powers, C., Richmond, L., Farag, C., Gross, R., Liang, T. W., & Grossman, M. (2010). Numerosity impairment in corticobasal syndrome. *Neuropsychology*, *24*, 476–492.

33. Roland, P. E., & Friberg, L. (1985). Localization of cortical areas activated by thinking. *J. Neurophysiol.*, *53*, 1219–1243.

34. Ogawa, S., Tank, D. W., Menon, R., Ellermann, J. M., Kim, S. G., Merkle, H., & Ugurbil, K. (1992). Intrinsic signal changes accompanying sensory stimulation: Functional brain mapping with magnetic resonance imaging. *Proc. Natl. Acad. Sci. U.S.A.*, *89*, 5951–5955.

35. Kwong, K. K., Belliveau, J. W., Chesler, D. A., Goldberg, I. E., Weisskoff, R. M., Poncelet, B. P., Kennedy, D. N., Hoppel, B. E., Cohen, M. S., & Turner, R., et al. (1992). Dynamic magnetic resonance imaging of human brain activity during primary sensory stimulation. *Proc. Natl. Acad. Sci. U.S.A.*, *89*, 5675–5679.

36. Logothetis, N. K. (2008). What we can do and what we cannot do with fMRI. *Nature*, *453*, 869–878.

37. Piazza, M., Izard, V., Pinel, P., Le Bihan, D., & Dehaene, S. (2004). Tuning curves for approximate numerosity in the human intraparietal sulcus. *Neuron*, *44*, 547–555.

38. Krekelberg, B., Boynton, G. M. & van Wezel, R. J. (2006). Adaptation: From single cells to BOLD signals. *Trends Neurosci.*, *29*, 250–256.

39. Jacob, S. N., & Nieder, A. (2009). Tuning to non-symbolic proportions in the human frontoparietal cortex. *Eur. J. Neurosci.*, *30*, 1432–1442.

40. Demeyere, N., Rotshtein, P., & Humphreys, G. W. (2014). Common and dissociated mechanisms for estimating large and small dot arrays: Value-specific fMRI adaptation. *Hum. Brain Mapp.*, *35*, 3988–4001.

41. Ansari, D., & Dhital, B. (2006). Age-related changes in the activation of the intraparietal sulcus during nonsymbolic magnitude processing: An event-related functional magnetic resonance imaging study. *J. Cogn. Neurosci.*, *18*, 1820–1828.

42. Castelli, F., Glaser, D. E., & Butterworth, B. (2006). Discrete and analogue quantity processing in the parietal lobe: A functional MRI study. *Proc. Natl. Acad. Sci. U.S.A.*, *103*, 4693–4698.

43. Roggeman, C., Santens, S., Fias, W., & Verguts, T. (2011). Stages of nonsymbolic number processing in occipitoparietal cortex disentangled by fMRI adaptation. *J. Neurosci.*, *31*, 7168–7173.

44. Santens, S., Roggeman, C., Fias, W., & Verguts, T. (2010). Number processing pathways in human parietal cortex. *Cereb. Cortex*, *20*, 77–88.

45. Eger, E., Michel, V., Thirion, B., Amadon, A., Dehaene, S., & Kleinschmidt, A. (2009). Deciphering cortical number coding from human brain activity patterns. *Curr. Biol.*, *19*, 1608–1615.

46. Eger, E., Pinel, P., Dehaene, S., & Kleinschmidt, A. (2015). Spatially invariant coding of numerical information in functionally defined subregions of human parietal cortex. *Cereb. Cortex*, *25*, 1319–1329.

47. Harvey, B. M., Klein, B. P., Petridou, N., & Dumoulin, S. O. (2013). Topographic representation of numerosity in the human parietal cortex. *Science*, *341*, 1123–1126.

48. Santens, Roggeman, Fias, & Verguts. Number processing pathways in human parietal cortex.

49. Dormal, V., Andres, M., Dormal, G., & Pesenti, M. (2010). Mode-dependent and mode-independent representations of numerosity in the right intraparietal sulcus. *Neuroimage*, *52*, 1677–1686.

50. Kansaku, K., Johnson, A., Grillon, M. L., Garraux, G., Sadato, N., & Hallett, M. (2006). Neural correlates of counting of sequential sensory and motor events in the human brain. *Neuroimage*, *31*, 649–660.

51. Damarla, S. R., Cherkassky, V. L., & Just, M. A. (2016). Modality-independent representations of small quantities based on brain activation patterns. *Hum. Brain Mapp.*, *37*, 1296–1307.

Chapter 7

1. Gazzaniga, M. S., Ivry, R. B., & Mangun, G. R. (2014). *Cognitive neuroscience: The biology of the mind* (4th ed). New York, NY: W. W. Norton.

2. Barlow, H. (1995). The neuron doctrine in perception. In M. S. Gazzaniga (Ed.), *The cognitive neurosciences* (pp. 415–435). Cambridge, MA: MIT Press.

3. DeCharms, R. C., & Zador, A. (2000). Neural representation and the cortical code. *Annu. Rev. Neurosci., 23*, 613–647.

4. Tsao, D. Y., Freiwald, W. A., Tootell, R. B., & Livingstone, M. S. (2006). A cortical region consisting entirely of face-selective cells. *Science, 311*, 670–674.

5. Nieder, A., Freedman, D. J., & Miller, E. K. (2002). Representation of the quantity of visual items in the primate prefrontal cortex. *Science, 297*, 1708–1711.

6. Nieder, A., & Miller, E. K. (2004). A parieto-frontal network for visual numerical information in the monkey. *Proc. Natl. Acad. Sci. U.S.A., 101*, 7457–7462.

7. Okuyama, S., Kuki, T., & Mushiake, H. (2015). Representation of the numerosity "zero" in the parietal cortex of the monkey. *Sci. Rep., 5*, 10059.

8. Quintana, J., Fuster, J. M., & Yajeya, J. (1989). Effects of cooling parietal cortex on prefrontal units in delay tasks. *Brain Res., 503*, 100–110.

9. Chafee, M. V., & Goldman-Rakic, P. S. (2000). Inactivation of parietal and prefrontal cortex reveals interdependence of neural activity during memory-guided saccades. *J. Neurophysiol., 83*, 1550–1566.

10. Duhamel, J. R., Colby, C. L., & Goldberg, M. E. (1998). Ventral intraparietal area of the macaque: Congruent visual and somatic response properties. *J. Neurophysiol., 79*, 126–136.

11. Colby, C. L., & Goldberg, M. E. (1999). Space and attention in parietal cortex. *Annu. Rev. Neurosci., 22*, 319–349.

12. Onoe, H., Komori, M., Onoe, K., Takechi, H., Tsukada, H., & Watanabe, Y. (2001). Cortical networks recruited for time perception: A monkey positron emission tomography (PET) study. *Neuroimage, 13*, 37–45.

13. Janssen, P., & Shadlen, M. N. (2005). A representation of the hazard rate of elapsed time in macaque area LIP. *Nat. Neurosci., 8*, 234–241.

14. Tudusciuc, O., & Nieder, A. (2007). Neuronal population coding of continuous and discrete quantity in the primate posterior parietal cortex. *Proc. Natl. Acad. Sci. U.S.A., 104*, 14513–14518.

15. Walsh, V. (2003). A theory of magnitude: Common cortical metrics of time, space and quantity. *Trends Cogn. Sci., 7*, 483–488.

16. Bueti, D., & Walsh, V. (2009). The parietal cortex and the representation of time, space, number and other magnitudes. *Philos. Trans. R. Soc. Lond. B. Biol. Sci., 364*, 1831–1840.

17. Rusconi, E., Walsh, V., & Butterworth, B. (2005). Dexterity with numbers: rTMS over left angular gyrus disrupts finger gnosis and number processing. *Neuropsychologia, 43*, 1609–1624.

18. Sawamura, H., Shima, K., & Tanji, J. (2002). Numerical representation for action in the parietal cortex of the monkey. *Nature, 415*, 918–922.

19. Thompson, R. F., Mayers, K. S., Robertson, R. T., & Patterson, C. J. (1970). Number coding in association cortex of the cat. *Science, 168*, 271–273.

20. Nieder, A., & Miller, E. K. (2003). Coding of cognitive magnitude: Compressed scaling of numerical information in the primate prefrontal cortex. *Neuron, 37*, 149–157.

21. Merten, K., & Nieder, A. (2009). Compressed scaling of abstract numerosity representations in adult humans and monkeys. *J. Cogn. Neurosci., 21*, 333–346.

22. Nieder & Miller, Coding of cognitive magnitude.

23. Nieder, A., & Merten, K. (2007). A labeled-line code for small and large numerosities in the monkey prefrontal cortex. *J. Neurosci., 27*, 5986–5993.

24. Pouget, A., Dayan, P., & Zemel, R. (2000). Information processing with population codes. *Nat. Rev. Neurosci., 1*, 125–132.

25. Tudusciuc, O., & Nieder, A. (2007). Neuronal population coding of continuous and discrete quantity in the primate posterior parietal cortex. *Proc. Natl. Acad. Sci. U.S.A., 104*, 14513–14518.

26. Piazza, M., Izard, V., Pinel, P., Le Bihan, D., & Dehaene, S. (2004). Tuning curves for approximate numerosity in the human intraparietal sulcus. *Neuron, 44*, 547–555.

27. Jacob, S. N., & Nieder, A. (2009). Tuning to non-symbolic proportions in the human frontoparietal cortex. *Eur. J. Neurosci., 30*, 1432–1442.

28. Kersey, A. J., & Cantlon, J. F. (2017). Neural Tuning to Numerosity Relates to Perceptual Tuning in 3–6-Year-Old Children. *J. Neurosci., 37*, 512–522.

29. Nieder, A. (2013). Coding of abstract quantity by "number neurons" of the primate brain. *J. Comp. Physiol. A Neuroethol. Sens. Neural Behav. Physiol., 199*, 1–16.

30. Tudusciuc & Nieder. Neuronal population coding of continuous and discrete quantity.

31. Sawamura, H., Shima, K., & Tanji J. (2010). Deficits in action selection based on numerical information after inactivation of the posterior parietal cortex in monkeys. *J. Neurophysiol., 104*, 902–910.

32. Nieder, A., Diester, I., & Tudusciuc, O. (2006). Temporal and spatial enumeration processes in the primate parietal cortex. *Science, 313*, 1431–1435.

33. Nieder, A. (2012). Supramodal numerosity selectivity of neurons in primate prefrontal and posterior parietal cortices. *Proc. Natl. Acad. Sci. U.S.A.*, *109*, 11860–11865.

34. Piazza, M., Mechelli, A., Price, C. J. & Butterworth, B. (2006). Exact and approximate judgements of visual and auditory numerosity: An fMRI study. *Brain Res.*, *1106*, 177–188.

35. Eger, E., Sterzer, P., Russ, M. O., Giraud, A. L. & Kleinschmidt A. (2003). A supramodal number representation in human intraparietal cortex. *Neuron*, *37*, 719–725.

36. Dehaene, S., & Cohen, L. (1995). Towards an anatomical and functional model of number processing. *Math. Cogn.*, *1*, 83–120.

37. Kumar, S., & Hedges, S. B. (1998). A molecular timescale for vertebrate evolution. *Nature*, *392*, 917–920.

38. Hedges, S. B. (2002). The origin and evolution of model organisms. *Nat. Rev. Genet.*, *3*, 838–849.

39. Nieder, A. (2016). The neuronal code for number. *Nat. Rev. Neurosci.*, *17*, 366–382.

40. Olkowicz, S., Kocourek, M., Lučan, R. K., Porteš, M., Fitch, W. T., Herculano-Houzel, S., & Němec, P. (2016). Birds have primate-like numbers of neurons in the forebrain. *Proc. Natl. Acad. Sci. U.S.A.*, *113*, 7255–7260.

41. Karten, H. J. (2015). Vertebrate brains and evolutionary connectomics: On the origins of the mammalian "neocortex." *Philos. Trans. R. Soc. Lond. B. Biol. Sci.*, *370*(1684), 20150060.

42. Dugas-Ford, J., & Ragsdale, C. W. (2015). Levels of homology and the problem of neocortex. *Annu. Rev. Neurosci.*, *38*, 351–368.

43. Jarvis, E. D., Güntürkün, O., Bruce, L., Csillag, A., & Karten, H., et al. (2005). Avian Brain Nomenclature Consortium. Avian brains and a new understanding of vertebrate brain evolution. *Nat. Rev. Neurosci.*, *6*, 151–159.

44. Butler, A. B., Reiner, A., & Karten, H. J. (2011). Evolution of the amniote pallium and the origins of mammalian neocortex. *Ann. N. Y. Acad. Sci.*, *1225*, 14–27.

45. Puelles, L., Kuwana, E., Puelles, E., Bulfone, A., Shimamura, K., Keleher, J., Smiga, S., & Rubenstein, J. L. (2000). Pallial and subpallial derivatives in the embryonic chick and mouse telencephalon, traced by the expression of the genes Dlx-2, Emx-1, Nkx-2.1, Pax-6, and Tbr-1. *J. Comp. Neurol.*, *424*, 109–138.

46. Reiner, A., Perkel, D. J., Bruce, L. L., Butler, A. B., & Csillag, A., et al. (2004). Avian Brain Nomenclature Forum. Revised nomenclature for avian telencephalon and some related brainstem nuclei. *J. Comp. Neurol.*, *473*, 377–414.

47. Güntürkün, O., & Bugnyar, T. (2016). Cognition without cortex. *Trends Cogn. Sci., 20*, 291–303.

48. Divac, I., & Mogensen, J. (1985). The prefrontal "cortex" in the pigeon catecholamine histofluorescence. *Neuroscience, 15*, 677–682.

49. Güntürkün, O. (2005). The avian "prefrontal cortex" and cognition. *Curr. Opin. Neurobiol., 15*, 686–693.

50. Nieder, A. (2017). Inside the corvid brain—probing the physiology of cognition in crows. *Curr. Opin. Behav. Sci., 16*, 8–14.

51. Schnupp, J. W., & Carr, C. E. (2009). On hearing with more than one ear: Lessons from evolution. *Nat. Neurosci., 12*, 692–697.

52. Ditz, H. M., & Nieder, A. (2016). Numerosity representations in crows obey the Weber–Fechner law. *Proc. Biol. Sci., 283*, 20160083.

53. Ditz, H. M., & Nieder, A. (2015). Neurons selective to the number of visual items in the corvid songbird endbrain. *Proc. Natl. Acad. Sci. U.S.A., 112*, 7827–7832.

54. Kutter, E. F., Bostroem, J., Elger, C. E., Mormann, F., & Nieder, A. (2018). Number neurons in the human brain. *Neuron, 100*, 753-761.

55. Menon, V. (2016). Memory and cognitive control circuits in mathematical cognition and learning. *Prog. Brain Res., 227*, 159–186.

56. Goldman-Rakic, P. S., Selemon, L. D., & Schwartz, M. L. (1984). Dual pathways connecting the dorsolateral prefrontal cortex with the hippocampal formation and parahippocampal cortex in the rhesus monkey. *Neuroscience, 12*, 719–743.

57. De Smedt, B., Holloway, I. D., & Ansari, D. (2011). Effects of problem size and arithmetic operation on brain activation during calculation in children with varying levels of arithmetical fluency. *Neuroimage, 57*, 771–781.

58. Supekar, K., Swigart, A. G., Tenison, C., Jolles, D. D., Rosenberg-Lee, M., Fuchs, L., & Menon, V. (2013). Neural predictors of individual differences in response to math tutoring in primary-grade school children. *Proc. Natl. Acad. Sci. U.S.A., 110*, 8230–8235.

59. Qin, S., Cho, S., Chen, T., Rosenberg-Lee, M., Geary, D. C., & Menon, V. (2014). Hippocampal-neocortical functional reorganization underlies children's cognitive development. *Nat. Neurosci., 17*, 1263–1269.

60. Kutter, Bostroem, Elger, Mormann, & Nieder. Number neurons in the human brain.

61. Aminoff, E. M., Kveraga, K., & Bar, M. (2013). The role of the parahippocampal cortex in cognition. *Trends Cogn. Sci., 17*, 379–390.

62. Kreiman, G., Koch, C., & Fried, I. (2000). Category-specific visual responses of single neurons in the human medial temporal lobe. *Nat. Neurosci.*, *3*, 946–953.

63. Mormann, F., Kornblith, S., Cerf, M., Ison, M. J., Kraskov, A., Tran, M., Knieling, S., Quian Quiroga, R., Koch, C., & Fried, I. (2017). Scene-selective coding by single neurons in the human parahippocampal cortex. *Proc. Natl. Acad. Sci. U.S.A.*, *114*, 1153–1158.

64. Mukamel, R., Ekstrom, A. D., Kaplan, J., Iacoboni, M., & Fried, I. (2010). Single-neuron responses in humans during execution and observation of actions. *Curr. Biol.*, *20*, 750–756.

65. Suzuki, W. A. (2009). Comparative analysis of the cortical afferents, intrinsic projections and interconnections of the parahippocampal region in monkeys and rats. In M. S. Gazzaniga (Ed.), *The cognitive neurosciences* (4th ed.) (pp. 659–674). Cambridge, MA: MIT Press.

66. Buckley, P. B., & Gillman, C. B. (1974). Comparisons of digits and dot patterns. *J. Exp. Psychol.*, *103*, 1131–1136.

67. Freedman, D. J., Riesenhuber, M., Poggio, T., & Miller, E. K. (2001). Categorical representation of visual stimuli in the primate prefrontal cortex. *Science*, *291*, 312–316.

68. Roy, J. E., Riesenhuber, M., Poggio, T., & Miller, E. K. (2010). Prefrontal cortex activity during flexible categorization. *J. Neurosci.*, *30*, 8519–8528.

69. Viswanathan, P., & Nieder, A. (2013). Neuronal correlates of a visual "sense of number" in primate parietal and prefrontal cortices. *Proc. Natl. Acad. Sci. U.S.A.*, *110*, 11187–11192.

70. Viswanathan, P., & Nieder, A. (2015). Differential impact of behavioral relevance on quantity coding in primate frontal and parietal neurons. *Curr. Biol.*, *25*, 1259–1269.

71. Park, J., DeWind, N. K., Woldorff, M. G., & Brannon, E. M. (2016). Rapid and Direct Encoding of Numerosity in the Visual Stream. *Cereb. Cortex*, *26*, 748–763.

72. Leibovich, T., Vogel, S. E., Henik, A., & Ansari D. (2016). Asymmetric processing of numerical and nonnumerical magnitudes in the brain: An fMRI study. *J. Cogn. Neurosci.*, *28*, 166–176.

73. Wagener, L., Loconsole, M., Ditz, H. M., & Nieder, A. (2018). Neurons in the endbrain of numerically naive crows spontaneously encode visual numerosity. *Curr. Biol.*, *28*, 1090–1094.

74. Burr, D., & Ross, J. (2008). A visual sense of number. *Curr. Biol.*, *18*, 425–428.

75. Ross, J., & Burr, D. C. (2010). Vision senses number directly. *J. Vis., 10,* 10.1–10.8.

76. Castaldi, E., Aagten-Murphy, D., Tosetti, M., Burr, D., & Morrone, M. C. (2016). Effects of adaptation on numerosity decoding in the human brain. *Neuroimage, 143,* 364–377.

77. Arrighi, R., Togoli, I., & Burr, D. C. (2014). A generalized sense of number. *Proc. Biol. Sci., 281*(1797).

78. Anobile, G., Arrighi, R., Togoli, I., & Burr, D. C. (2016). A shared numerical representation for action and perception. *eLife, 5,* e16161.

79. Danzig, T. (1930). *Number: The language of science.* New York, NY: Free Press.

80. Meck, W. H., & Church, R. M. (1983). A mode control model of counting and timing processes. *J. Exp. Psychol. An. Behav. Proc., 9,* 320–334.

81. Dehaene, S., & Changeux, J. P. (1993). Development of elementary numerical abilities: A neural model. *J. Cogn. Neurosci., 5,* 390–407.

82. Verguts, T., & Fias, W. (2004). Representation of number in animals and humans: A neural model. *J. Cogn. Neurosci., 16,* 1493–1504.

83. Roitman, J. D., Brannon, E. M., & Platt, M. L. (2007). Monotonic coding of numerosity in macaque lateral intraparietal area. *PLoS Biol., 8,* e208.

84. Nieder, A. (2017). Evolution of cognitive and neural solutions enabling numerosity judgements: Lessons from primates and corvids. *Philos. Trans. R. Soc. Lond. B. Biol. Sci., 373*(1740).

85. Viswanathan & Nieder. Neuronal correlates of a visual "sense of number."

86. Wagener, Loconsole, Ditz, & Nieder. Neurons in the endbrain of numerically naive crows.

87. Stoianov, I., & Zorzi, M. (2012). Emergence of a "visual number sense" in hierarchical generative models. *Nat. Neurosci., 15,* 194–196.

88. Yamins, D. L., & DiCarlo, J. J. (2016). Using goal-driven deep learning models to understand sensory cortex. *Nat. Neurosci., 19,* 356–365.

89. Nasr, K., Viswanathan, P., & Nieder A. (2019). Number detectors spontaneously emerge in a deep neural network designed for visual object recognition. *Sci. Adv., 5,* eaav7903.

90. Fuster, J. M., & Alexander, G. E. (1971). Neuron activity related to short-term memory. *Science, 173,* 652–654.

91. Miller, E. K. (2013). The "working" of working memory. *Dialogues Clin. Neurosci., 15,* 411–418.

92. Shadlen, M. N., & Gold, J. I. (2004). The neurophysiology of decision-making as a window on cognition. In M. S. Gazzaniga (Ed.), *The cognitive neurosciences* (3rd ed.) (pp. 1229–1241). Cambridge, MA: MIT Press.

93. Selemon, L. D., & Goldman-Rakic, P. S. (1988). Common cortical and subcortical targets of the dorsolateral prefrontal and posterior parietal cortices in the rhesus monkey: Evidence for a distributed neural network subserving spatially guided behavior. *J. Neurosci.*, *8*, 4049–4068.

94. Grieve, K. L., Acuña, C., & Cudeiro, J. (2000). The primate pulvinar nuclei: Vision and action. *Trends Neurosci.*, *23*, 35–39.

95. Goldman-Rakic, P. S. (1988). Topography of cognition: Parallel distributed networks in primate association cortex. *Annu. Rev. Neurosci.*, *11*, 137–156.

96. Dehaene, S., & Changeux, J. P. (2011). Experimental and theoretical approaches to conscious processing. *Neuron*, *70*, 200–227.

97. Nieder, Diester, & Tudusciuc. Temporal and spatial enumeration processes in the primate parietal cortex.

98. Nieder. Supramodal numerosity selectivity of neurons in primate prefrontal and posterior parietal cortices.

99. MacLeod, C. M. (2007). The concept of inhibition in cognition. In D. S. Gorfein & C. M. MacLeod (Eds.), *Inhibition in cognition* (pp. 3–23). Washington, DC: American Psychological Association.

100. Jacob, S. N., & Nieder, A. (2014). Complementary roles for primate frontal and parietal cortex in guarding working memory from distractor stimuli. *Neuron*, *83*, 226–237.

101. Postle, B. R. (2006). Working memory as an emergent property of the mind and brain. *Neuroscience*, *139*, 23–38.

102. Lara, A. H., & Wallis, J. D. (2015). The role of prefrontal cortex in working memory: A mini review. *Front. Syst. Neurosci.*, *9*, 173.

103. Malmo, R. B. (1942). Interference factors in delayed response in monkeys after removal of frontal lobes. *J. Neurophysiol.*, *5*, 295–308.

104. Chao, L. L., & Knight, R. T. (1998). Contribution of human prefrontal cortex to delay performance. *J. Cogn. Neurosci.*, *10*, 167–177.

105. Menon. Memory and cognitive control circuits in mathematical cognition and learning.

106. Corbetta, M., & Shulman, G. (2002). Control of goal-directed and stimulus-driven attention in the brain. *Nat. Rev. Neurosci.*, *3*, 201–215.

107. Bressler, S. L., & Menon, V. (2010). Large-scale brain networks in cognition: Emerging methods and principles. *Trends Cogn. Sci.*, *14*, 277–290.

108. Raichle, M. E., Macleod, A. M., Snyder, A. Z., Powers, W. J., Gusnard, D. A., & Shulman, G. L. (2001). A default mode of brain function. *Proc. Natl. Acad. Sci. U.S.A.*, *98*, 676–682.

109. Rivera, S. M., Reiss, A. L., Eckert, M. A., & Menon, V. (2005). Developmental changes in mental arithmetic: Evidence for increased functional specialization in the left inferior parietal cortex. *Cereb. Cortex*, *15*, 1779–1790.

Chapter 8

1. Powell, A., Shennan, S., & Thomas, M. G. (2009). Late Pleistocene demography and the appearance of modern human behavior. *Science*, *324*, 1298–1301.

2. Barton, R. N. E., & d'Errico, F. (2012). North African origins of symbolically mediated behaviour and the Aterian. In S. Elias (Ed.), *Origins of human innovation and creativity developments in quaternary science*, Vol. 16 (pp. 23–34). Amsterdam, the Netherlands: Elsevier.

3. Neubauer, S., Hublin, J. J., & Gunz, P. (2018). The evolution of modern human brain shape. *Sci. Adv.*, *4*, eaao5961.

4. Klein, R. G. (2000). Archeology and the evolution of human behavior. *Evol. Anthropol.*, *9*, 17–36.

5. Abramiuk, M. A. (2012). *The foundations of cognitive archaeology*. Cambridge, MA: MIT Press.

6. Malafouris, L. (2013). *How things shape the mind: A theory of material engagement.* Cambridge, MA: MIT Press.

7. Malafouris. *How things shape the mind.*

8. Overmann, K. A. (2013). Material scaffolds in number and time. *Camb. Archaeol. J.*, *23*, 19–39.

9. Peirce, C. (1955). *Philosophical writings of Peirce*. J. Buchler (Ed.). New York, NY: Dover.

10. Deacon, T. (1997). *The symbolic species: The co-evolution of language and the human brain*. London: Norton.

11. Deacon, T. W. (1996). Prefrontal cortex and symbol learning: Why a brain capable of language evolved only once. In B. M. Velichkovsky & D. M. Rumbaugh (Eds.), *Communicating meaning: The evolution and development of language* (pp. 103–138). Hillsdale, NJ: Erlbaum.

12. Wiese, H. (2003). *Numbers, language, and the human mind.* Cambridge, UK: Cambridge Univ. Press.

13. Flegg, G. (1983). *Numbers: Their history and meaning.* New York, NY: Schocken Books.

14. d'Errico, F., Backwell, L., Villa, P., Degano, I., Lucejko, J. J., Bamford, M. K., Higham, T. F. G., Colombini, M. P., & Beaumont, P. B. (2012). Early evidence of San material culture represented by organic artifacts from Border Cave, South Africa. *Proc. Natl. Acad. Sci. U.S.A., 109,* 13214–13219.

15. Saxe, G. B. (2014). *Cultural development of mathematical ideas: Papua New Guinea studies.* New York, NY: Cambridge University Press.

16. Saxe, G. B. (1981). Body parts as numerals: A developmental analysis of numeration among the Oksapmin in Papua New Guinea. *Child Dev., 52,* 306–316.

17. Menninger, K. (1969). *Number words and number symbols.* Cambridge, MA: MIT Press

18. Flegg, G. (1989). *Numbers through the ages.* London, UK: Macmillan.

19. Wiese, H. (2007). The co-evolution of number concepts and counting words. *Lingua, 117,* 758–772.

20. Chrisomalis, S. (2010). *Numerical notation—A comparative history.* Cambridge, UK: Cambridge University Press.

21. Butterworth, B. (1999). *The mathematical brain.* London, UK: Macmillan.

22. Eccles, P. J. (2007). *An introduction to mathematical reasoning: Lectures on numbers, sets, and functions.* New York, NY: Cambridge University Press.

23. Wiese, H. (2003). Iconic and non-iconic stages in number development: The role of language. *Trends Cogn. Sci., 7,* 385–390.

24. Fayol, M., Barrouillet, P., & Marinthe, C. (1998). Predicting arithmetical achievement from neuropsychological performance: A longitudinal study. *Cognition, 68,* 63–70.

25. Andres, M., Michaux, N., & Pesenti, M. (2012). Common substrate for mental arithmetic and finger representation in the parietal cortex. *Neuroimage, 62,* 1520–1528.

26. Gerstmann, J. (1940). Syndrome of finger agnosia, disorientation for right and left, agraphia, and acalculia. *Arch Neurol Psychiatry, 44,* 398–408.

27. Butterworth. *The mathematical brain.*

28. Fuson, K. C. (1992). Relationships between counting and cardinality from age 2 to age 8. In J. Bideaud, C. Meljac, & J.-P. Fischer (Eds.), *Pathways to number: Children's developing numerical abilities* (pp. 127–149). Hillsdale, NJ: Erlbaum.

29. Terrace, H. S., Son, L. K., & Brannon, E. M. (2003). Serial expertise of rhesus macaques. *Psychol. Sci.*, *14*, 66–73.

30. Gelman, R., & Gallistel, C. R. (1978). *The child's understanding of number.* Cambridge, MA: Harvard University Press.

31. Wynn, K. (1990). Children's understanding of counting. *Cognition*, *36*, 155–193.

32. Carey, S. (2009). *The origin of concepts.* Oxford, UK: Oxford University Press.

33. Núñez, R. E. (2017). Is there really an evolved capacity for number? *Trends Cogn. Sci.*, *21*, 409–424.

34. Cheney, D. L., & Seyfarth, R. M. (1990). *How monkeys see the world. Inside the mind of another species.* Chicago, IL: University of Chicago Press.

35. Templeton, C. N., Greene, E., & Davis, K. (2005). Allometry of alarm calls: Black-capped chickadees encode information about predator size. *Science*, *308*, 1934–1937.

36. Pepperberg, I. M. (1987). Evidence for conceptual quantitative abilities in the African grey parrot: Labeling of cardinal sets. *Ethology*, *75*, 37–61.

37. Pepperberg, I. M. (1994). Numerical competence in an African grey parrot (*Psittacus erithacus*). *J. Comp. Psychol.*, *108*, 36–44.

38. Xia, L., Emmerton, J., Siemann, M., & Delius, J. D. (2001). Pigeons (*Columba livia*) learn to link numerosities with symbols. *J. Comp. Psychol.*, *115*, 83–91.

39. Diester, I., & Nieder, A. (2010). Numerical values leave a semantic imprint on associated signs in monkeys. *J. Cogn. Neurosci.*, *22*, 174–183.

40. Xia, L., Siemann, M., & Delius J. D. (2000). Matching of numerical symbols with number of responses by pigeons. *Anim. Cogn.*, *3*, 35–43.

41. Beran, M. J., & Rumbaugh, D. M. (2001). "Constructive" enumeration by chimpanzees (*Pan troglodytes*) on a computerized task. *Anim. Cogn.*, *4*, 81–89.

42. Pepperberg. Evidence for conceptual quantitative abilities in the African parrot.

43. Boysen, S. T., & Bernston, G. G. (1989). Numerical competence in a chimpanzee. *J. Comp. Psychol.*, *103*, 23–31.

44. Matsuzawa, T. (1985). Use of numbers by a chimpanzee. *Nature*, *315*, 57–59.

45. Carey. *The origin of concepts.*

46. Hauser, M. D., Chomsky, N., & Fitch, W. T. (2002). The faculty of language: What is it, who has it, and how did it evolve? *Science*, *298*, 1554–1555.

Chapter 9

1. Cipolotti, L., Butterworth, B., & Denes, G. (1991). A specific deficit for numbers in a case of dense acalculia. *Brain, 114*, 2619–2637.

2. Cipolotti, Butterworth, & Denes. A specific deficit for numbers in a case of dense acalculia.

3. Cipolotti, Butterworth, & Denes. A specific deficit for numbers in a case of dense acalculia.

4. Henschen, S. E. (1925). Clinical and anatomical contributions on brain pathology. *Arch. Neurol. Psychiatry, 13*, 226–249.

5. Berger, H. (1929). Ueber das Elektroenkephalogramm des Menschen. *Arch. Psychiatr. Nervenkr., 87*, 527–570.

6. Berger, H (1926). Über Rechenstörungen bei Herderkrankungen des Großhirns. *Arch. Psychiatr. Nervenkr., 78*, 238–263.

7. Kahn, H. J., & Whitaker H. A. (1991). Acalculia: A historical review of localization. Brain and *Cognition, 17*, 102–115.

8. Goldstein, K. (1948). *Language and language disturbances.* New York, NY: Grune & Stratton, p. 133.

9. Hécaen, H., Angelergues, R., & Houillier, S. (1961). Les variétés cliniques des acalculies au cours des lésions rétrorolandique: Approche statistique du problème. *Rev. Neurol. (Paris), 105*, 85–103.

10. Dehaene, S., Piazza, M., Pinel, P., & Cohen, L. (2003). Three parietal circuits for number processing. *Cogn. Neuropsychol., 20*, 487–506.

11. Ansari, D. (2008). Effects of development and enculturation on number representation in the human brain. *Nat. Rev. Neurosci., 9*, 278–291.

12. Arsalidou, M., & Taylor, M. J. (2011). Is 2 + 2 = 4? Meta-analyses of brain areas needed for numbers and calculations. *Neuroimage, 54*, 2382–2393.

13. Notebaert, K., Nelis, S., & Reynvoet, B. (2011). The magnitude representation of small and large symbolic numbers in the left and right hemisphere: an event-related fMRI study. *J. Cogn. Neurosci., 3*, 622–630.

14. Notebaert, Nelis, & Reynvoet. The magnitude representation of small and large symbolic numbers.

15. Holloway, I. D., Battista, C., Vogel, S. E., & Ansari, D. (201 3) Semantic and perceptual processing of number symbols: Evidence from a cross-linguistic fMRI adaptation study. *J. Cogn. Neurosci., 25*, 388–400.

16. Jacob, S. N., & Nieder, A. (2009). Notation-independent representation of fractions in the human parietal cortex. *J. Neurosci., 29*, 4652–4657.

17. Piazza, M., Pinel, P., Le Bihan, D., & Dehaene, S. (2007). A magnitude code common to numerosities and number symbols in human intraparietal cortex. *Neuron, 53*, 293–305.

18. Cohen Kadosh, R., Cohen Kadosh, K., Kaas, A., Henik, A., & Goebel, R. (2007). Notation-dependent and -independent representations of numbers in the parietal lobes. *Neuron, 53*, 307–314.

19. Arsalidou & Taylor. Is 2 + 2 = 4?

20. Eger, E., Sterzer, P., Russ, M. C., Giraud, A.-L., & Kleinschmidt, A. (2003). A supramodal number representation in human intraparietal cortex. *Neuron, 37*, 719–725.

21. Nieder, A. (2009). Prefrontal cortex and the evolution of symbolic reference. *Curr. Opin. Neurobiol., 19*, 99–108.

22. Diester, I., & Nieder, A. (2007). Semantic associations between signs and numerical categories in the prefrontal cortex. *PLoS Biol., 5*, e294.

23. Kaufmann, L., Koppelstaetter, F., Siedentopf, C., Haala, I., & Haberlandt, E., Zimmerhackl, L. B., Felber, S., & Ischebeck, A. (2006). Neural correlates of the number-size interference task in children. *Neuroreport, 17*, 587–591.

24. Cantlon, J. F., Libertus, M. E., Pinel, P., Dehaene, S., Brannon, E. M., & Pelphrey, K. A. (2009). The neural development of an abstract concept of number. *J. Cogn. Neurosci., 21*, 2217–2229.

25. Ansari, D., Garcia, N., Lucas, E., Hamon, K., & Dhital, B. (2005). Neural correlates of symbolic number processing in children and adults. *Neuroreport, 16*, 1769–1773.

26. Rivera, S. M., Reiss, A. L., Eckert, M. A., & Menon, V. (2005). Developmental changes in mental arithmetic: Evidence for increased functional specialization in the left inferior parietal cortex. *Cereb. Cortex, 15*, 1779–1790.

27. Miller, E. K., & Cohen, J. D. (2001). An integrative theory of prefrontal cortex function. *Annu. Rev. Neurosci., 24*, 167–202.

28. Tanaka, K. (1996). Inferotemporal cortex and object vision. *Annu. Rev. Neurosci., 19*, 109–139.

29. Nieder, A., & Miller, E. K. (2004). A parieto-frontal network for visual numerical information in the monkey. *Proc. Natl. Acad. Sci. U.S.A., 101*, 7457–7462.

30. Rainer, G., Rao, S. C., & Miller, E. K. (1999). Prospective coding for objects in primate prefrontal cortex. *J. Neurosci., 19*, 5493–5505.

31. Fuster, J. M., Bodner, M., & Kroger, J. K. (2000). Cross-modal and cross-temporal association in neurons of frontal cortex. *Nature, 405*, 347–351.

32. Tomita, H., Ohbayashi, M., Nakahara, K., Hasegawa, I., & Miyashita, Y. (1999). Top-down signal from prefrontal cortex in executive control of memory retrieval. *Nature, 401*, 699–703.

33. Rivera, Reiss, Eckert, & Menon. Developmental changes in mental arithmetic.

34. Kutter, E. F., Bostroem, J. Elger, C. E., Mormann, F., & Nieder, A. (2018). Number neurons in the human brain. *Neuron, 100*, 753-761.

35. Buckley, P. B., & Gillman, C. B. (1974). Comparisons of digits and dot patterns. *J. Exp. Psychol., 103*, 1131–1136.

36. Verguts, T., & Fias, W. (2004). Representation of number in animals and humans: A neural model. *J. Cogn. Neurosci., 16*, 1493–1504.

37. Szkudlarek, E., & Brannon, E. M. (2017). Does the approximate number system serve as a foundation for symbolic mathematics? *Lang. Learn. Dev., 13*, 171–190.

38. Dehaene, Piazza, Pinel, & Cohen. Three parietal circuits for number processing.

39. Allison, T., McCarthy, G., Nobre, A., Puce, A., & Belger, A. (1994). Human extrastriate visual cortex and the perception of faces, words, numbers, and colors. *Cereb. Cortex, 4*, 544–554.

40. Martin, A. (2007). The representation of object concepts in the brain. *Annu. Rev. Psychol., 58*, 25–45.

41. Dehaene, S., & Cohen, L. (2011). The unique role of the visual word form area in reading. *Trends Cogn. Sci., 15*, 254–262.

42. Starrfelt, R., & Behrmann, M. (2011). Number reading in pure alexia: A review. *Neuropsychologia, 49*, 2283–2298.

43. Roux, F. E., Lubrano, V., Lauwers-Cances, V., Giussani, C., & Démonet, J. F. (2008). Cortical areas involved in Arabic number reading. *Neurology, 70*, 210–217.

44. Shum, J., Hermes, D., Foster, B. L., Dastjerdi, M., Rangarajan, V., Winawer, J., Miller, K. J., & Parvizi, J. (2013). A brain area for visual numerals. *J. Neurosci., 33*, 6709–6715.

45. Freiwald, W. A., & Tsao, D. Y. (2010). Functional compartmentalization and viewpoint generalization within the macaque face-processing system. *Science, 330*, 845–851.

46. Pinsk, M. A., DeSimone, K., Moore, T., Gross, C. G., & Kastner, S. (2005). Representations of faces and body parts in macaque temporal cortex: A functional MRI study. *Proc. Natl. Acad. Sci. U.S.A., 102*, 6996–7001.

47. Abboud, S., Maidenbaum, S., Dehaene, S., & Amedi, A. (2015). A number-form area in the blind. *Nat. Commun.*, *6*, 6026.

48. Abboud, Maidenbaum, Dehaene, & Amedi, A number-form area in the blind.

49. Dehaene, S., & Cohen, L. (2007). Cultural recycling of cortical maps. *Neuron*, *56*, 384–398.

50. Reich, L., Szwed, M., Cohen, L., & Amedi, A. (2011). A ventral visual stream reading center independent of visual experience. *Curr. Biol.*, *21*, 363–368.

51. Amalric, M., & Dehaene S. (2016). Origins of the brain networks for advanced mathematics in expert mathematicians. *Proc. Natl. Acad. Sci. U.S.A.*, *113*, 4909–4917.

52. Dehaene-Lambertz, G., Monzalvo, K., & Dehaene S. (2018). The emergence of the visual word form: Longitudinal evolution of category-specific ventral visual areas during reading acquisition. *PLoS Biol.*, *16*, e2004103.

53. Srihasam, K., Mandeville, J. B., Morocz, I. A., Sullivan, K. J., & Livingstone, M. S. (2012). Behavioral and anatomical consequences of early versus late symbol training in macaques. *Neuron*, *73*, 608–619.

54. Srihasam, K., Vincent, J. L., & Livingstone, M. S. (2014). Novel domain formation reveals proto-architecture in inferotemporal cortex. *Nat. Neurosci.*, *17*, 1776–1783.

Chapter 10

1. Pica, P., Lemer, C., Izard, V., & Dehaene, S. (2004). Exact and approximate arithmetic in an Amazonian indigene group. *Science*, *306*, 499–503.

2. Wynn, K. (1992). Addition and subtraction by human infants. *Nature*, *358*, 749–750.

3. McCrink, K., & Wynn, K. (2004). Large-number addition and subtraction by 9-month-old infants. *Psychol. Sci.*, *15*, 776–781.

4. Flombaum, J. I., Junge, J. A., & Hauser, M. D. (2005). Rhesus monkeys (*Macaca mulatta*) spontaneously compute addition operations over large numbers. *Cognition*, *97*, 315–325.

5. Cantlon, J. F., & Brannon, E. M. (2007). Basic math in monkeys and college students. *PLoS Biol.*, *5*, e328.

6. Okuyama, S., Iwata, J., Tanji, J., & Mushiake, H. (2013). Goal-oriented, flexible use of numerical operations by monkeys. *Anim. Cogn.*, *16*, 509–518.

7. Cantlon, J. F., Merritt, D. J., & Brannon, E. M. (2016). Monkeys display classic signatures of human symbolic arithmetic. *Anim. Cogn.*, *19*, 405–415.

Notes

8. Wilson, M. L., Hauser, M. D., & Wrangham, R. W. (2001). Does participation in intergroup conflict depend on numerical assessment, range location, or rank for wild chimpanzees? *Anim. Behav., 61*, 1203–1216.

9. Bongard, S., & Nieder, A. (2010). Basic mathematical rules are encoded by primate prefrontal cortex neurons. *Proc. Natl. Acad. Sci. U.S.A., 107*, 2277–2282.

10. Vallentin, D., Bongard, S., & Nieder A. (2012). Numerical rule coding in the prefrontal, premotor, and posterior parietal cortices of macaques. *J. Neurosci., 32*, 6621–6630.

11. Eiselt, A. K., & Nieder, A. (2013). Representation of abstract quantitative rules applied to spatial and numerical magnitudes in primate prefrontal cortex. *J. Neurosci., 33*, 7526–7534.

12. Dehaene, S., & Changeux, J. P. (1991). The Wisconsin card sorting test: Theoretical analysis and modeling in a neuronal network. *Cereb. Cortex, 1*, 62–79.

13. Seamans, J. K., & Yang, C. R. (2004). The principal features and mechanisms of dopamine modulation in the prefrontal cortex. *Prog. Neurobiol., 74*, 1–58.

14. Jacob, S. N., Ott, T., & Nieder, A. (2013). Dopamine regulates two classes of primate prefrontal neurons that represent sensory signals. *J. Neurosci., 33*, 13724–13734.

15. Ott, T., & Nieder A. (2019). Dopamine and cognitive control in prefrontal cortex. Trends in Cognitive Sciences (in press).

16. Ott, T., Jacob, S. N., & Nieder A. (2014). Dopamine receptors differentially enhance rule coding in primate prefrontal cortex neurons. *Neuron, 84*, 1317–1328.

17. Miller, E. K. (2013). The "working" of working memory. *Dialogues Clin. Neurosci., 15*, 411–418.

18. Fuster, J. (2008). *The prefrontal cortex* (4th ed). London, UK: Elsevier.

19. Miller, E., & Cohen, J. (2001). An integrative theory of prefrontal cortex function. *Annu. Rev. Neurosci., 24*, 167–202.

20. Luria, A. R. (1966). *Higher cortical functions in man*. London, UK: Tavistock.

21. Shallice, T. Evans, M. E. (1978). The involvement of the frontal lobes in cognitive estimation. *Cortex, 14*, 294–303.

22. Smith, M. L., & Milner, B. (1984). Differential effects of frontal-lobe lesions on cognitive estimation and spatial memory. *Neuropsychologia, 22*, 697–705.

23. Della Sala, S., MacPherson, S. E., Phillips, L. H., Sacco, L., & Spinnler, H. (2004). The role of semantic knowledge on the cognitive estimation task—Evidence from Alzheimer s disease and healthy adult aging. *J. Neurol., 251*, 156–164.

24. Revkin, S. K., Piazza, M., Izard, V., Zamarian, L., Karner, E., & Delazer, M. (2008). Verbal numerosity estimation deficit in the context of spared semantic representation of numbers: A neuropsychological study of a patient with frontal lesions. *Neuropsychologia, 46,* 2463–2475.

25. Domahs, F., Benke, T., & Delazer, M. (2011). A case of "task-switching acalculia." *Neurocase, 17,* 24–40.

26. Roland, P. E., Friberg, L (1985). Localization of cortical areas activated by thinking. *J. Neurophysiol., 53,* 1219–1243.

27. Dehaene, S., Tzourio, N., Frak, V., Raynaud, L., Cohen, L., Mehler, J., & Mazoyer, B. (1996). Cerebral activations during number multiplication and comparison: A PET study. *Neuropsychologia, 34,* 1097–1106.

28. Sakurai, Y., Momose, T., Iwata, M., Sasaki, Y., & Kanazawa, I. (1996). Activation of prefrontal and posterior superior temporal areas in visual calculation. *J. Neurol. Sci., 139,* 89–94.

29. Burbaud, P., Degreze, P., Lafon, P., Franconi, J. M., Bouligand, B., Bioulac, B., Caille, J. M., & Allard, M. (1995). Lateralization of prefrontal activation during internal mental calculation: a functional magnetic resonance imaging study. *J. Neurophysiol., 74,* 2194–2200.

30. Rueckert, L., Lange, N., Partiot, A., Appollonio, I., Litvan, I., Le Bihan, D., & Grafman, J. (1996). Visualizing cortical activation during mental calculation with functional MRI. *Neuroimage, 3,* 97–103.

31. Chochon, F., Cohen, L., van de Moortele, P. F., & Dehaene, S. (1999). Differential contributions of the left and right inferior parietal lobules to number processing. *J. Cogn. Neurosci., 11,* 617–630.

32. Dehaene, S., Spelke, E., Pinel, P., Stanescu, R., & Tsivkin, S. (1999). Sources of mathematical thinking: Behavioral and brain-imaging evidence. *Science, 284,* 970–974.

33. Gruber, O., Indefrey, P., Steinmetz, H., & Kleinschmidt, A. (2001). Dissociating neural correlates of cognitive components in mental calculation. *Cereb. Cortex, 11,* 350–359.

34. Arsalidou, M., & Taylor, M. J. (2011). Is 2 + 2 = 4? Meta-analyses of brain areas needed for numbers and calculations. *Neuroimage, 54,* 2382–2393.

35. Dehaene, S., & Cohen, L. (1995). Towards an anatomical and functional model of number processing. *Math. Cogn., 1,* 83–120.

36. Nuerk, H.-C., Weger, U., & Willmes, K. (2001). Decade breaks in the mental number line? Putting the tens and units back in different bins. *Cognition, 82,* B25–B33.

37. Chochon, Cohen, van de Moortele, & Dehaene. Differential contributions of the left and right inferior parietal lobules to number processing.

38. Shum, J., Hermes, D., Foster, B. L., Dastjerdi, M., Rangarajan, V., Winawer, J., Miller, K. J., & Parvizi, J. (2013). A brain area for visual numerals. *J. Neurosci.*, *33*, 6709–6715.

39. Anderson, S. W., Damasio, A. R., & Damasio, H. (1990). Troubled letters but not numbers. *Brain*, *113*, 749–766.

40. Cipolotti, L. (1995). Multiple routes for reading words, why not numbers? Evidence from a case of Arabic numeral dyslexia. *Cogn. Neuropsychol.*, *12*, 313–342.

41. Arsalidou & Taylor. Is 2+2=4?

42. Campbell, J. I. D., & Xue, Q. (2001). Cognitive arithmetic across cultures. *J. Exp. Psychol.*, *130*, 299–315.

43. Grabner, R. H., Ansari, D., Koschutnig, K., Reishofer, G., Ebner, F., & Neuper, C. (2009). To retrieve or to calculate? Left angular gyrus mediates the retrieval of arithmetic facts during problem solving. *Neuropsychologia*, *47*, 604–608.

44. Singer, H. D., & Low, A. A. (1933). Acalculia: A clinical study. *Arch Neurol Psychiatry*, *29*, 467–498.

45. Dagenbach, D., & McCloskey, M. (1992). The organization of arithmetic facts in memory: Evidence from a brain-damaged patient. *Brain Cogn.*, *20*, 345–366.

46. McNeil, J. E., & Warrington, E. K. (1994). A dissociation between addition and subtraction with written calculation. *Neuropsychologia*, *32*, 717–728.

47. Presenti, M., Seron, X., & Van Der Linden, M. (1994). Selective impairment as evidence for mental organisation of arithmetical facts: BB, A case of preserved subtraction? *Cortex*, *30*, 661–671.

48. Van Harskamp, N. J., & Cipolotti, L. (2001). Selective impairment for addition, subtraction and multiplication. Implications for the organisation of arithmetical facts. *Cortex*, *37*, 363–388.

49. Benson, F., & Weir, W. S. (1972). Acalculia: Acquired anarithmetria. *Cortex*, *8*, 465–472.

50. Delazer, M., & Benke, T. (1997). Arithmetic facts without meaning. *Cortex*, *33*, 697–710.

51. Lee, K. M. (2000). Cortical areas differentially involved in multiplication and subtraction: A functional magnetic resonance imaging study and correlation with a case of selective acalculia. *Ann. Neurol.*, *48*, 657–661.

52. Lampl, Y., Eshel, Y., Gilad, R., & Sarova-Pinhas, I. (1994). Selective acalculia with sparing of the subtraction process in a patient with left parietotemporal haemorrhage. *Neurology, 44*, 1759–1761.

53. Dehaene, S., & Cohen, L. (1997). Cerebral pathways for calculation: Double dissociation between rote verbal and quantitative knowledge of arithmetic. *Cortex, 33*, 219–250.

54. Baldo, J. V., & Dronkers, N. F. (2007). Neural correlates of arithmetic and language comprehension: A common substrate? *Neuropsychologia, 45*, 229–235.

55. Selimbeyoglu, A., & Parvizi, J. (2010). Electrical stimulation of the human brain: Perceptual and behavioral phenomena reported in the old and new literature. *Front. Hum. Neurosci., 4*, 46.

56. Penfield, W., & Boldrey, E. (1937). Somatic motor and sensory representation in the cerebral cortex of man as studied by electrical stimulation. *Brain, 60*, 389–443.

57. Penfield, W., & Jasper, H. H. (1954). *Epilepsy and the Functional Anatomy of the Human Brain*. London, UK: J. & A. Churchill.

58. Duffau, H., Denvil, D., Lopes, M., Gasparini, F., Cohen, L., Capelle, L., & Van Effenterre, R. (2002). Intraoperative mapping of the cortical areas involved in multiplication and subtraction: An electrostimulation study in a patient with a left parietal glioma. *J. Neurol. Neurosurg. Psychiatry, 73*, 733–738.

59. Pu, S., Li, Y. N., Wu, C. X., Wang, Y. Z., Zhou, X. L., & Jiang, T. (2011). Cortical areas involved in numerical processing: An intraoperative electrostimulation study. *Stereotact. Funct. Neurosurg., 89*, 42–47.

60. Duffau, Denvil, Lopes, Gasparini, Cohen, Capelle, & Van Effenterre. Intraoperativemapping of the cortical areas involved in multiplication and subtraction.

61. Whalen, J., McCloskey, M., Lesser, R. P., & Gordon, B. (1997). Localizing arithmetic processes in the brain: Evidence from a transient deficit during cortical stimulation. *J. Cogn. Neurosci., 9*, 409–417.

62. Pu, Li, Wu, Wang, Zhou, & Jiang. Cortical areas involved in numerical processing.

63. Kurimoto, M., Asahi, T., Shibata, T., Takahashi, C., Nagai, S., Hayashi, N., Matsui, M., & Endo, S. (2006). Safe removal of glioblastoma near the angular gyrus by awake surgery preserving calculation ability. *Neurol. Med. Chir. (Tokyo), 46*, 46–50.

64. Arsalidou & Taylor. Is 2 + 2 = 4?

65. Göbel, S. M., Rushworth, M. F., & Walsh, V. (2006). Inferior parietal rTMS affects performance in an addition task. *Cortex, 42*, 774–781.

66. Salillas, E., Semenza, C., Basso, D., Vecchi, T., & Siegal, M. (2012). Single pulse TMS induced disruption to right and left parietal cortex on addition and multiplication. *Neuroimage, 59*, 3159–3165.

67. Andres, M., Pelgrims, B., Michaux, N., Olivier, E., & Pesenti, M. (2011). Role of distinct parietal areas in arithmetic: An fMRI-guided TMS study. *Neuroimage, 54*, 3048–3056.

68. Yu, X., Chen, C., Pu, S., Wu, C., Li, Y., Jiang, T., & Zhou, X. (2011). Dissociation of subtraction and multiplication in the right parietal cortex: Evidence from intraoperative cortical electrostimulation. *Neuropsychologia, 49*, 2889–2895.

69. Della Puppa, A., De Pellegrin, S., d'Avella, E., Gioffrè, G., Munari, M., Saladini. M., Salillas, E., Scienza, R., & Semenza, C. (2013). Right parietal cortex and calculation processing: Intraoperative functional mapping of multiplication and addition in patients affected by a brain tumor. *J. Neurosurg., 119*, 1107–1111.

70. Della Puppa, A., De Pellegrin, S., Rossetto, M., Rustemi, O., Saladini, M., Munari. M., & Scienza, R. (2015). Intraoperative functional mapping of calculation in parietal surgery. New insights and clinical implications. *Acta Neurochir. (Wien), 157*, 971–977.

71. Gelman, R., & Butterworth, B. (2005). Number and language: How are they related? *Trends Cogn. Sci., 9*, 6–10.

72. Dahmen, W., Hartje, W., Büssing, A., & Sturm, W. (1982). Disorders of calculation in aphasic patients—spatial and verbal components. *Neuropsychologia, 20*, 145–153.

73. Delazer, M., Girelli, L., Semenza, C., & Denes, G. (1999). Numerical skills and aphasia. *J. Int. Neuropsychol. Soc., 5*, 213–221.

74. Henschen, S. E. (1925). Clinical and anatomical contributions on brain pathology. *Arch. Neurol. Psychiatry, 13*, 226–249.

75. Rossor, M. N., Warrington, E. K., & Cipolotti, L. (1995). The isolation of calculation skills. *J. Neurol., 242*, 78–81.

76. Rossor, Warrington, & Cipolotti. The isolation of calculation skills.

77. Butterworth, B., Cappelletti, M., & Kopelman, M. (2001). Category specificity in reading and writing: The case of number words. *Nat. Neurosci., 4*, 784–786.

78. Cappelletti, M., Butterworth, B., & Kopelman, M. (2001). Spared numerical abilities in a case of semantic dementia. *Neuropsychologia, 39*, 1224–1239.

79. Varley, R. A., Klessinger, N. J., Romanowski, C. A., & Siegal, M. (2005). Agrammatic but numerate. *Proc. Natl. Acad. Sci. U.S.A., 102*, 3519–3524.

80. Cappelletti, M., Butterworth, B., & Kopelman, M. (2012). Numeracy skills in patients with degenerative disorders and focal brain lesions: A neuropsychological investigation. *Neuropsychology, 26*, 1–19.

81. Rath, D., Domahs, F., Dressel, K., Claros-Salinas, D., Klein, E., Willmes, K., & Krinzinger, H. (2015). Patterns of linguistic and numerical performance in aphasia. *Behav. Brain Funct., 11*, 2.

82. Luchelli, F., & De Renzi, E. (1993). Primary dyscalculia after a medial frontal lesion of the left hemisphere. *J. Neurol. Neurosurg. Psychiatry, 56*, 304–307.

83. Warrington, E. K. (1982). The fractionation of arithmetical skills: A single study. *Q. J. Exp. Psychol, 34*, 31–51.

84. Anderson, S. W., Damasio, A. R., & Damasio, H. (1990). Troubled letters but not numbers. Domain specific cognitive impairments following focal damage in frontal cortex. *Brain, 113*, 749–766.

85. Luchelli & De Renzi. Primary dyscalculia after a medial frontal lesion of the left hemisphere.

86. Dehaene & Cohen. Cerebral pathways for calculation.

87. Varley, R. A., Klessinger, N. J., Romanowski, C. A., & Siegal, M. (2005). Agrammatic but numerate. *Proc. Natl. Acad. Sci. U.S.A., 102*, 3519–3524.

88. Varley, Klessinger, Romanowski, & Siegal. Agrammatic but numerate.

89. Baldo & Dronkers. Neural correlates of arithmetic and language comprehension.

90. Dronkers, N. F., Wilkins, D. P., Van Valin, R. D. Jr, Redfern, B. B., & Jaeger, J. J. (2004). Lesion analysis of the brain areas involved in language comprehension. *Cognition, 92*, 145–177.

91. Klessinger, N., Szczerbinski, M., & Varley, R. (2007). Algebra in a man with severe aphasia. *Neuropsychologia, 45*, 1642–1648.

92. Roux, F. E., Boukhatem, L., Draper, L., Sacko, O., & Démonet, J. F. (2009). Cortical calculation localization using electrostimulation. *J. Neurosurg., 110*, 1291–1299.

93. Fedorenko, E., & Varley, R. (2016). Language and thought are not the same thing: Evidence from neuroimaging and neurological patients. *Ann. N. Y. Acad. Sci., 1369*, 132–153.

94. Wagner, R. (1860). *Vorstudien zu einer wissenschaftlichen Morphologie und Physiologie des menschlichen Gehirns als Seelenorgan / 1: Über die typischen Verschiedenheiten der Windungen der Hemisphären und über die Lehre vom Hirngewicht, mit besonderer Rücksicht auf die Hirnbildung intelligenter Männer.* Göttingen, Germany: Dieterich.

95. Finger, S. (2001). *Origins of neuroscience. A history of explorations into brain function.* Oxford, UK: Oxford University Press.

96. Wagner. *Vorstudien zu einer wissenschaftlichen Morphologie und Physiologie des menschlichen Gehirns als Seelenorgan.*

97. Broca, P. (1861). Sur le volume et la forme du cerveau suivant les individus et les races. *Bull. Soc. d'Anthrop. (Paris), 2*, 139–204.

98. Spitzka, E. A. (1907). A study of the brains of six eminent scientists and scholars belonging to the American anthropometric society, together with a description of the skull of Professor E. D. Cope. *Trans. Am. Philos. Soc., 21*(4), 175–308.

99. Spitzka, E. A. (1907). A study of the brains of six eminent scientists and scholars.

100. Mall, F. P. (1909). On several anatomical characters of the human brain, said to vary according to race and sex, with especial reference to the weight of the frontal lobe. *Am. J. Anat., 9*, 1–32.

101. Schweizer, R., Wittmann, A., & Frahm, J. (2014). A rare anatomical variation newly identifies the brains of C. F. Gauss and C. H. Fuchs in a collection at the University of Gottingen. *Brain, 137*, e269.

102. Bodanis, D. (2000). *E = mc2: A biography of the world's most famous equation.* New York, NY: Walker.

103. Paterniti, M. (2000). *Driving Mr. Albert: A trip across America with Einstein's brain.* New York, NY: Dial Press.

104. Burrell, B. D. (2015). Genius in a jar. *Scientific American, 313*, 83–87.

105. Paterniti. *Driving Mr. Albert.*

106. Diamond, M. C., Scheibel, A. B., Murphy, G. M., & Harvey, T. (1985). On the brain of a scientist: Albert Einstein. *Exp. Neurol., 88*, 198–204.

107. Diamond, M. C., Scheibel, A. B., Murphy, G. M., & Harvey. On the brain of a scientist.

108. Anderson, B., & Harvey, T. (1996). Alterations in cortical thickness and neuronal density in the frontal cortex of Albert Einstein. *Neurosci. Lett., 210*, 161–164.

109. Witelson, S. F., Kigar, D. L., & Harvey T. (1999). The exceptional brain of Albert Einstein. *Lancet, 353*, 2149–2153.

110. Falk, D., Lepore, F. E., & Noe, A. (2013). The cerebral cortex of Albert Einstein: A description and preliminary analysis of unpublished photographs. *Brain, 136*, 1304–1327.

111. Men, W., Falk, D., Sun, T., Chen, W., Li, J., Yin, D., Zang, L., & Fan, M. (2014). The corpus callosum of Albert Einstein's brain: Another clue to his high intelligence? *Brain, 137*, e268.

112. Hines, T. (2014). Neuromythology of Einstein's brain. *Brain Cogn., 88*, 21–25.

113. Sextilliarden, Trillionen: Rüdiger Gamm hat nichts als Zahlen im Kopf. *Handelsblatt*, June 11, 2004. https://www.handelsblatt.com/archiv/sextilliarden-trillionen-ruediger-gamm-hat-nichts-als-zahlen-im-kopf/2341536.html?ticket=ST-632046-CAgpRrES225zdOvzneYw-ap2.

114. Adam, D. (2000). He's a nurtural. *Nature*, December 28, 2000. doi:10.1038/news001228-5

115. Pesenti, M., Zago, L., Crivello, F., Mellet, E., Samson, D., Duroux, B., Seron, X., Mazoyer, B., & Tzourio-Mazoyer, N. (2001). Mental calculation in a prodigy is sustained by right prefrontal and medial temporal areas. *Nat. Neurosci.*, 4, 103–107.

116. Amalric, M., & Dehaene, S. (2016). Origins of the brain networks for advanced mathematics in expert mathematicians. *Proc. Natl. Acad. Sci. U.S.A.*, 113, 4909–4917.

Chapter 11

1. Galton, F. (1880). Visualised numerals. *Nature*, 21, 252–256.

2. Seron, X., Pesenti, M., Noel, M. P., Deloche, G., & Cornet, J. A. (1992). Images of numbers, or "When 98 is upper left and 6 sky blue." *Cognition*, 44, 159–196.

3. Rickmeyer, K. (2001). "Die Zwölf liegt hinter der nächsten Kurve und die Sieben ist pinkrot": Zahlenraumbilder und bunte Zahlen. *J. Mathematik-Didaktik*, 22, 51–71.

4. Restle, F. (1970). Speed of adding and comparing numbers. *J. Exp. Psychol.*, 91, 191–205.

5. Dehaene, S., Bossini, S., & Giraux, P. (1993). The mental representation of parity and number magnitude. *J. Exp. Psychol. Gen.*, 122, 371–396.

6. Fischer, M. H., Castel, A. D., Dodd, M. D., & Pratt, J. (2003). Perceiving numbers causes spatial shifts of attention. *Nat. Neurosci.*, 6, 555–556.

7. Zorzi, M., Priftis, K., & Umilta, C. (2002). Brain damage: Neglect disrupts the mental number line. *Nature*, 417, 138–139.

8. Vuilleumier, P., Ortigue, S., & Brugger, P. (2004). The number space and neglect. *Cortex*, 40, 399–410.

9. Shaki, S., & Fischer, M. H. (2008). Reading space into numbers: A crosslinguistic comparison of the SNARC effect. *Cognition*, 108, 590–599.

10. Dehaene, Bossini, & Giraux. The mental representation of parity and number magnitude.

11. de Hevia, M. D., Veggiotti, L., Streri, A., & Bonn, C. D. (2017). At Birth, Humans Associate "Few" with Left and "Many" with Right. *Curr. Biol.*, 27, 3879–3884.e2.

12. Drucker, C. B., & Brannon, E. M. (2014). Rhesus monkeys (Macaca mulatta) map number onto space. *Cognition, 132*, 57–67.

13. Rugani, R., Vallortigara, G., Priftis, K., & Regolin, L. (2015). Animal cognition. Number-space mapping in the newborn chick resembles humans' mental number line. *Science, 347*, 534–536.

14. McCrink, K., Dehaene, S., & Dehaene-Lambertz, G. (2007). Moving along the number line: Operational momentum in nonsymbolic arithmetic. *Percept. Psychophys., 69*, 1324–1333.

15. McCrink, K., & Wynn, K. (2009). Operational momentum in large-number addition and subtraction by 9-month-olds. *J. Exp. Child Psychol., 103*, 400–408.

16. McCrink, K., & Hubbard, T. (2017). Dividing Attention Increases Operational Momentum. *J. Numer. Cogn., 3*, 230–245.

17. Kahneman, D. (2013). *Thinking, fast and slow*. New York, NY: Farrar, Straus, and Giroux.

18. Knops, A., Thirion, B., Hubbard, E. M., Michel, V., & Dehaene, S. (2009). Recruitment of an area involved in eye movements during mental arithmetic. *Science, 324*, 1583–1585.

19. Koenigs, M., Barbey, A. K., Postle, B. R., & Grafman, J. (2009). Superior parietal cortex is critical for the manipulation of information in working memory. *J. Neurosci., 29*, 14980–14986.

20. Knops, Thirion, Hubbard, Michel, & Dehaene. Recruitment of an area involved in eye movements during mental arithmetic.

21. Knops, Thirion, Hubbard, Michel, & Dehaene. Recruitment of an area involved in eye movements during mental arithmetic.

22. Colby, C. L., & Goldberg, M. E. (1999). Space and attention in parietal cortex. *Annu. Rev. Neurosci., 22*, 319–349.

23. Snyder, L. H., Batista, A. P., & Andersen, R. A. (2000). Intention-related activity in the posterior parietal cortex: A review. *Vision Res., 40*, 1433–1441.

24. Duhamel, J. R., Bremmer, F., Benhamed, S., & Graf, W. (1997). Spatial invariance of visual receptive fields in parietal cortex neurons. *Nature, 389*, 845–858.

25. Duhamel, J. R., Colby, C. L., & Goldberg, M. E. (1998). Ventral intraparietal area of the macaque: Congruent visual and somatic response properties. *J. Neurophysiol., 79*, 126–136.

26. Taira, M., Georgopolis, A. P., Murata, A., & Sakata, H. (1990). Parietal cortex neurons of the monkey related to the visual guidance of hand movement. *Exp. Brain Res., 79*, 155–166.

27. Murata, A., Gallese, V., Luppino, G., Kaseda, M., & Sakata, H. (2000). Selectivity for the shape, size, and orientation of objects for grasping in neurons of monkey parietal area AIP. *J. Neurophysiol., 83*, 2580–2601.

28. Orban, G. A., Van Essen, D., & Vanduffel, W. (2004). Comparative mapping of higher visual areas in monkeys and humans. *Trends Cogn. Sci., 8*, 315–324.

29. Simon, O., Mangin, J. F., Cohen, L., Le Bihan, D., & Dehaene, S. (2002). Topographical layout of hand, eye, calculation, and language-related areas in the human parietal lobe. *Neuron, 33*, 475–487.

30. Culham, J. C., & Kanwisher, N. G. (2001). Neuroimaging of cognitive functions in human parietal cortex. *Curr. Opin. Neurobiol., 11*, 157–163.

31. Hubbard, E. M., Piazza, M., Pinel, P., & Dehaene, S. (2005). Interactions between number and space in parietal cortex. *Nat. Rev. Neurosci., 6*, 435–448.

32. Henik, A., & Tzelgov, J. (1982). Is three greater than five: The relation between physical and semantic size in comparison tasks. *Mem. Cognit., 10*, 389–395.

33. Walsh, V. (2003). A theory of magnitude: Common cortical metrics of time, space and quantity. *Trends Cogn. Sci., 7*, 483–488.

34. Bueti, D., & Walsh, V. (2009). The parietal cortex and the representation of time, space, number and other magnitudes. *Philos. Trans. R. Soc. Lond. B. Biol. Sci., 364*(1525), 1831–1840.

35. Tudusciuc, O., & Nieder, A. (2009). Contributions of primate prefrontal and posterior parietal cortices to length and numerosity representation. *J. Neurophysiol., 101*, 2984–2994.

36. Genovesio, A., Tsujimoto, S., & Wise, S. P. (2011). Prefrontal cortex activity during the discrimination of relative distance. *J. Neurosci., 31*, 3968–3980.

37. Vallentin, D., & Nieder, A. (2008). Behavioural and prefrontal representation of spatial proportions in the monkey. *Curr. Biol., 18*, 1420–1425.

Chapter 12

1. Frege, G. (1980). *The foundations of arithmetic: A logico-mathematical enquiry into the concept of number* (J. L. Austin, Trans). Evanston, IL: Northwestern University Press (Originallly published 1884).

2. Gelman, R., & Gallistel, C. R. (1978). *The child's understanding of number.* Cambridge, MA: Harvard University Press.

3. Carey, S. (2009). *The origin of concepts.* Oxford, UK: Oxford University Press.

4. LeCorre, M., Brannon, E. M., Van de Walle, G. A., & Carey, S. (2006). Re-visiting the competence/performance debate in the acquisition of the counting principles. *Cogn. Psychol.*, *52*, 130–169.

5. Le Corre, M., & Carey, S. (2007). One, two, three, four, nothing more: An investigation of the conceptual sources of the verbal counting principles. *Cognition*, *105*, 395–438.

6. Carey. *The origin of concepts*.

7. Dedekind, R. (1901). Essays in the theory of numbers, 1. Continuity of irrational numbers, 2. The nature and meaning of numbers (W. W. Beman, Trans.). (Originally published in 1872 and 1888, respectively).

8. Izard, V., Pica, P., Spelke, E. S., & Dehaene, S. (2008). Exact equality and successor function: Two key concepts on the path towards understanding exact numbers. *Philos. Psychol.*, *21*, 491–505.

9. Spelke, E. S., & Kinzler, K. D. (2007). Core knowledge. *Dev. Sci.*, *10*, 89–96.

10. Piazza, M. (2010). Neurocognitive start-up tools for symbolic number representations. *Trends Cogn. Sci.*, *14*, 542–551.

11. Gallistel, C. R., & Gelman, R. (1992). Preverbal and verbal counting and computation. *Cognition*, *44*, 43–74.

12. Huntley-Fenner, G. (2001). Children's understanding of number is similar to adults' and rats': Numerical estimation by 5–7-year olds. *Cognition*, *78*, B27–B40.

13. Lipton, J. S., & Spelke, E. (2005). Preschool children's mapping of number words to nonsymbolic numerosities. *Child Dev.*, *76*, 978–988.

14. Carey. *The origin of concepts*.

15. Hauser, M. D., Chomsky, N., & Fitch, W. T. (2002). The faculty of language: What is it, who has it, and how did it evolve? *Science*, *298*, 1569–1579.

16. Spelke, E. S. (2017). Core knowledge, language, and number. *Lang. Learn. Dev.*, *13*, 147–170.

17. Moyer, R. S., & Landauer, T. K. (1967). Time required for judgements of numerical inequality. *Nature*, *215*, 1519–1520.

18. Buckley P. B., & Gillman C. B. (1974). Comparisons of digits and dot patterns. *J. Exp. Psychol.*, *103*, 1131–1136.

19. Dehaene, S., Dupoux, E., & Mehler, J. (1990). Is numerical comparison digital? Analogical and symbolic effects in two-digit number comparison. *J. Exp. Psychol. Hum. Percept. Perform.*, *16*, 626–641.

20. Koechlin, E., Naccache, N., Block, E., & Dehaene, S. (1999). Primed numbers: Exploring the modularity of numerical representations with masked and unmasked priming. *J. Exp. Psychol. Hum. Percept. Perform.*, *25*, 1882–1905.

21. Halberda, J., Mazzocco, M. M. M., & Feigenson, L. (2008). Individual differences in non-verbal number acuity correlate with maths achievement. *Nature*, *455*, 665–668.

22. Wynn, K. (1992). Addition and subtraction by human infants. *Nature*, *358*, 749–750.

23. Gilmore, C. K., McCarthy, S. E., & Spelke, E. S. (2007). Symbolic arithmetic knowledge without instruction. *Nature*, *447*, 589–591.

24. Park, J., & Brannon, E. M. (2013). Training the approximate number system improves math proficiency. *Psychol. Sci.*, *24*, 2013–2019.

25. Piazza, M., Pica, P., Izard, V., Spelke, E. S., & Dehaene, S. (2013). Education enhances the acuity of the nonverbal approximate number system. *Psychol. Sci.*, *24*, 1037–1043.

26. Piazza, Pica, Izard, Spelke, & Dehaene. Education enhances the acuity of the nonverbal approximate number system.

27. Park & Brannon. Training the approximate number system improves math proficiency.

28. Iuculano, T., Tang, J., Hall, C. W. B., & Butterworth, B. (2008). Core information processing deficits in developmental dyscalculia and low numeracy. *Dev. Sci.*, *11*, 669–680.

29. Holloway, I. D., & Ansari, D. (2009). Mapping numerical magnitudes onto symbols: The numerical distance effect and individual differences in children's mathematics achievement. *J. Exp. Child Psychol.*, *103*, 17–29.

30. Sasanguie, D., Defever, E., Maertens, B., & Reynvoet, B. (2014). The approximate number system is not predictive for symbolic number processing in kindergarteners. *Q. J. Exp. Psychol.*, *67*, 271–280.

31. Chen, Q., & Li, J. (2014). Association between individual differences in non-symbolic number acuity and math performance: A meta-analysis. *Acta Psychol. (Amst.)*, *148*, 163–172.

32. Fazio, L. K., Bailey, D. H., Thompson, C. A., & Siegler, R. S. (2014). Relations of different types of numerical magnitude representations to each other and to mathematics achievement. *J. Exp. Child Psychol.*, *123*, 53–72.

33. Schneider, M., Beeres, K., Coban, L., Merz, S., Susan Schmidt, S., Stricker, J., & De Smedt, B. (2017). Associations of non-symbolic and symbolic numerical magnitude processing with mathematical competence: A meta-analysis. *Dev. Sci.*, *20*, e12372.

34. Cantlon, J. F., Brannon, E. M., Carter, E. J., & Pelphrey, K. A. (2006). Functional imaging of numerical processing in adults and 4-y-old children. *PLoS Biol., 4*, e125.

35. Hyde, D. C., Boas, D. A., Blair, C., & Carey, S. (2010). Near-infrared spectroscopy shows right parietal specialization for number in pre-verbal infants. *Neuroimage, 53*, 647–652.

36. Izard, V., Dehaene-Lambertz, G., & Dehaene, S. (2008). Distinct cerebral pathways for object identity and number in human infants. *PLoS Biol., 6*, e11.

37. Piazza, M., Izard, V., Pinel, P., Le Bihan, D., & Dehaene, S. (2004). Tuning curves for approximate numerosity in the human intraparietal sulcus. *Neuron, 44*, 547–555.

38. Kutter, E. F., Bostroem, J., Elger, C. E., Mormann, F., & Nieder, A. (2018). Number neurons in the human brain. *Neuron, 100*, 753–761.

39. Nieder, A. (2016). The neuronal code for number. *Nat. Rev. Neurosci., 17*, 366–382.

40. Kersey, A. J., & Cantlon, J. F. (2017). Neural Tuning to Numerosity Relates to Perceptual Tuning in 3–6-Year-Old Children. *J. Neurosci., 37*, 512–522.

41. Kaufmann, L., Wood, G., Rubinsten, O., & Henik, A. (2011). Meta-analyses of developmental fMRI studies investigating typical and atypical trajectories of number processing and calculation. *Dev. Neuropsychol., 36*, 763–787.

42. Butterworth, B. (2005). The development of arithmetical abilities. *J. Child Psychol. Psychiatry, 46*, 3–18.

43. Kaufmann, L., Vogel, S. E., Wood, G., Kremser, C., Schocke, M., Zimmerhackl, L.-B., & Koten, J. W. (2008). A developmental fMRI study of nonsymbolic numerical and spatial processing. *Cortex, 44*, 376–385.

44. Ansari, D., Garcia, N., Lucas, E., Hamon, K., & Dhital, B. (2005). Neural correlates of symbolic number processing in children and adults. *Neuroreport, 16*, 1769–1773.

45. Kaufmann, Vogel, Wood, Kremser, Schocke, & Zimmerhackl. A developmental fMRI study of nonsymbolic numerical and spatial processing.

46. Cantlon, J. F., Libertus, M. E., Pinel, P., Dehaene, S., Brannon, E. M., & Pelphrey, K. A. (2009). The neural development of an abstract concept of number. *J. Cogn. Neurosci., 21*, 2217–2229.

47. Holloway, I. D., & Ansari, D. (2010). Developmental specialization in the right intraparietal sulcus for the abstract representation of numerical magnitude. *J. Cogn. Neurosci., 22*, 2627–2637.

48. Rivera, S. M., Reis, A. L., Eckert, M. A., & Menon, V. (2005). Developmental changes in mental arithmetic: Evidence for increased functional specialization in the left inferior parietal cortex. *Cereb. Cortex, 15*, 1779–1790.

49. Kucian, K., von Aster, M., Loenneker, T., Dietrich, T., & Martin, E. (2008). Development of neural networks for exact and approximate calculation: A fMRI study. *Dev. Neuropsychol.*, *33*, 447–473.

50. De Smedt, B., Holloway, I. D., & Ansari, D. (2011). Effects of problem size and arithmetic operation on brain activation during calculation in children with varying levels of arithmetical fluency. *Neuroimage*, *57*, 771–781.

51. Supekar, K., Swigart, A. G., Tenison, C., Jolles, D. D., Rosenberg-Lee, M., Fuchs, L., & Menon, V. (2013). Neural predictors of individual differences in response to math tutoring in primary-grade school children. *Proc. Natl. Acad. Sci. U.S.A.*, *110*, 8230–8235.

52. Qin, S., Cho, S., Chen, T., Rosenberg-Lee, M., Geary, D. C., & Menon, V. (2014). Hippocampal-neocortical functional reorganization underlies children's cognitive development. *Nat. Neurosci.*, *17*, 1263–1269.

53. Menon, V. (2016). Memory and cognitive control circuits in mathematical cognition and learning. *Prog. Brain Res.*, *227*, 159–186.

54. Kucian, von Aster, Loenneker, Dietrich, & Martin. Development of neural networks for exact and approximate calculation.

55. Diester, I., & Nieder, A. (2007). Semantic associations between signs and numerical categories in the prefrontal cortex. *PLoS Biol.*, *5*, e294.

56. Piazza, M., Pinel, P., Le Bihan, D., & Dehaene, S. (2007). A magnitude code common to numerosities and number symbols in human intraparietal cortex. *Neuron*, *53*, 293–305.

57. Holloway, I. D., Price, G. R., & Ansari, D. (2010). Common and segregated neural pathways for the processing of symbolic and nonsymbolic numerical magnitude: An fMRI study. *Neuroimage*, *49*, 1006–1017.

58. Arsalidou, M., & Taylor, M. J. (2011). Is 2 + 2 = 4? Meta-analyses of brain areas needed for numbers and calculations. *Neuroimage*, *54*, 2382–2393.

59. Eger, E., Michel, V., Thirion, B., Amadon, A., Dehaene, S., & Kleinschmidt, A. (2009). Deciphering cortical number coding from human brain activity patterns. *Curr. Biol.*, *19*, 1608–1615.

60. Cappelletti, M., Barth, H., Fregni, F., Spelke, E. S., & Pascual-Leone, A. (2007). rTMS over the intraparietal sulcus disrupts numerosity processing. *Exp. Brain Res.*, *179*, 631–642.

61. Lemer, C., Dehaene, S., Spelke, E., & Cohen, L. (2003). Approximate quantities and exact number words: Dissociable systems. *Neuropsychologia*, *41*, 1942–1958.

62. Ashkenazi, S., Henik, A., Ifergane, G., & Shelef, I. (2008). Basic numerical processing in left intraparietal sulcus (IPS) acalculia. *Cortex, 44*, 439–448.

63. Koss, S., Clark, R., Vesely, L., Weinstein, J., Powers, C., Richmond, L., Farag, C., Gross, R., Liang, T. W., & Grossman, M. (2010). Numerosity impairment in corticobasal syndrome. *Neuropsychology, 24*, 476–492.

64. Naccache, L., & Dehaene, S. (2001). The priming method: Imaging unconscious repetition priming reveals an abstract representation of number in the parietal lobes. *Cereb. Cortex, 11*, 966–974.

65. Notebaert, K., Pesenti, M., & Reynvoet, B. (2010). The neural origin of the priming distance effect: Distance-dependent recovery of parietal activation using symbolic magnitudes. *Hum. Brain Mapp., 31*, 669–677.

66. Cohen Kadosh, R., Cohen Kadosh, K., Kaas, A., Henik, A., & Goebel, R. (2007). Notation-dependent and -independent representations of numbers in the parietal lobes. *Neuron, 53*, 307–314.

67. Eger, E., Sterzer, P., Russ, M. C., Giraud, A.-L., & Kleinschmidt, A. (2003). A supramodal number representation in human intraparietal cortex. *Neuron, 37*, 719–725.

68. Piazza, M., Mechelli, A., Price, C. J., & Butterworth, B. (2006). Exact and approximate judgements of visual and auditory numerosity: An fMRI study. *Brain Res., 1106*, 177–188.

69. Cavdaroglu, S., Katz, C., & Knops, A. (2015). Dissociating estimation from comparison and response eliminates parietal involvement in sequential numerosity perception. *Neuroimage, 116*, 135–148.

70. Damarla, S. R., Cherkassky, V. L., & Just, M. A. (2016). Modality-independent representations of small quantities based on brain activation patterns. *Hum. Brain Mapp., 37*, 1296–1307.

71. Cohen Kadosh, R., & Walsh, V. (2009). Numerical representation in the parietal lobes: Abstract or not abstract? *Behav. Brain Sci., 32*, 313–328.

72. Cohen Kadosh & Walsh. Numerical representation in the parietal lobes.

73. Nieder. The neuronal code for number.

74. Kutter, Bostroem, Elger, Mormann, & Nieder. Number neurons in the human brain.

Chapter 13

1. Butterworth, B. (2017). The implications for education of an innate numerosity-processing mechanism. *Philos. Trans. R. Soc. Lond. B. Biol. Sci., 373*(1740).

2. Butterworth, B., & Kovas, Y. (2013). Understanding neurocognitive developmental disorders can improve education for all. *Science, 340*, 300–305.

3. Shalev, R. S. (2007). Prevalence of developmental dyscalculia. In D. B. Berch & M. M. M. Mazzocco (Eds.), *Why is math so hard for some children? The nature and origins of mathematical learning difficulties and disabilities* (pp. 151–172). Baltimore, MD: Paul H. Brookes Publishing.

4. Gabrieli, J. D. (2009). Dyslexia: A new synergy between education and cognitive neuroscience. *Science, 325*, 280–283.

5. Beddington, J., Cooper, C. L., Field, J., Goswami, U., Huppert, F. A., Jenkins, R., Jones, H. S., Kirkwood, T. B., Sahakian, B. J., & Thomas, S. M. (2008). The mental wealth of nations. *Nature, 455*, 1057–1060.

6. Parsons, S., & Bynner, J. (2005). *Does Numeracy Matter More?* London, UK: National Research and Development Centre for Adult Literacy and Numeracy, Institute of Education.

7. Butterworth, B., Varma, S., & Laurillard, D. (2011). Dyscalculia: From brain to education. *Science, 332*, 1049–1053.

8. Gross-Tsur, V., Manor, O., & Shalev, R. S. (1996). Developmental dyscalculia: Prevalence and demographic features. *Dev. Med. Child Neurol., 38*, 25–33.

9. Landerl, K., & Moll, K. (2010). Comorbidity of learning disorders: Prevalence and familial transmission. *J. Child. Psychol. Psychiatry, 51*, 287–294.

10. Donlan, C. (2007). Mathematical development in children with specific language impairments. In D. B. Berch & M. M. M. Mazzocco (Eds.), *Why is math so hard for some children? The nature and origins of mathematical learning difficulties and disabilities* (pp. 151–172). Baltimore, MD: Paul H. Brookes Publishing.

11. Monuteaux, M. C., Faraone, S. V., Herzig, K., Navsaria, N., & Biederman, J. (2005). ADHD and dyscalculia: Evidence for independent familial transmission. *J. Learn. Disabil., 38*, 86–93.

12. Landerl, K., Bevan, A., & Butterworth, B. (2004). Developmental dyscalculia and basic numerical capacities: A study of 8–9-year-old students. *Cognition, 93*, 99–125.

13. Cappelletti, M., Chamberlain, R., Freeman, E. D., Kanai, R., Butterworth, B., Price, C. J., & Rees, G. (2014). Commonalities for numerical and continuous quantity skills at temporo-parietal junction. *J. Cogn. Neurosci., 26*, 986–999.

14. Landerl, Bevan, & Butterworth. Developmental dyscalculia and basic numerical capacities.

15. Piazza, M., Facoetti, A., Trussardi, A. N., Berteletti, I., Conte, S., Lucangeli, D., Dehaene, S., & Zorzi, M. (2010). Developmental trajectory of number acuity reveals a severe impairment in developmental dyscalculia. *Cognition, 116*, 33–41.

16. Piazza, Facoetti, Trussardi, Berteletti, Conte, Lucangeli, Dehaene, & Zorzi. Developmental trajectory of number acuity.

17. Reeve, R., Reynolds, F., Humberstone, J., Butterworth, B. (2012). Stability and change in markers of core numerical competencies. *J. Exp. Psychol. Gen.*, *141*, 649–666.

18. Gilmore, C. K., McCarthy, S. E., & Spelke, E. S. (2007). Symbolic arithmetic knowledge without instruction. *Nature*, *447*, 589–591.

19. Halberda, J., Mazzocco, M. M. M., & Feigenson, L. (2008). Individual differences in non-verbal number acuity correlate with maths achievement. *Nature*, *455*, 665–668.

20. Lyons, I. M., Price, G. R., Vaessen, A., Blomert, L., & Ansari, D. (2014). Numerical predictors of arithmetic success in grades 1–6. *Dev. Sci.*, *17*, 714–726.

21. Chen, Q., & Li, J. (2014). Association between individual differences in non-symbolic number acuity and math performance: A meta-analysis. *Acta Psychol. (Amst.)*, *148*, 163–172.

22. Fazio, L. K., Bailey, D. H., Thompson, C. A., & Siegler, R. S. (2014). Relations of different types of numerical magnitude representations to each other and to mathematics achievement. *J. Exp. Child Psychol.*, *123*, 53–72.

23. Schneider, M., Beeres, K., Coban, L., Merz, S., Susan Schmidt, S., Stricker, J., & De Smedt, B. (2017). Associations of non-symbolic and symbolic numerical magnitude processing with mathematical competence: A meta-analysis. *Dev. Sci.*, *20*(3), e12372.

24. Li, Y., Hu, Y., Wang, Y., Weng, J., & Chen, F. (2013). Individual structural differences in left inferior parietal area are associated with schoolchildrens' arithmetic scores. *Front. Hum. Neurosci.*, *7*, 844.

25. Price, G. R., Wilkey, E. D., Yeo, D. J., & Cutting, L. E. (2016). The relation between 1st grade grey matter volume and 2nd grade math competence. *Neuroimage*, *124*, 232–237.

26. Lubin, A., Rossi, S., Simon, G., Lanoë, C., Leroux, G., Poirel, N., Pineau, A., & Houdé, O. (2013). Numerical transcoding proficiency in 10-year-old schoolchildren is associated with gray matter inter-individual differences: A voxel-based morphometry study. *Front. Psychol.*, *4*, 197.

27. Isaacs, E. B., Edmonds, C. J., Lucas, A., & Gadian, D. G. (2001). Calculation difficulties in children of very low birthweight: A neural correlate. *Brain*, *124*, 1701–1707.

28. Rotzer, S., Kucian, K., Martin, E., von Aster, M., Klaver, P., & Loenneker, T. (2008). Optimized voxel-based morphometry in children with developmental dyscalculia. *Neuroimage*, *39*, 417–422.

29. Ranpura, A., Isaacs, E., Edmonds, C., Rogers, M., Lanigan, J., Singhal, A., Clayden, J., Clark, C., & Butterworth, B. (2013). Developmental trajectories of grey and white matter in dyscalculia. *Trends Neurosci. Educ.*, *2*, 56–64.

30. Le Bihan, D., Mangin, J. F., Poupon, C., Clark, C. A., Pappata, S., Molko, N., & Chabriat, H. (2001). Diffusion tensor imaging: Concepts and applications. *J. Magn. Reson. Imaging*, *13*, 534–546.

31. Rykhlevskaia, E., Uddin, L. Q., Kondos, L., & Menon, V. (2009). Neuroanatomical correlates of developmental dyscalculia: Combined evidence from morphometry and tractography. *Front. Hum. Neurosci.*, *3*, 51.

32. Landerl, Bevan, & Butterworth. Developmental dyscalculia and basic numerical capacities.

33. Butterworth, Varma, & Laurillard. Dyscalculia.

34. Fias, W., Menon, V., & Szucs, D. (2014). Multiple components of developmental dyscalculia. *Trends Neurosci. Educ.*, *2*, 43–47.

35. Kaufmann, L., Wood, G., Rubinsten, O., & Henik, A. (2011). Meta-analyses of developmental fMRI studies investigating typical and atypical trajectories of number processing and calculation. *Dev. Neuropsychol.*, *36*, 763–787.

36. Price, G. R., Holloway, I., Räsänen, P., Vesterinen, M., & Ansari, D. (2007). Impaired parietal magnitude processing in developmental dyscalculia. *Curr. Biol.*, *17*, R1042–R1043.

37. Menon, V. (2016). Working memory in children's math learning and its disruption in dyscalculia. *Curr. Opin. Behav. Sci.*, *10*, 125–132.

38. Geary, D. C. (2013). Early foundations for mathematics learning and their relations to learning disabilities. *Curr. Dir. Psychol. Sci.*, *22*, 23–27.

39. Fias, W., Menon, V., & Szucs, D. (2013). Multiple components of developmental dyscalculia. *Trends Neurosci. Educ.*, *2*, 43–47.

40. Bull, R., & Lee, K. (2014). Executive functioning and mathematics achievement. *Child Dev. Perspect.*, *8*, 36–41.

41. Rivera, S. M., Reiss, A. L., Eckert, M. A., & Menon, V. (2005). Developmental changes in mental arithmetic: Evidence for increased functional specialization in the left inferior parietal cortex. *Cereb. Cortex*, *15*, 1779–1790.

42. Rotzer, S., Loenneker, T., Kucian, K., Martin, E., Klaver, P., & von Aster, M. (2009). Dysfunctional neural network of spatial working memory contributes to developmental dyscalculia. *Neuropsychologia*, *47*, 2859–2865.

43. Szucs, D., Devine, A., Soltesz, F., Nobes, A., & Gabriel, F. (2013). Developmental dyscalculia is related to visuo-spatial memory and inhibition impairment. *Cortex*, *49*, 2674–2688.

44. Plomin, R., DeFries, J. C., Knopik, V. S., & Neiderhiser, J. M. (2012). *Behavioral genetics* (6th ed.). New York, NY: Macmillan.

45. Izard, V., Sann, C., Spelke, E. S., & Streri, A. (2009). Newborn infants perceive abstract numbers. *Proc. Natl. Acad. Sci. U.S.A.*, *106*, 10382–10385.

46. Kovas, Y., Haworth, C., Dale, P., & Plomin, R. (2007). The genetic and environmental origins of learning abilities and disabilities in the early school years. *Monogr. Soc. Res. Child Dev.*, *72*, 1–144.

47. Alarcón, M., Defries, J., Gillis Light, J., & Pennington, B. (1997). A twin study of mathematics disability. *J. Learn. Disabil.*, *30*, 617–623.

48. Kovas, Haworth, Dale, & Plomin. The genetic and environmental origins of learning abilities and disabilities.

49. Kovas, Y., & Plomin, R. (2006). Generalist genes: Implications for the cognitive sciences. *Trends Cogn. Sci.*, *10*, 198–203.

50. Kovas, Haworth, Dale, & Plomin. The genetic and environmental origins of learning abilities and disabilities.

51. Kovas, Y., & Plomin, R. (2007). Learning abilities and disabilities: Generalist genes, specialist environments. *Curr. Dir. Psychol. Sci.*, *16*, 284–288.

52. Tosto, M. G., Petrill, S. A., Halberda, J., Trzaskowski, M., Tikhomirova, T. N., Bogdanova, O. Y., Ly, R., Wilmer, J. B., Naiman, D. Q., Germine, L., Plomin, R., & Kovas, Y. (2014). Why do we differ in number sense? Evidence from a genetically sensitive investigation. *Intelligence*, *43*, 35–46.

53. Hettema, J. M., Annas, P., Neale, M. C., Kendler, K. S., & Fredrikson, M. (2003). A twin study of the genetics of fear conditioning. *AMA Arch. Gen. Psychiatry.*, *60*, 702–708.

54. Lindberg, S. M., Hyde, J. S., Petersen, J. L., & Linn, M. C. (2010). New trends in gender and mathematics performance: A meta-analysis. *Psychol. Bull.*, *136*, 1123–1135.

55. Kovas, Y., Haworth, C. M. A., Petrill, S. A., & Plomin, R. (2007). Mathematical ability of 10-year-old boys and girls. *J. Learn. Disabil.*, *40*, 554–567.

56. Tosto, Petrill, Halberda, Trzaskowski, Tikhomirova, Bogdanova, Ly, Wilmer, Naiman, Germine, Plomin, & Kovas. Why do we differ in number sense?

57. Bruandet, M., Molko, N., Cohen, L., & Dehaene, S. (2004). Cognitive characterization of dyscalculia in Turner syndrome. *Neuropsychologia*, *42*, 288–298.

58. Molko, N., Cachia, A., Riviere, D., Mangin, J. F., Bruandet, M., Le Bihan, D., Cohen, L., & Dehaene S. (2003). Functional and structural alterations of the intraparietal sulcus in a developmental dyscalculia of genetic origin. *Neuron*, *40*, 847–858.

59. Molko, N., Cachia, A., Riviere, D., Mangin, J. F., Bruandet, M., LeBihan, D., Cohen, L., & Dehaene, S. (2004). Brain anatomy in Turner syndrome: Evidence for impaired social and spatial-numerical networks. *Cereb. Cortex, 14*, 840–850.

60. Rivera, S. M., Menon, V., White, C. D., Glaser, B., & Reiss, A. L. (2002). Functional brain activation during arithmetic processing in females with fragile X syndrome is related to FMR1 protein expression. *Hum. Brain Mapp., 16*, 206–218.

61. Kovas, Haworth, Dale, & Plomin. The genetic and environmental origins of learning abilities and disabilities.

62. Kovas, Haworth, Dale, & Plomin. The genetic and environmental origins of learning abilities and disabilities.

63. Räsänen, P., Salminen, J., Wilson, A. J., Aunio, P., & Dehaene, S. (2009). Computer-assisted intervention for children with low numeracy skills. *Cogn. Dev., 24*, 450–472.

64. Iuculano, T., Rosenberg-Lee, M., Richardson, J., Tenison, C., Fuchs, L., Supekar, K., & Menon, V. (2015). Cognitive tutoring induces widespread neuroplasticity and remediates brain function in children with mathematical learning disabilities. *Nat. Commun., 6*, 8453.

Chapter 14

1. Wiese, H. (2003). *Numbers, language, and the human mind.* Cambridge, UK: Cambridge University Press.

2. Carey, S. (2009). *The origin of concepts.* Oxford, UK: Oxford University Press.

3. Boyer, C. B. (1944). Zero: The symbol, the concept, the number. *Nat. Math. Mag., 18*, 323–330.

4. Dantzig, T. (1930). *Number: The language of science.* New York, NY: Free Press.

5. Eccles, P. J. (2007). *An introduction to mathematical reasoning: Lectures on numbers, sets, and functions.* Cambridge, UK: Cambridge University Press.

6. Dantzig, *Number.*

7. Ifrah, G. (2000). *Universal history of numbers: From prehistory to the invention of the computer.* Hoboken, NJ: John Wiley & Sons.

8. Ifrah, G. (1985). *From one to zero: A universal history of numbers.* New York, NY: Viking.

9. Boyer, Zero.

10. Houston, S., Mazariegos, O. C., & Stuart, D. (Eds.). (2001). *The decipherment of ancient Maya writing.* Norman, OK: University of Oklahoma Press

11. Ifrah. *Universal history of numbers.*

12. Ifrah. *Universal history of numbers.*

13. Colebrooke, H. T. (1817). *Algebra, with arithmetic and mensuration, from the Sanscrit of Brahmegupta and Bháscara* (p. 339). London: John Murray.

14. Colebrooke, *Algebra.*

15. Plofker, K., Keller, A., Hayashi, T., Montelle, C., & Wujastyk, D. (2017). The Bakhshālī manuscript: A response to the Bodleian Library's radiocarbon dating. *History of science in South Asia*, 5.1, 134–150.

16. Menninger, K. (1969). *Number words and number symbols.* Cambridge, MA: MIT Press

17. Hockney, M. (2012). *The god equation.* Miami, FL: Hyper Reality Books.

18. Schimmel, A. (1993). *The mystery of numbers.* Oxford, UK: Oxford University Press

19. Bellos, A. (2010). *Alex's adventures in numberland.* London, UK: Bloomsbury.

20. Fry, H. (2016). We couldn't live without "zero"—but we once had to. BBC Future, 6 Dec 2016. (http://www.bbc.com/future/story/20161206-we-couldnt-live-without-zero-but-we-once-had-to)

21. Kaplan, R. (2000). *The nothing that is: A natural history of zero.* Oxford, UK: Oxford University Press.

22. Tammet, D. (2012). *Thinking in numbers: How maths illuminates our lives.* London, UK: Hodder & Stoughton.

23. Rotman, B. (1987). *Signifying nothing: The semiotics of zero.* London, UK: Macmillan Press

24. Recorde, R. (1543). *The grounde of artes.* London, UK: Reynold Wolff

25. Blank, P. (2006). *Shakespeare and the mismeasure of man.* Ithaca, NY: Cornell University Press.

26. Devlin, K. (2002). The most beautiful equation. *Wabash Magazine*, Winter/Spring 2002.

27. Feynman, R., Leighton, R. B., & Sands, M. (1963). *The Feynman lectures on physics*, Vol. I (p. 22). Amsterdam, the Netherlands: Addison-Wesley Longman.

28. Hockney. *The god equation.*

29. Eccles, P. J. *An introduction to mathematical reasoning.*

30. Leibniz, G. (1679). *De progressione Dyadica.*

31. Wynn, K., & Chiang, W. C. (1998). Limits to infants' knowledge of objects: The case of magical appearance. *Psychol. Sci.*, *9*, 448–455.

32. Liszkowski, U., Schäfer, M., Carpenter, M., & Tomasello, M. (2009). Prelinguistic infants, but not chimpanzees, communicate about absent entities. *Psychol. Sci.*, *20*, 654–660.

33. Wellman, H. M., & Miller, K. F. (1986). Thinking about nothing: Development of concepts of zero. *Br. J. Dev. Psychol.*, *4*, 31–42.

34. Wellman & Miller. Thinking about nothing.

35. Bialystok, E., & Codd, J. (2000). Representing quantity beyond whole numbers: Some, none, and part. *Can. J. Exp. Psychol.*, *54*, 117–128.

36. Wellman & Miller. Thinking about nothing.

37. Merritt, D. J., & Brannon, E. M. (2013). Nothing to it: Precursors to a zero concept in preschoolers. *Behav. Process.*, *93*, 91–97.

38. Wellman & Miller. Thinking about nothing.

39. Brysbaert, M. (1995). Arabic number reading—On the nature of the numerical scale and the origin of phonological recoding. *J. Exp. Psychol. Gen.*, *124*, 434–452.

40. Fias, W. (2001). Two routes for the processing of verbal numbers: Evidence from the SNARC effect. *Psychol. Res.*, *12*, 415–423.

41. Nuerk, H.-C., Iversen, W., & Willmes, K. (2004). Notational modulation of the SNARC and the MARC (linguistic markedness of response codes) effect. *Q. J. Exp. Psychol. Hum. Exp. Psychol.*, *57*, 835–863.

42. Wheeler, M., & Feghali, I. (1983). Much ado about nothing: Preservice elementary school teachers' concept of zero. *J. Res. Math. Educ.*, *14*, 147–155.

43. De Lafuente, V., & Romo, R. (2005). Neuronal correlates of subjective sensory experience. *Nat. Neurosci.*, *8*, 1698–1703.

44. Merten, K., & Nieder, A. (2012). Active encoding of decisions about stimulus absence in primate prefrontal cortex neurons. *Proc. Natl. Acad. Sci. U.S.A.*, *109*, 6289–6294.

45. Pepperberg, I. M. (1988). Comprehension of "absence" by an African grey parrot: Learning with respect to questions of same/different. *J. Exp. Anal. Behav.*, *50*, 553–564.

46. Pepperberg, I. M., & Gordon, J. D. (2005). Number comprehension by a grey parrot (*Psittacus erithacus*), including a zero-like concept. *J. Comp. Psychol.*, *119*, 197–209.

47. Pepperberg, I. M. (2006). Grey parrot (*Psittacus erithacus*) numerical abilities: Addition and further experiments on a zero-like concept. *J. Comp. Psychol., 120*, 1–11.

48. Boysen, S. T., & Berntson, G. G. (1989). Numerical competence in a chimpanzee (*Pan troglodytes*). *J. Comp. Psychol., 103*, 23–31.

49. Olthof, A., Iden, C. M., & Roberts, W. A. (1997). Judgments of ordinality and summation of number symbols by squirrel monkeys (*Saimiri sciureus*). *J. Exp. Psychol. Anim. Behav. Proc., 23*, 325–339.

50. Matsuzawa T. (1985). Use of numbers by a chimpanzee. *Nature, 315*, 57–59.

51. Biro, D., & Matsuzawa, T. (2001). Use of numerical symbols by the chimpanzee (*Pan troglodytes*): Cardinals, ordinals, and the introduction of zero. *Anim. Cogn., 4*, 193–199.

52. Biro, D., & Matsuzawa, T. (1999). Numerical ordering in a chimpanzee (*Pan troglodytes*): Planning, executing, and monitoring. *J. Comp. Psychol., 113*, 178–185.

53. Merritt, D. J., Rugani, R., & Brannon, E. M. (2009). Empty sets as part of the numerical continuum: Conceptual precursors to the zero concept in rhesus monkeys. *J. Exp. Psychol. Gen., 138*, 258–269.

54. Ramirez-Cardenas, A., Moskaleva, M., & Nieder, A. (2016). Neuronal representation of numerosity zero in the primate parieto-frontal number network. *Curr. Biol., 26*, 1285–1294.

55. Okuyama, S., Iwata, J., Tanji, J., & Mushiake, H. (2013). Goal-oriented, flexible use of numerical operations by monkeys. *Anim. Cogn., 16*, 509–518.

56. Howard, S. R., Avarguès-Weber, A., Garcia, J. E., Greentree, A. D., & Dyer, A. G. (2018). Numerical ordering of zero in honey bees. *Science, 360*, 1124–1126.

57. Merten & Nieder. Active encoding of decisions about stimulus absence.

58. De Lafuente, V., & Romo, R. (2006). Neural correlate of subjective sensory experience gradually builds up across cortical areas. *Proc. Natl. Acad. Sci. U.S.A., 103*, 14266–14271.

59. Merten & Nieder. Active encoding of decisions about stimulus absence.

60. Ramirez-Cardenas, Moskaleva, & Nieder. Neuronal representation of numerosity zero.

61. Schultz, W., Dayan, P., & Montague, P. R. (1997). A neural substrate of prediction and reward. *Science, 275*, 1593–1599.

62. Schultz, W. (2007). Multiple dopamine functions at different time courses. *Annu. Rev. Neurosci., 30*, 259–288.

63. Law, C.-T., & Gold, J. I. (2009). Reinforcement learning can account for associative and perceptual learning on a visual-decision task. *Nat. Neurosci.*, *12*, 655–663.

64. Rombouts, J., Bohte, S., & Roelfsema, P. (2012). Neurally plausible reinforcement learning of working memory tasks. *Adv. Neural Inf. Process. Syst.*, *25*, 1880–1888.

65. Engel, T. A., Chaisangmongkon, W., Freedman, D. J., & Wang, X. J. (2015). Choice-correlated activity fluctuations underlie learning of neuronal category representation. *Nat. Commun.*, *6*, 6454.

66. Hockney. *The god equation.*

Index

Abboud, Sami, 190
Absolute numerosity judgment, 35, 52, 60
Acalculia, 102, 103, 169, 179, 206, 207, 219, 260
Action potential, 116–121, 128, 188, 301
Adaptation, 10, 27, 147, 192
Adaptive value, 27, 40, 64, 70, 72, 75
Addition, 58, 142, 193, 200, 208, 209, 212–224, 237, 253, 254, 258, 279
African grey parrot, 41, 171, 173, 295
Agnosia, 169
Agraphia, 169
Agrillo, Christian, 36, 38
AIP. *See* Anterior intraparietal area (AIP)
Alarm call, 41, 69, 170, 171
Algebra, 222, 294, 295
Allele, 26, 27
Alzheimer's disease, 101
Amalric, Marie, 231, 232
Amedi, Amir, 190
Amphibians, 23, 25, 38–40, 48, 66, 67
Amygdala, 98, 99, 141
Analogous trait, 24
Angelfish, 36
Angular gyrus, 93, 98, 102, 105, 154, 155, 169, 207–210, 214–217, 219, 222, 223, 228, 242, 260, 271, 272
ANS. *See* Approximate number system (ANS)

Ansari, Daniel, 258, 260
Ant, 46, 65
Anterior intraparietal area (AIP), 97, 98, 241, 242
Anterior temporal cortex, 123
Anti-realism, 4. *See also* Non-Platonism
Apes, 43, 67, 295
Aphasia, 177, 220–224
Approximate number system (ANS), 14, 15, 36, 47, 52, 59, 61, 62, 80–85, 172, 195, 199, 210, 237, 250–254, 270, 279
Archeology, 159, 160, 161
Arithmetic, 3, 101, 102, 103, 155, 170, 193, 198, 214, 219, 220, 258, 268
Arsalidou, Marie, 180, 209, 210, 212, 217
Arthropods, 22, 23, 44–48
Artifacts, 161, 164
Association cortex, 95, 96, 98–101, 152, 159, 240
Association neuron, 184, 186
Attention, 124, 179, 206, 214, 235, 238, 241, 242, 258, 259
Attention-deficit hyperactivity disorder (ADHD), 268

Baboons, 43
Babylonians, 286, 287, 291
Bacteria, 64, 65
Bakhshali manuscript, 288

Baldo, Juliana, 214, 221
Barlow, Horace Basil, 118
Base-2 system, 168, 292
Base-10 system, 168, 288, 291
Base-20 system, 168, 286
Base-60 system, 286, 291
Bears, 43
Bees. *See* Honeybees
Beetles, 23, 46, 72
Behavioral interference, 242, 243
Behavioral training, 32, 34–36, 60, 81, 82, 146
Berger, Hans, 179
Bianchi, Leonardo, 95
Bilaterian, 23, 25
Binary nomenclature, 23, 24
Bioluminescence, 65
Birds, 17, 23, 24, 25, 30–32, 40–42, 48, 57, 59–62, 69, 70, 73–75, 137, 138, 140, 171
Bisazza, Angelo, 36
Blind subjects, 190, 191
Blood flow, 101, 105–107
Blood oxygen level–dependent signal, 106–109, 111, 112, 129, 136, 238, 255, 257, 261, 262, 278
Body parts as numerical signs, 165, 166
BOLD signal. *See* Blood oxygen level–dependent signal
Bongard, Sylvia, 201
Booth, Julie, 87, 88
Boysen, Sarah, 296
Brahmagupta, 288
Brain damage, 94, 101, 102, 104, 177–180, 213, 214, 219, 221–224, 235. *See also* Lesions
Brannon, Elizabeth M., 11, 49–52, 54, 58–60, 198, 199, 293, 297,
Broca, Paul, 94, 226
Broca's area, 94, 222
Brodmann, Korbinian, 95
Brood parasitism, 73–75
Budgerigars, 31

Burr, David, 17, 147
Bush babies, 42
Butterworth, Brian, 265, 269

Calculation, 28, 84, 98, 101, 102, 105, 163, 167, 170, 177–180, 193–200, 207–210, 212, 213, 216, 217, 219, 220, 238, 239, 240, 242, 253, 259, 268
Cambrian explosion, 22, 44, 45
Cantlon, Jessica F., 52, 129, 198, 199, 256, 257
Cappelletti, Marinella, 268
Capuchin monkeys, 43. *See also* Monkeys
Cardinality, 11, 12, 14, 17, 56, 81, 123, 129, 134, 147, 162, 165, 166, 169, 171, 172, 248, 249, 251, 257, 279
Cardinality principle, 248
Carey, Susan, 174, 249, 250
Carnivora, 42
Cats, 43
Cavefish, 38
Central sulcus, 92, 95, 98
Cephirum, 289
Cerebellum, 92, 181, 212
Cerebral cortex, 92–96, 101, 104, 146
Cerebral hemispheres, 92
Cerebrovascular damage, 101, 103
Cerebrum, 92, 93
Cetacea, 42
Changeux, Jean-Pierre, 148
Chickadees, 41, 69, 70, 171
Chickens, 41
Chiffre, 289
Children, 18, 35, 80, 87, 88, 129, 155, 161, 168, 174, 182, 185, 247–251, 254–259, 268, 269 292–295
Chimpanzees, 42, 43, 71, 96, 173, 174, 201, 295
Chittka, Lars, 46, 66
Chomsky, Noam, 251

Chordates, 22, 23, 25
Church, Russell M., 148
Cicadas, 46
Cifra, 289
Cingulate cortex, 99, 181, 260, 272
Cipolotti, Lisa, 177
"Clever Hans," 28–30
  "Clever Hans" effect, 29
Cognitive control, 99, 151, 152, 155, 205, 206, 258, 274. *See also* Executive function
Cognitive inhibition, 153, 206
Cohen Kadosh, Roi, 261
Cohen, Laurent, 136, 210
Colombo, Michael, 60
Confusion effect, 69
Conservation-of-number task, 247
Convergent evolution, 24, 48, 137, 138, 140
Cooper, Robert G., 78, 82
Coots, 40, 73
Core knowledge, 8, 249–251
Correct responses, 13, 120, 182
Corvids, 60, 61 138
Counting, 7, 11, 13–15, 28, 32, 39, 52, 57, 61, 77, 82, 84, 85, 87, 102, 103, 126, 136, 155, 164–170, 172–175, 177–180, 182, 193, 215, 237, 247–252, 265–268, 279, 284, 285, 292, 293
Cowbirds, 40, 73–75
Coyotes, 67
Crocodiles, 39
Cross-modal number representation, 58, 59, 79, 112, 113
Crows, 35, 37, 40, 41, 60, 61, 67, 137–140, 144, 146, 150, 151
  carrion crows, 60, 61
Cultural recycling, 191, 204
Culture, 13, 14, 82, 83, 87, 159, 160, 161, 163, 165–168, 170, 211, 235, 236, 285–292, 303, 305
Cuneiform numerals, 167

Dacke, Marie, 46, 66
Damarla, Saudamini, 112, 262
Dantzig, Tobias, 7, 147
Darwin, Charles, 25. *See also* Theory of evolution
Davis, Phillip J., 3, 5
De Hevia, Maria Dolores, 236
Deacon, Terrence W., 163
Decision making, 45, 48, 55, 65, 68, 70, 72, 74, 95, 127, 131, 193, 203–205, 210, 300–302
Dedekind, Richard, 249
Dedekind–Peano axioms, 249
Dehaene, Stanislas, 8, 108, 109, 128, 136, 148, 180, 208, 210, 231, 232, 233, 238, 240, 242, 259, 261
Dehaene-Lambertz, Ghislaine, 237
Delayed association task, 172, 182
Delayed matching-to-sample task, 52, 53, 56, 57, 60, 120, 132, 135, 145, 152, 153, 182, 243
Della Puppa, Alessandro, 218
Deoxyribonucleic acid (DNA), 26
Deuterostomes, 23, 25, 137
Development, 8, 14, 18, 23, 77, 88, 91, 92, 159–163, 185, 190, 227, 238, 248–259, 285, 292, 295, 304. *See also* Ontogeny
Devlin, Keith, 291
Diagnostic and Statistical Manual of Mental Disorders (DMS), 266
Diamond, Marian C., 228
Diester, Ilka, 134, 182
Diffusion tensor imaging, 272
Dillusion effect, 69
Distractors, 153, 154
Ditz, Helen M., 60, 140
Division, 212–224
Dogs, 43, 67
Dolphins, 43
Domain-general capability, 268, 269
Domain-specific capability, 268
Dopamine, 138, 205, 206, 304

Dormal, Valérie, 111
Dronkers, Nina, 214, 221, 222
Dumoulin, Serge, 111
Dyer, Adrian G., 47, 298
Dyscalculia, 253, 265–280
Dyslexia, 188, 267

Earth history, 21
Education, 87, 200, 213, 252, 265, 267, 270, 279, 289
Eger, Evelyn, 109, 112, 136, 260, 262
Eggs, 73–75
Egypt, 167, 168, 286
Einstein, Albert, 7, 227–229
Eiselt, Anne-Kathrin, 204
Electrical activity (of neurons), xiv, 107, 117, 118, 188. See also Action potential
Electrical stimulation, 94, 188, 215–224
Electrocorticography (ECoG), 188, 211
Electroencephalography (EEG), 107, 179
Elephants, 43, 67
Empty set, 283–307
Endbrain, 92, 138, 140, 146, 160
Endothermy, 40, 42
Engel, Tatiana, 304
Entorhinal cortex, 98, 141
Epilepsy, 141, 142, 186, 188, 215
Epistemology, 9, 28. See also Theory of knowledge
Equinumerosity, 247–249
Euler, Leonhard, 291
Euler's identity, 291, 305
Evolution, 5, 21, 24, 25, 26, 27, 28, 45, 72, 75, 96, 137, 140, 159–163, 172, 190, 204, 237, 241, 254, 277, 280
Exact equality. See Equinumerosity
Executive function, 154, 206–209. See also Cognitive control

Faces, 190
Fact-retrieval strategy, 212–216, 272
Falk, Dean, 228

Fechner, Gustav Theodor, 54, 55
Fechner's law, 55, 61, 87, 88, 128, 129, 140, 151
Feigenson, Lisa, 252
Ferrier, David, 94, 95
Feynman, Richard, 291
Fias, Wim, 149
Fibonacci, 288
Finger counting, 112, 258
Fingers, 161, 163–166, 168, 169
Fish, 23, 25, 36–38, 48, 57, 69, 137
Fissure, 92, 94
Fitness, 27, 28, 63
Flechsig, Paul Emil, 95, 96
fMRI. See Functional magnetic resonance imaging (fMRI)
Foraging, 39, 45, 46, 66, 67
Fractions, 16, 170, 180
Fragile X syndrome, 278
Freed-Brown, Grace, 74
Frege, Gottlob, 247
Friberg, Lars, 105, 207
Fritsch, Gustav Theodor, 94
Frogs, 39, 67
Frontal lobe, 93, 96, 98, 99, 101, 103, 105, 112, 113, 124, 137, 141, 146, 177, 179, 180, 202, 206, 207, 215, 220, 221, 223, 231, 240, 255, 273
Fuchs, Conrad Heinrich, 227
Functional brain imaging, 105, 106, 128, 142, 148, 180–182, 188, 192, 207, 210, 241, 255, 260, 274, 279
Functional magnetic resonance imaging (fMRI), 105–108, 111, 136, 147, 190–192, 208, 209, 212, 216, 217, 230, 231, 238, 239, 255, 260, 273, 278, 280. See also Functional brain imaging
  adaptation, 108, 109, 128, 129, 180, 255, 257, 259
Fusiform gyrus, 181, 211, 260
Fuster, Joaquin M., 152

Gallistel, Charles Ransom, 62, 248, 250
Galton, Francis, 233
Gamm, Rüdiger, 230, 231
Gauss, Carl Friedrich, 225–227
Gauss function, 54, 55
Gazzaniga, Michael, 117
Geiger, Karl, 46, 66
Gelman, Rochel, 62, 248, 250
Gender differences, 277, 278
Genes, 24, 26, 27, 63, 64, 72, 75, 274–279
Genetic code, 27, 275
Genetic variation, 10, 26, 160
Geniuses, 7, 48, 226–230
Genome, 26, 275
Genotype, 26, 27, 64
Gerstmann's syndrome, 169
Gilmore, Camilla, 253
Give-a-number task, 248
Global workspace, 152–154
Gödel, Kurt, 7, 8, 227
Goldstein, Kurt, 102, 179
Gordon, Peter, 83
Gorillas, 43
Grey matter, 270, 271
Gruber, Oliver, 208
Guppies, 36
Gyrus, 92, 97

Habituation, 78, 81, 108, 128, 255
  protocol, 78
Halberda, Justin, 252
Hardy, Godfrey Harold, 4
Harvey, Ben, 111
Harvey, Thomas, 228
Hécaen, Henry, 179
Hedonic value, 33, 35, 67, 296
Henschen, Salomon Eberhard, 102, 179, 219
Herculano-Houzel, Suzana, 91
Hieroglyphic numerals, 167
High cultures, 160

Higher brain functions, 95, 96, 210,
Hines, Terence, 229
Hippocampus, 98, 141, 240, 258, 259
Hitzig, Eduard, 94, 95
Hockney, Mike, 288, 291, 306
Hominids, 10, 42, 43
Homologous trait, 24
Homo sapiens, 6, 23, 43, 63, 159, 160
Honeybees, 23, 37, 45–48, 57, 65, 66, 298
Horses, 28, 29, 30, 43
Howard, Scarlett, 47, 298
Human history, 159–168, 284, 285–292, 304
Human revolution, 160
Hunting, 67–69
Huntington's disease, 101
Hyenas, 43, 71

Icon, 161–165, 168–170, 293
Imprinting, 41, 42
Inactivation of neurons, 131
Incas, 165
Index, 161–163, 165–169, 171, 182
Indians, 168, 286
Indigenous people, 14, 82–84, 165, 193–195, 253. *See also* Incas, Munduruku, Pirahã
Infants, xiv, 14, 18, 32, 77–82, 163, 195–200, 237, 238, 249, 250, 252, 292
Inferior frontal gyrus (F3), 93, 102, 105, 113, 181, 185, 212, 214, 215, 222, 258, 260, 272, 274
Inferior frontal lobe, 94, 102, 208, 209, 212, 214, 222
Inferior parietal lobule (IPL), 93, 98, 105, 113, 123, 169, 181, 210, 214, 215, 222, 255, 260
Inferior temporal gyrus (ITG), 93, 112, 188, 190, 191, 211, 232

Insects, 23, 25, 44–48, 137
Insula, 181, 260, 274
Intelligence, 60, 178, 206, 225, 227, 253, 268, 269, 276, 278
Intervention, 279, 280
Intraparietal sulcus (IPS), 92, 96–98, 102–104, 109, 112, 113, 123, 124, 129, 134–136, 147, 149, 155, 180–182, 185, 191, 202, 214, 215, 217, 223, 231, 232, 239–243, 255, 257–263, 272–274, 278, 279
IPL. See Inferior parietal lobule (IPL)
IPS. See Intraparietal sulcus (IPS)
Isaacs, Elizabeth, 271
Ishango bone, 164
ITG. See Inferior temporal gyrus (ITG)
Iuculano, Teresa, 279
Izard, Veronique, 79, 80

Jackdaws, 30, 31, 41
Jacob, Simon N., 109, 129, 153
Jordan, Kelly, 58, 59
Just, Marcel, 112, 262
Just-noticeable difference, 16, 55

Kasner, Edward, 4
Kaufmann, Liane, 257
Kersey, Alyssa, 256, 257
Kleinschmidt, Andreas, 109, 112, 136, 208, 260, 262
Klessinger, Nicolai, 222
Knops, André, 238, 240
Knots, 161, 166
Koehler, Otto, 30–32, 34, 40, 49, 60
Konrad Lorenz, 30
Kovas, Yulia, 276–278
Kutter, Esther, 141
Kwong, Ken, 106

Labeled-line code, 127, 150, 151, 186, 243
Lanchester's square law, 71
Landauer, Thomas K., 252

Language, 17, 18, 32, 83, 94, 95, 98, 102, 161, 163, 179, 198, 214, 218, 219–224, 236, 242, 250, 251, 255, 268–270, 276, 289
Last Universal Common Ancestor (LUCA), 21
Lateral intraparietal area (LIP), 97, 109, 241, 242
Learning disorder, 266, 267, 274, 275, 278, 279
Lebombo bone, 164
Leibniz, Gottfried Wilhelm, 292
Lemurs, 42, 43
Lesions, 95, 102–104, 131, 154, 188, 206, 207, 214, 222–224, 240, 260. See also Brain damage
Lewandowsky, Max, 102
Linear number scale, 87
Lions, 43, 70
LIP. See Lateral intraparietal area (LIP)
Livio, Mario, 4
Lizards, 39
Locke, John, 6
Logarithmic number scale, 53, 55, 56, 61, 85–88, 128, 129, 140, 151, 210, 257, 295
Log-normal distribution, 55
Long-term memory, 98, 99, 171, 219, 230, 231, 267
Luria, Alexander R., 206

Macaques, 42, 43, 96, 170, 198. See also Monkeys
Mafia, 74
Magpies, 31
Malafouris, Lambros, 161
Mall, Franklin Paine, 227
Mammals, 23, 25, 30, 40, 42–44, 48, 57, 59, 60, 62, 70, 137, 138, 140, 163, 171, 277
Many eyes effect, 69
Matching-to-sample task, 30
Material engagement theory, 161

Mathematicians, 4, 5, 7, 8, 30, 42, 191, 201, 224, 225, 230–232, 247, 249, 288–292, 306
Mathematics, 3–10, 91, 95, 101, 102, 168, 178, 193, 207, 221–232, 249, 252, 253, 266, 275–278, 283, 286, 289, 291, 305, 306
Mating, 63, 64, 72
Matsuzawa, Tetsuro, 174, 296
Maya, 167, 286, 287
McComb, Karen, 70
McCrink, Koleen, 196, 237
Mechner, Francis, 43, 44, 49, 54
Meck, Warren H., 148
Medial intraparietal area (MIP), 97, 241
Medial temporal lobe (MTL), 97, 98, 141–145, 186, 187
Menon, Vinod, 155, 258, 279
Mental number line, 47, 55, 87, 61, 86–88, 109, 140, 233–236, 238, 294, 297, 298, 306
Merritt, Dustin, 293
Merten, Katharina, 56, 300
Mesopotamia, 160, 167, 286
Microelectrodes, 116, 118, 120, 142
Middle Ages, 289
Middle frontal gyrus (F2), 93, 210, 212, 223, 255, 258, 260, 272
Middle temporal gyrus, 93, 208, 222
Miller, Earl K., 52, 119, 132
Miller, Kevin, 294
MIP. *See* Medial intraparietal area (MIP)
Monkeys, 9, 10, 25, 35, 37, 43–45, 49–61, 67–100, 119–124, 128, 131, 132, 135, 138–140, 144, 149–155, 184, 190, 199, 200, 237, 240–244, 259, 263, 285, 295, 298–302. *See also* Rhesus monkeys
  brains of, 94, 96–100, 119–129
Mormann, Florian, 141, 186
Mosquitofish, 36, 38

Mice, 67
Moyer, Robert S, 252
MTL. *See* Medial temporal lobe (MTL)
Multiplication, 163, 207–210, 212–224
Munduruku, 83–87, 193–195, 253
Muscimol, 131
Mushiake, Hajime, 123, 199, 298

Naccache, Lionel, 261
Nasr, Khaled, 150
Natural selection, 10, 27, 28, 64, 75, 190
Navigation, 65
NCL. *See* Nidopallium caudolaterale (NCL)
Negative numbers, 170, 214, 283, 289
Neocortex, 92, 137
Neolithic revolution, 160
Neural model, 128, 148–151
Neural network, 149–151
Neural tube, 91, 92
Neurodegenerative disease, 101, 103, 219
Neuromodulator, 205, 206
Neurons, xiv, 18, 36, 45, 48, 60, 91, 92, 101, 105–108, 115–119, 182, 201–204, 228, 240–244, 262, 264, 299–306
Neurosurgery, 94, 118, 120, 141, 188, 215–217, 224
Newborns, xiv, 77, 79–82, 95, 151, 236, 292
Newman, James Roy, 4
Nidopallium caudolaterale (NCL), 138, 139, 146
Non-human primate. *See* Monkeys, Apes
Non-Platonism, 4. *See also* Anti-realism
Nothing, 283–306
Null, 289
Numbers
  bisection of, 103
  cardinal, 5, 6–8, 11, 12, 28, 54, 58, 60, 150, 167, 173, 175, 248–252

Numbers (cont.)
  format of, 13, 49, 80, 132, 134, 143, 150, 153, 180–182, 184, 186, 187, 210, 211, 221, 239, 240, 257–263
  modality of, 14, 38, 58, 59, 78, 80, 112, 126, 135, 136, 149, 150, 181, 190, 191, 259, 262
  nominal, 12
  non-symbolic representation of, 10, 12–18, 21, 32, 33, 43, 56–61, 103, 111, 129, 142–145, 161, 170, 180–187, 193, 198, 200, 210, 237–240, 250–263, 275, 277
  notation of, 17, 136, 143, 144, 159–170, 180, 180–185, 210, 259, 285, 261, 285–293
  number codes, xiv, 127, 136, 140, 144, 148–151, 172
  number form area, 188–192, 211, 232
  number instinct, xv, 7, 8, 10, 87, 146, 147, 232. *See also* Numbers: number sense
  number line. *See* Mental number line
  number neurons, xii, 36, 108, 115–155, 185–187, 242, 243, 263, 302, 303
  number sense, 7, 8, 148, 266, 277. *See also* Numbers: number instinct
  ordinal, 11, 12, 35, 58, 169, 174, 248, 297
  preferred, 120, 121, 127, 135
  symbolic representation of, 13–18, 22, 33, 43, 98, 129, 136, 143–145, 169, 171, 173, 180–186, 191, 193–198, 217, 218, 249–261, 270, 293, 295
  symbols for, 17, 18, 77, 83, 87, 103, 136, 156–163, 167–173, 179–190, 195, 220, 240, 250–254, 260, 261, 270, 299, 306
  theory of, xiii, 18, 161, 219, 288, 299, 305

Numeral, 13, 17, 35, 77, 88, 103, 104, 143, 168, 170–174, 180, 183, 184, 186–192, 212, 235, 239, 258, 288, 289, 290
Numerical distance effect, 15, 43, 56, 61, 62, 103, 127, 128, 140, 145, 172, 186, 187, 252, 294, 297, 298, 302
Numerical size effect, 15, 16, 56, 61, 127, 128, 140, 145, 172, 252
Numerosity, map for, 109, 110

Object tracking system (OTS), 14, 15, 61, 62, 81, 82, 250
Occipital lobe, 93, 96, 188, 211
Ogawa, Seiji, 106
Oksapmin (Papua New Guinea), 165
Okuyama, Sumito, 198
One-to-one correspondence, 247–251
One-to-one principle, 248
Ontogeny, xiii, 8, 237, 259, 285, 303. *See also* Development
Operant conditioning, 30, 32, 34, 35
Operational momentum, 237, 238
Orangutans, 43
Ordinality, 11, 12, 58, 174, 248
Ott, Torben, 205
Overmann, Karenleigh, 161
Oxygen revolution, 22

Paleoanthropology, 159, 160
Paleolithic age, 160, 164
Pallium, 92, 138
Parahippocampal gyrus, 98, 141, 144, 272
Parietal lobe, 93, 96, 98, 99, 102, 103, 112, 113, 126, 137, 141, 146, 177, 208–210, 215–223, 240, 255, 258, 274, 280
Parieto-frontal number network, 144, 255, 256, 262, 271, 278
Parieto-occipital region, 103, 179
Parieto-temporal junction, 98, 102, 214

Parkinson's disease, 101
Parrots, 31. *See also* African grey parrots
Parvizi, Josef, 188, 211
Patients, xiv, 94, 96, 101–104, 141, 142, 177–180, 186, 188, 206–223, 235, 260, 261
Peano, Guiseppe, 249
Peddington, Bruce, 276
Peirce, Charles Sanders, 162
Pepperberg, Irene, 171, 295
Perissodactyla, 42
Peritz, Georg, 102
Pesenti, Mauro, 111, 230
PET. *See* Positron emission tomography (PET)
PFC. *See* Prefrontal cortex (PFC)
Pfungst, Oskar, 29
Phenotype, 26, 27, 64
Phrenology, 225
Phylogeny, 8, 21, 24–26, 36, 64, 237, 259, 305
Phylum, 23–25
Piaget, Jean, 247
Piazza, Manuela, 108, 109, 128, 180, 259, 268
Pica, Pierre, 83–85, 193–195
Pigeon, 31, 41, 60, 171–173
Pirahã, 83, 85
Place value system. *See* Positional notation system
Platonism, 4, 6. *See also* Realism
Platt, Michael, 149
Playback study, 70–72
Positional notation system, 168, 285–293
Positron emission tomography (PET), xiv, 105, 106, 207, 208, 230
Postcentral gyrus, 93, 98, 257, 258
Posterior parietal cortex (PPC), 98, 99, 103, 108, 109, 123, 124, 126, 136, 152–155, 179–181, 185, 187, 190, 210, 217, 222, 223, 238, 240–243, 257, 259, 260, 273

PPC. *See* Posterior parietal cortex (PPC)
Precentral gyrus, 93, 98, 109, 112, 113, 181, 214, 255, 260
Predation, 64–70
Prefrontal cortex (PFC), 93, 95, 98–100, 109, 120–124, 129, 135, 136, 138, 145, 146, 149, 152–155, 180–185, 187, 201, 205–212, 223, 228, 232, 239, 243, 258, 259, 263, 273, 274, 280, 300–304
Premotor cortex, 98, 99, 101, 112, 120, 121, 136, 208, 239, 274
Presentation format
  sequential, 14, 57, 58, 80, 111, 112, 132, 133, 153
  simultaneous, 14, 49, 57, 80, 111, 112, 132, 133, 153
Primary motor cortex, 98, 99
Primary sensory cortex, 100
Primates, 10, 16, 42, 43, 45, 49, 59–62, 92, 96, 100, 124, 127, 129, 137, 138, 140, 141, 144, 146, 169, 182, 185, 190, 200, 204, 259, 285, 296, 298, 302. *See also* Monkeys, Apes
Proboscidea, 42
Procedural strategy, 212–216
Prodigy, 224, 230
Proportions, 243
Protostomes, 23, 25, 137
Pylyshyn, Zenon, 82

Quine, Willard V. O., 9
Quipu, 166. *See also* Knots
Quorum sensing, 64, 65

Raccoons, 43
Ramirez-Cardenas, Araceli, 297, 302
Ratio dependency, 16, 52, 59, 62, 65, 80, 81
Rats, 43, 44, 49
Ravens, 31
Reaction time, 13, 236
Reading, 212

Realism, 4, 6–9. *See also* Platonism
Receptive field, 240, 241
Recorde, Robert, 290
Recursion, 175, 249, 251
Reeve, Robert, 269
Relative numerosity judgment, 34, 35, 52
Representation
 cross-modal, 58, 59, 79, 112, 113
 mental, 12–14, 18, 43, 118, 153, 187
 multimodal, 58, 59, 100, 124, 149, 241
 neuronal, 18, 118, 119
 supramodal, 135, 136, 181, 262
Reptiles, 23, 25, 39–42, 48, 137
Rhesus monkeys, 43, 49–61, 139, 172, 192, 198, 201, 237, 243, 295, 295–300
Rivera, Susan, 155, 258, 274
Robins, 40, 67
Rodentia, 42
Roitman, Jamie D., 149
Roland, Per, 105, 207
Roman numerals, 167, 168
Rotzer, Stephanie, 272
Roux, Franck-Emmanuel, 223, 224
Rule following, 206, 210
Rules
 greater-than/less-than, 47, 48, 200–204, 298
 mathematical, 18, 200–205, 207, 208, 220, 294
 numerical, 35, 51, 198, 205, 206

Salamanders, 37–39, 67
Saxe, Geoffrey B., 165
Scorpions, 72
Sea lions, 43
Semantics, 17, 219, 220
 Semantic association, 182–184
Sex chromosomes, 277, 278
Shadlen, Michael, 152

Sieger, Robert, 87, 88
Signs, 35, 160–163, 171, 172, 182, 285, 292
Simon, Olivier, 242
Simpson, George Gaylord, 9, 10
Single-cell/single-neuron recording, 117, 118, 123, 124, 129, 141, 148, 154, 182, 186, 205, 256, 263
Skinner, B. F., 34, 49
Snakes, 39
SNARC effect, 233–235
Space, 126, 233–244
Spatial neglect, 235, 240
Spelke, Elizabeth S., 79, 251, 253
Sperm competition, 72, 73
Spiders, 44–46, 67
Spitzka, Edward Anthony, 226, 227
SPL. *See* Superior parietal lobule (SPL)
Spontaneous choice test, 32–36, 38, 41, 43, 62, 81, 236
Squirrel monkeys, 43, 295
Squirrels, 31
Srinivasan, Mandyam, 46, 66
Stable order principle, 248
Stadelmann, Ernst, 102
Staianov, Ivilin, 150
Starkey, Prentice, 78, 81
Statistical classifier, 109, 112, 130, 131, 144, 186, 239, 243, 260, 262
Stimulus-response function, 119
Stone Age, 160
Streri, Arlette, 79
Structural magnetic resonance imaging, 271
Subitizing, 15, 103
Subtraction, 142, 193, 195, 200, 208, 209, 212–224, 237, 238, 253, 254, 258, 279
Successor function. *See* Successor relation
Successor relation, 170, 174, 175, 249–251
Sulci, 92, 93

Summation coding, 148–151
Sunya, 288
Superior frontal gyrus (F1), 93, 105, 113, 208, 210, 212, 255, 260
Superior parietal lobule (SPL), 93, 98, 109, 111, 113, 123, 126, 131, 181, 214, 219, 238, 239, 240, 255, 260
Superior temporal gyrus, 222
Supramarginal gyrus, 93, 98, 154, 155, 214–216, 219, 222, 272, 274
Survival, xiii, xv, 5, 9, 10, 24, 27, 28, 64, 75
Sustained activity, 152, 304
Swordtails (fish), 36
Sylvian fissure, 98
Symbol grounding, 254
Symbolic reference, 163
Symbols, 13, 77, 103, 159, 160, 161, 169, 170, 172, 180, 182, 186, 212
Symbol system for numbers, xiii, 17, 18, 103, 159, 167, 169, 170, 184, 219, 220, 223, 250, 269, 288
Synapses, 115–118, 131
Syntax, 17, 163, 219, 220

Tally marks, 164
Tanji, Jun, 126, 131
Taxonomy, 23
Taylor, Margot, 180, 209, 210, 212, 217
Telencephalon, 92, 138
Temporal lobe, 93, 96, 98, 99, 119, 123, 137, 138, 177, 214, 220, 221, 231, 280
Terrace, Herbert S., 49–51, 54
Territory defense, 70
Tetrapods, 25, 38, 42
Theory of evolution, 9, 10, 25–28, 63
Theory of knowledge, 9. *See also* Epistemology
Theory of magnitude, 126
Theory of optimal foraging, 66
Thompson, Richard F., 126

Thomson, Paul, 9
Thorndike, Edward L., 34
Tinbergen, Nikolaas, xiii, 30
TMS. *See* Transcranial magnetic stimulation (TMS)
Toads, 39
Transcranial magnetic stimulation (TMS), xiv, 217, 218, 260
Tree of life, 21, 24, 25, 32, 48, 75, 137, 298
Triple-code model, 136, 210–212
Tudusciuc, Oana, 130, 134, 243
Tuning curve, 119, 121, 122, 125–133, 139, 143, 144, 149, 151, 154, 183, 243, 257
Turner's syndrome, 278
Turtles, 39
Twins, 275, 278, 279
study of, 275, 276

Vallentin, Daniela, 203, 243
Vallortigara, Giorgio, 41, 42
Van Essen, David, 96
Varley, Rosemary, 221, 222
Ventral intraparietal area (VIP), 97, 98, 123, 124, 126, 134, 135, 145, 146, 149, 154, 184, 204, 241–243, 302–304
Ventral occipito-temporal cortex (VOT), 188, 273
Verguts, Tom, 149
Vertebrates, 23, 25, 36, 38–40, 44, 45, 48, 66, 72, 92, 118, 138, 146
Vervet monkeys, 170
Violation of expectation, 195, 198
VIP. *See* Ventral intraparietal area (VIP)
Visual word form area, 191, 192
Viswanathan, Pooja, 145
Vollmer, Gerhard, 9
von Frisch, Karl, 30
von Linné, Carl, 23
von Osten, Wilhelm, 28–30
Voxel-based morphometry, 271

Wagener, Lysann, 146
Wagner, Rudolph F. J. H., 225, 227
Walsh, Vincent, 126, 242
Wang, Xiao-Jing, 304
Weber, Ernst Heinrich, 16, 55
Weber fraction, 16, 17
Weber's law, 16, 52, 57, 59, 62, 65, 80, 81, 83, 128, 129, 140, 151, 172, 173, 195, 252
Wellman, Henrey, 294
Whales, 43
White, David J., 74
White matter, 270, 271
Wilson, Michael, 71
Wolf bone, 164
Wolves, 68, 69
Working memory, 47, 48, 95, 99, 106, 120–124, 151–155, 205–210, 219, 230, 231, 239, 258, 259, 267, 268, 273, 274, 304
Writing, 212
Wynn, Karen, 195–197, 253

Zero, 48, 168, 170, 283–306
Ziffer, 289
Zorzi, Marco, 150
Zuse, Konrad, 292